Models of the Atomic Nucleus

Norman D. Cook

Models of the Atomic Nucleus

With Interactive Software

Springer

Professor Norman D. Cook
Department of Informatics
Kansai University
Takatsuki
Osaka 569-1095
Japan
E-mail: cook@res.kutc.kansai-u.ac.jp

Library of Congress Control Number: 2005934039

ISBN-10 3-540-28569-5 Springer Berlin Heidelberg New York
ISBN-13 978-3-540-28569-4 Springer Berlin Heidelberg New York

This work is subject to copyright. All rights are reserved, whether the whole or part of the material is concerned, specifically the rights of translation, reprinting, reuse of illustrations, recitation, broadcasting, reproduction on microfilm or in any other way, and storage in data banks. Duplication of this publication or parts thereof is permitted only under the provisions of the German Copyright Law of September 9, 1965, in its current version, and permission for use must always be obtained from Springer. Violations are liable for prosecution under the German Copyright Law.

Springer is a part of Springer Science+Business Media
springer.com
© Springer-Verlag Berlin Heidelberg 2006
Printed in The Netherlands

The use of general descriptive names, registered names, trademarks, etc. in this publication does not imply, even in the absence of a specific statement, that such names are exempt from the relevant protective laws and regulations and therefore free for general use.

Typesetting: by the author and TechBooks using a Springer LATEX macro package
Cover design: *design & production* GmbH, Heidelberg

Printed on acid-free paper   SPIN: 11427988   56/TechBooks   5 4 3 2 1 0

Preface

Models of the Atomic Nucleus is a largely non-technical introduction to nuclear theory – an attempt to explain one of the most important objects in natural science in a way that makes nuclear physics as comprehensible as chemistry or cell biology. Unlike most other scientific fields, the popularization of nuclear physics has not generally been successful because many fundamental issues remain controversial and, even after more than 70 years of study, a "unified theory" of nuclear structure has not yet been established. Nevertheless, despite some widely-acknowledged unfinished business, the theme developed in this book is that each of the many models in use in nuclear theory provides a partial perspective on the nucleus that can be integrated into a coherent whole within the framework of a *lattice* of nucleons. The proposed unification itself is not uncontroversial, but, at the very least, the lattice representation of nuclear structure illustrates how a complex physical object such as the nucleus can be understood as simultaneously exhibiting the properties of a gas, a liquid, a molecular cluster and a solid.

The book is divided into three parts. Part I (Chaps. 1–4) introduces the main topics that must be addressed in any discussion of nuclear physics. My intention has been to present a brief, but even-handed summary of the basic models that have been devised to explain nuclear phenomena and to review the theoretical ideas that have occupied the minds of theorists for many decades. These chapters are historical and are intended to convey what the major insights and conceptual challenges of nuclear physics have been thus far.

Part II (Chaps. 5–8) is a more detailed look at four topics that have not found satisfactory resolution within any of the current models of the nucleus – arguably the most basic nuclear properties that theory must eventually explain. These include topics concerning the size, shape and density of nuclei, and the related problem of the distance that nucleons travel in the nucleus before they interact with other nucleons (the so-called "mean-free-path" problem). Other issues discussed here are the nuclear force, the non-existence of the so-called super-heavy nuclei, and the phenomena of nuclear fission. In this part of the book, no attempt is made to resolve problems; on the contrary, I merely point to: (i) their existence, (ii) the long history of related debate, and (iii) the lack of any widely acknowledged solutions. An attempt at resolving these problems on the basis of a lattice model is the burden of Part III, but before we get to the stage of discussing solutions, we must be agreed that indeed there are problems that need to be solved.

Finally, Part III (Chaps. 9–10) addresses the questions of nuclear structure within the framework of a specific lattice model. If the lattice were simply one more model of the nucleus, the entire book would be nothing but another, incomplete and inherently controversial approach to nuclear phenomena, but the lattice – a relative latecomer to nuclear theory – is of special interest for two reasons. First, it can be readily demonstrated that the lattice reproduces the main features of the other, well-established nuclear models within it. Secondly, because it is a lattice with well-defined geometry, it allows for explicit visualization of nuclear structure through computer graphics.

As surprising as it may at first seem, the lattice model reproduces the properties that motivated the development of the gaseous-phase (shell), liquid-phase (liquid-drop) and molecule-like (cluster) models. Moreover, the strongest argument for focusing on a lattice approach to nuclear structure is that the lattice exhibits the known symmetries of quantum numbers that are an essential part of the conventional description of the nucleus. In brief, the known quantum mechanics of nucleon states has a (relatively simple) geometrical structure that lends itself to (relatively simple) graphical display, within which the diverse models of nuclear structure can also be found. It should be noted that the lattice model does not, in one easy swoop, solve all of the conceptual mysteries of quantum theory. Specifically, the particle/wave nature of all elementary particles and the fact that the nucleons themselves appear to have substructure remain unsolved problems at the subnuclear level, and these problems intrinsic to the nature of the nucleon persist in the lattice model. Be that as it may, the lattice view of nuclear structure does suggest how certain of the classic paradoxes in the realm of nuclear structure can be resolved in a coherent fashion at the level of nuclear structure theory.

The lattice model – its properties, symmetries and substructures – is discussed in detail in Chaps. 9 and 10. Whatever the final verdict on the model may eventually be, these chapters show that, while gaseous, liquid and cluster models have been on center-stage since the early days of nuclear physics, one specific lattice configuration shows a remarkable correspondence with nuclear properties, but was entirely overlooked until the 1970s. Explicating that model and showing how it is consistent with quantum mechanics is the main message of Chap. 9 and using the model to predict nuclear properties is the main topic of Chap. 10.

In the Appendix, the *Nuclear Visualization Software (NVS)* on CD-ROM is described. The graphics software is not by any means a complete explication of the nuclear realm, but it is arguably one gateway to a basic understanding insofar as it provides an intuitive, graphical display of quantum mechanics at the nuclear level and allows one to view nuclei in a manner consistent with the various nuclear models. Viewing modes that can be selected in the software include all of the important models of nuclear structure theory: (1) the shell model, (2) the liquid-drop model, (3) the alpha-cluster model, and two varieties of lattice model, [(4) the fcc model and (5) the scp model], as well as (6) the boson model and (7) the quark model. Nuclear physicists unfamiliar with the lattice approach may, on first exposure, find the visualization to be counter-intuitive, but closer examination will show that the visual displays of nuclei are consistent with experimental facts, and indeed allow for a rather straight-forward understanding of the known properties of nuclei at the low-energies of nuclear structure physics. For both research and teaching purposes, thinking in terms of a lattice has untapped potential that the *NVS* program may help to unleash.

In order to understand the core strengths of nuclear theory as well as the puzzles that remain, I urge all readers both to read the text summarizing what is known and debated about the nucleus, *and* to see how those issues lend themselves to visualization, as can be experienced using the software. In other words, the usefulness of the visualization of the nucleus is itself *not* an abstract argument, but rather is something that any reader can experience and evaluate in relation to the facts. The heuristic value of a graphical representation of the many models of nuclear structure is so obvious that I have no doubt that most readers will both enjoy the visual displays and learn more easily the essential concepts of nuclear physics while using the software, rather than through textbook discussions alone. The more difficult question concerning the possible unification of diverse nuclear models within the lattice approach, as discussed in Chaps. 9 and 10, is a theoretical issue that is likely to remain controversial for some time.

Several versions of the software tailored to various operating systems are on the enclosed CD-ROM – one for Linux-based computers, one for Windows computers, and one more for Macintosh computers. With access to any of these popular platforms, the reader can experience the visual message that lies behind the text. Finally, for those interested in graphics programming or in adding functionality to the present *NVS* program, the source code written entirely in C can also be found on the CD-ROM.

Philadelphia *Norman D. Cook*
September, 2005

Contents

Part I Fundamentals

1 Introduction 3
   1.1 The Essence of Nuclear Physics – Energy 3
   1.2 The Possible Unification of Nuclear Models 5

2 Atomic and Nuclear Physics 9
   2.1 The Atom 11
   2.2 The Nucleus 14
   2.3 Electron Shells in Atomic Physics 19
   2.4 Nucleon Shells in Nuclear Physics 27
   2.5 Summary 38

3 A Brief History of Nuclear Theory 41

4 Nuclear Models 55
   4.1 The Collective Models 56
   4.2 The Cluster Models 65
   4.3 The Independent-Particle Models 72
   4.4 Other Models 77
   4.5 Summary 84

Part II Long-Standing Problems

5 The Mean Free Path of Nucleons in Nuclei 87
   5.1 Avoiding the Issue 94
   5.2 The Persisting Problem of the MFP 97
   5.3 The Weisskopf Solution 108
   5.4 Exclusion Principle "Correlations" 114
   5.5 Further Doubts 118

	5.6	What is the Pauli Exclusion Principle? 121
	5.7	Summary 122
6	**The Nuclear Size and Shape** 123	
	6.1	The Nuclear Density 123
	6.2	The Nuclear Skin 135
	6.3	The Nuclear Radius 137
	6.4	Summary 139
7	**The Nuclear Force and Super-Heavy Nuclei** 141	
	7.1	The Nuclear Force 142
	7.2	Super-Heavy Nuclei? 144
	7.3	Summary 149
8	**Nuclear Fission** 151	
	8.1	Basic Facts of Fission 152
	8.2	The History of Nuclear Fission 158
	8.3	Textbook Treatment of Asymmetric Fission 159
	8.4	The Empirical Data on Fission Fragments 164
	8.5	Adjusting the Nuclear Potential-Well to Produce Asymmetry . 168
	8.6	What Needs to be Explained? 172
	8.7	Summary 174

Part III The Lattice Model

9	**The Lattice Model: Theoretical Issues** 177	
	9.1	The Independent-Particle Model Again 177
	9.2	Reproduction of the Independent-Particle Model in an fcc Lattice 182
	9.3	Symmetries of the Unit Cube of the fcc Lattice 194
	9.4	The Lattice-Gas Model 204
	9.5	Conclusions 209
10	**The Lattice Model: Experimental Issues** 211	
	10.1	Nuclear Size and Shape 214
	10.2	The Alpha-Particle Texture of Nuclei 222
	10.3	Nuclear Spin in the Lattice Model 225
	10.4	The Coulomb Force and Super-Heavy Nuclei 233
	10.5	Nuclear Binding Energies 235
	10.6	Fission of a Lattice 239
	10.7	Conclusions 246

Contents

A The "Nuclear Visualization Software" 247
    A.1 A Brief User's Manual 248
    A.2 Literature References to the Lattice Models 266
    A.3 Installation Notes 270
    A.4 Keyboard Shortcuts 271
    A.5 Keyboard Templates 272
    A.6 Nuclear Model Definitions 273

References 275

Name Index 283

Subject Index 287

Part I

Fundamentals

The technological developments of the modern world leave little room for skepticism concerning the validity of the basic discoveries of reductionist science. Lively debates concerning the ultimate meaning and the appropriate application of scientific discoveries are ongoing, but there is no doubt about what the fundamental pieces of physical reality are. In the realm of atomic physics, protons, neutrons and electrons have been well characterized, and their various properties are known with great precision. While it is universally accepted that quantum mechanics is the theory that best describes these particles, there is a surprising absence of consensus concerning how protons and neutrons interact with one another inside of stable nuclei. This book is concerned with the various ways in which the nucleus, as a complex system of nucleons, can be modeled. Many of the conceptual issues that nuclear physicists have dealt with over the past 100 years will therefore be discussed, but, before addressing the controversial issues, it is best to review what is well-known and fully agreed upon at the nuclear level. Only then will it be possible to understand the significance of the continuing debates and the possibility for resolution of old problems.

1
Introduction

1.1 The Essence of Nuclear Physics – Energy

The bottom line in the study of nuclear physics is energy, and it is for this reason that the field of nuclear physics is not a topic of irrelevant academic speculation, but is a matter of real-world concern. Because of its constructive and destructive power, issues concerning the release of energy from the nucleus are fundamental to both domestic and international politics – and this hard fact is not likely to change in the foreseeable future. From a purely intellectual perspective as well, nuclear physics has a strong claim to being a topic central to a scientific view of the world because it is the ultimate origin of most forms of *physical energy*. In lighting a candle, starting a car engine or switching on a computer, the source of energy is not obvious, but in nearly every case the fact that energy is available at all can be traced back to the extremely powerful nuclear events in the sun at some time in the history of our solar system. Prior to the 20th Century, the different types of energy – biological, chemical, mechanical or electrical – were studied in separate academic disciplines, but modern science has shown that the seemingly different forms of energy have common roots that lie in the field known as nuclear physics.

The general argument for thinking that nuclear physics is intellectually, practically and politically relevant is today almost a matter of common sense. And yet, despite its unparalleled importance, the study of nuclear physics has not in fact entered the mainstream curriculum at either the high school or the university level; it is still considered to be an area for specialists only – and most normally-intelligent people have come to dismiss the entire topic in the same way that "rocket science" was once considered to be beyond the reach of normal folk. Reinforcing this neglect of the nuclear realm, it is strange but true that most bookstores across America do not carry any books – popular or technical – on nuclear physics. There are of course many books on atomic physics, particle physics, and on what might be described as "quantum philosophy", but most bookstores and all of the major chain stores carry no books on the nucleus itself. None. Straight-forward descriptions of the physical object that is the cause of so much geopolitical anxiety and the source of one-fourth of our daily energy needs are not to be found.

The current neglect of nuclear physics in the education of most people worldwide is highly regrettable. Intellectually, without an understanding of nuclear physics, the core concept of physical energy cannot be properly understood and, practically, without the benefit of even a rough idea about the nucleus, the awesome power that is released in nuclear reactions has led to an uninformed skepticism and an emotional revulsion to all things nuclear – an attitude that is not warranted by the facts. While caution is of course required when applying nuclear technology, the growing energy needs of the world demand the cheap and potentially-clean energy that only the nucleus can provide. The disasters of Chernobyl, Hiroshima and Three Mile Island are useful reminders and permanent landmarks in nuclear history, but a future world that allows currently impoverished nations to participate in truly modern human civilization will require a massive increase in clean energy that only nuclear technology can provide. To arrive at that future, the anti-nuclear ideology that is currently fashionable across the political spectrum simply must be overcome.

The safety precautions and waste disposal issues that were once huge problems have of course been studied by nuclear engineers, and a direct comparison of the (economic, environmental and political) merits and demerits of all energy technologies shows the overwhelming case for pursuing safe nuclear energy. Most significantly, the radioactive-waste problems that came to be understood in the 1940s and 1950s have been largely resolved, and the prospect of a genuinely "green" nuclear technology is no longer a pipedream. Unfortunately, public opinion has already turned decidedly negative, and optimistic pronouncements from those within the nuclear power industry have had little effect. Despite the known and insoluble political and environmental problems inherent to coal- and oil-based sources of energy, the use of both coal and oil continues to grow, while nuclear reactors are no longer being designed and built in the U.S., Germany and Italy have outlawed such developments, Britain, Canada and New Zealand have declared moratoriums on nuclear energy, and almost everywhere truly democratic decision-making would mean a non-nuclear future – quite unrelated to the merits of the argument. The only alternative to an (undemocratic, energy-rich) nuclear future or a (democratic, but energy-poor) non-nuclear future lies in an educational revolution where not only the costs and benefits of various energy sources are properly understood, but also where there is an intellectual appreciation of the nucleus itself as a part of the natural world.

In brief, nuclear physics may be "as difficult as rocket science", but it is no more deserving of revulsion by the common man than is airplane travel. There are dangers inherent to air travel and dangers inherent to tapping the power of the nucleus, but they are known and worth confronting for the benefits they can bring. In spite of its currently ambiguous status, the field of nuclear physics is not an occult brotherhood in need of hiding from public scrutiny nor an elite club for geniuses only. And, despite some loose ends still in need of further study, the basics of nuclear theory do not require a PhD in mathematics to understand and are not beyond the intellectual reach of many high school students. Moreover, contrary to some popular misconceptions, an understanding of nuclear structure does not demand that we sort out the perennial philosophical debates concerning the directionality of time, the meaning of electrostatic charge, or the interpretation of the uncertainty relations. To be sure, those are fascinating metaphysical topics worthy of study, but the atomic nucleus is a physical object in the natural world and can be understood in a manner that basic concepts in chemistry, biology and brain science are widely understood by the intelligent layman.

Today, education concerning nuclear physics is in an appalling state. On the one hand, discussion of the politics of nuclear issues is to be heard almost daily on the evening news. On the other hand, a visit to any typical bookstore in the US will reveal a total lack of interest in this academic field. Even university bookstores rarely have any texts on the nucleus, and, if you find a book on nuclear physics, check the copyright date! Books written in the 1950s or 1960s and reprinted *in unaltered form* are on offer to students in the 21st Century! (Today in the bookstores, hot off the press: Blatt and Weisskopf, *Theoretical Nuclear Physics*, 1952; Landau and Smorodinsky, *Lectures on Nuclear Theory*, 1959; de-Shalit and Talmi, *Nuclear Shell Theory*, 1963. Of course, specialist texts continue to be written, but they are not generally available in the bookstores.)

1.2 The Possible Unification of Nuclear Models

The curious state of nuclear physics today is a result of many complex factors, political as well as educational, but "curious" it is. It can be summarized briefly by noting that a great deal is known about the technology of nuclear energy, and yet our understanding of the nucleus itself is seemingly quite incomplete. More than 30 (!) nuclear models – based on strikingly different assumptions – are currently employed (Greiner & Maruhn, 1998). Each provides some insight into nuclear structure or dynamics, but none can claim to be more than a partial truth, often in conflict with the partial truths offered by other models.

The strengths and weaknesses of the nuclear models are of course acknowledged and debated in the specialist journals, but they are not often directly compared in order to highlight the contradictions they embody. Instead, their diverse starting assumptions are stated and explained, but justified simply by the fact that the different models are indeed all useful and, in total, contribute to the astounding successes of nuclear technology. As a consequence, the opinion that these contradictory models are somehow "complementary" is now the orthodox view in nuclear theory. Perhaps reluctantly, most physicists have come to the conclusion that, through the use of a variety of models, the diverse topics within nuclear structure physics have been elucidated as thoroughly as the human mind is capable of doing. Given the wealth and precision of experimental data on virtually every stable and unstable nucleus, and given the capability of explaining most of the data within one model or another, there is little expectation that a rethinking of nuclear theory is necessary. That is the current state of nuclear structure theory.

It is rather disconcerting, however, to discover that some of the remaining unanswered questions include truly basic issues, such as the phase state of nuclear matter (Is it a liquid, a gas, or a solid?), the nature of the nuclear force (Is it strong with effects extending only to nearest-neighbor nucleons, or is it weak and extending long-range to all nucleons in the nucleus?), and the nature of the nucleons themselves (Are they probability waves, point-particles or space-occupying objects?). Clearly, these are *not* minor issues, but the different models of nuclear theory require the use of vastly different model parameters that explicitly or implicitly assume the nucleus, the nucleon and the nuclear force to have certain characteristics, but not others.

In spite of the fact that these basic questions in nuclear theory have not been definitively answered, the focus of most theoretical work is now at the sub-nuclear level, i.e., particle physics – where it is hoped that extremely high-energy experimental work will provide solutions concerning the nuclear force and, tangentially, resolve old problems in nuclear theory. Physicists who have remained within (relatively low-energy) nuclear structure physics now work primarily on exotic nuclei – nuclei with extreme properties and, invariably, very short half-lives. Whatever the reality may be regarding those second-order issues, the traditional models in nuclear structure physics were devised to explain the first-order issues – nuclear sizes, spins, binding energies, fission dynamics, etc., of the stable and semi-stable nuclei. That is where significant problems stubbornly remain, and where answers are still needed – before we can hope to understand the short-lived exotic nuclei. While only future research will reveal what extremely high-energy particle physics might contribute to the low-energy issues of nuclear physics, there are certain topics in nuclear structure theory that need to be elucidated at the level of nuclear structure *per se*.

The view propounded here is that the core assumptions of each of the dominant models in nuclear structure theory are indeed correct. Alpha-clustering of nucleons in the nuclear interior and on the nuclear surface has been experimentally shown and lies at the heart of the cluster models. At the same time, the reality of nearest-neighbor interactions among all bound nucleons is the fundamental assumption of the liquid-drop model, and this model is essential for explanation of the experimental data on nuclear densities, nuclear radii and nuclear binding energies. Simultaneously, the quantum mechanical description of each nucleon based on the Schrödinger wave-equation is the essence of the (gaseous-phase) independent-particle model. The unique quantal state of each nucleon as described in this model is the basis for an understanding of other nuclear properties – most impressively, the nuclear spins of all 2000+ known isotopes and their many thousands of excited states.

As contradictory as it may first seem, each of these three main classes of cluster, liquid and gaseous model has demonstrable strengths. Unfortunately, theoretical elaboration of any one of these models on its own leads to blatant contradictions with the other models, so that a final decision concerning which model alone is the one-and-only correct view simply cannot be made. What I argue in the following chapters is that a lattice model contains within it the essential properties of these models of the nuclear realm, and can be shown to reproduce the semi-solid cluster structures, the liquid-state nearest-neighbor effects and the gaseous-state quantal description of nuclei that have made these diverse models so useful. In reproducing these properties, I claim that the lattice constitutes the basis for a unified view that has the strengths of the traditional models, but does not insist on the paradoxical properties, contradictory ideas or unrealistic parameters that have been the source of countless theoretical headaches for over 50 years.

Final pronouncements about the necessity of multiple models and the role of the lattice in nuclear structure theory may not yet be possible, but already several conclusions can be made about the uneven successes of nuclear modeling over the past half century. The first, with which most nuclear physicists would agree, is that fundamental problems have remained unsolved for many years and that nuclear structure theory has not yet reached a successful completion. The capabilities of the nuclear power industry reflect the strengths of a sophisticated and mature technology, but theoretical nuclear physics is not by any means a "closed chapter" of scientific endeavor, and does not approach the coherence and unanimity of opinion that has already been achieved in, for example, chemistry or molecular biology.

The second conclusion, that many physicists would perhaps prefer *not* to acknowledge, but that is fully justified from the published literature, is that several core problems – problems that, historically speaking, were once at the center of theoretical debates – have come to be treated as rather minor issues or have been elevated to the dubious status of "inevitable paradoxes" in the quantum world. Instead of frank discussion of an incomplete understanding, these problem areas are given cursory mention in textbooks, and often-times dismissed altogether as historical curiosities despite the fact that theoretical and experimental efforts have not resolved the basic issues.

The third conclusion is that the lattice model of nuclear structure provides a new means to address and perhaps resolve these old issues in nuclear theory. What I consider to be modest, but real successes in that direction are presented in the final chapters. Discussion of the unsolved problems and neglected topics in nuclear structure theory is necessarily controversial, but the lattice model unambiguously achieves an unprecedented visualization of nuclear phenomena within which the various viewpoints on nuclear structure and dynamics can literally be seen. The potentially-unifying framework provided by the lattice suggests that the "multiple models" of nuclear physics are, contrary to current opinion, not an insoluble paradox inherent to the nuclear realm, but rather a temporary stage in the evolution of nuclear theory. By reconfiguring the entire discussion of nuclear structure within the framework of a specific, dynamic lattice, the century-old paradoxes of quantum mechanics do not automatically disappear, but they do remain confined to our understanding of the individual particles (inherently probabilistic wave/particle entities), while nuclear structure itself becomes somewhat less mysterious.

Finally, before entering into the maze of nuclear theory, let me state what I believe is the essential starting point for study in this field. The theoretical and experimental efforts that have led to the elucidation of the atomic nucleus and the nature of physical energy in the universe are, in total, perhaps as great as any intellectual endeavor in the history of human thought. For those of us who have come after the major developments in both atomic and nuclear physics, serious study is required simply to understand what has already been accomplished, and a real appreciation can be had only by active participation in the deciphering of this complex many-body system. While some genuine humility is therefore appropriate when reviewing the field of nuclear theory, it is essential to distinguish between real accomplishments and unresolved problems, between proven facts and seductive suggestions, and to be unafraid of calling a joker "a joker" when we examine the cards! If our central concern is to achieve an understanding of the complex physical system that is the ultimate source of virtually all physical energy, the issues of academic egos, professorial personality quirks and university politics should not divert us from our main intellectual task. Unquestionably, the quantum level presents new and difficult challenges, but it is a mistake to throw controversial interpretations of the uncertainty principle at every problem that arises and declare that paradox is the final answer. Particularly for a system that has been studied with as much care, precision and ingenuity as the atomic nucleus, there is no reason to declare that progress has already come to an end. On the contrary, the remaining problems in nuclear theory should be pursued until puzzles are solved, enigmas unraveled and paradoxes no longer paradoxical. Only then will we be able to declare that we understand the atomic nucleus.

2
Atomic and Nuclear Physics

Two important areas where the concepts of quantum theory have been successfully applied are atomic physics (concerned primarily with the interactions between nuclei and electrons) and low-energy nuclear physics (concerned primarily with the interactions among protons and neutrons). Despite some strong theoretical connections, the practical issues in atomic and nuclear physics can be dealt with separately because the amount of energy required to bring about nuclear effects is generally many thousands of times greater than that required for electronic effects. For this reason, *atomic* physicists can work in an energy range where nuclear reactions do not occur, and *nuclear* physicists can work in an energy range where electron reactions occur, but are usually negligible in comparison with the much stronger nuclear phenomena. This separation of electronic and nuclear effects means that atomic and nuclear physicists are normally housed in different university departments, but there are, nevertheless, many conceptual links and similarities in the theoretical techniques employed in both realms.

Most importantly, in both nuclear and atomic physics, quantum mechanics is the theoretical framework for most quantitative work. Quantum mechanics is, above all else, a theory of the discreteness of physical quantities. Whereas classical mechanics assumed that all quantities of mass and energy are continuous, experimental findings obtained near the end of the 19th Century suggested that, at the atomic level, there is quantization of some physical quantities into integral units, and quantities in-between do not occur. The theoretical efforts that followed the first experimental indications of discrete units of mass and energy led eventually to atomic theory with quantum mechanics at its heart – a comprehensive, self-consistent theory of the microphysical world, where indeed quantal jumps and a certain discreteness of physical quantities are the general rule. The development of that theory was an unanticipated conceptual revolution at the turn of the century, but is today fully established. Although the discreteness of the physical world at the quantal level is not evident at the macroscopic level of classical physics, both classical macroscopic physics and quantal microscopic physics are correct, when applied to their respective realms. Today, it is no longer a controversial issue that both kinds of physics have their own applications, formulated in somewhat different terms. Both are correct in the very real sense that both can be used to understand natural phenomena and to predict physical events in the material world.

The validity of classical mechanics is demonstrated again and again every time a rocket is launched, a skateboarder wiggles down a sidewalk or a 50 story building refuses to come tumbling down. Similarly, classical electromagnetic theory may not be "easy," but there are no more macroscopic electromagnetic mysteries; it is a known world and, as a consequence, heavily exploited in countless practical ways. When we arrive at the microscopic realm of quantum mechanics, however, we run into a world where much is known, but little is understood. At least, quantum theory is not "understood" in the concrete way that we understand that a brick dropped on a toe will hurt more than a wad of cotton. Within the framework of classical mechanics, an analysis of force, mass and acceleration of the brick and the cotton wad would lead to technical conclusions that fully support our common sense. But in quantum mechanics, technical conclusions stand alone, and loose analogies and imperfect models based on common sense notions are never more than suggestions. The visceral understanding that we normally have with everyday objects is simply missing in the quantum world.

It is this *lack* of a visceral understanding of quantum mechanics that makes it an interesting and difficult intellectual puzzle and why it is the source of endless speculation along the lines of "What does quantum mechanics mean for our understanding of the universe, consciousness or human existence?" In contrast, there is really nothing of interest to discuss about the workings of a classical clockwork mechanism. It works, and the cause-and-effect throughout the entire system can be traced to whatever level of precision we wish to pursue. But quantum mechanics is different. The concepts of causality that we know (and feel in our bones!) from classical mechanics don't seem to work at the quantal level. We are left in an abstract cerebral world with its own logic and rules, but with only weak connections to familiar dynamics, gut feelings, and common-sense cause-and-effect.

The puzzles of the atomic level motivated a huge intellectual effort in the philosophy of science over the entire 20th Century. Since quantal systems – atoms or nuclei – cannot be individually measured nearly all knowledge of these small systems is obtained from experiments dealing with huge numbers of similar systems. This makes all of quantum mechanics inherently probabilistic. Our knowledge and intuitions from classical physics might still sometimes be relevant to aspects of the world of quantum physics, but we do not know how quantal causality will relate to classical causality. This dilemma has not yet been resolved by philosophical discussion over the course of a century – with illustrious figures such as Bohr and Einstein defending diametrically opposing views. Despite the reality of certain unresolved, perhaps unsolvable, issues in metaphysics, we can nonetheless expect to find logical consistency within the quantum world, and for that reason it is essential to study both the atomic and the nuclear realms where the ideas of quantum physics have been applied with success. Let us begin with a brief review of electron physics and the issues of atomic structure, and then examine the related, but more controversial nuclear realm.

2.1 The Atom

Some modern textbooks treat quantum theory as a recent revolution in scientific thinking, but that is simply no longer the case. For more than five generations of scientific inquiry, it has been known that the basic building block of all matter on earth is the atom; and for almost as long, the quantal nature of atomic phenomena has been studied.

The atom contains a centrally-located nucleus and a covering of electron "clouds". Although more than 99% of the atomic *volume* is due to the electron clouds, more than 99% of the atomic *mass* is contained in the protons and neutrons (nucleons) locked inside the nucleus. The density of matter in the nucleus and in the electron clouds is thus very different, but the nucleus and the electron periphery also differ with regard to their electric charges. Each electron contains one unit of negative charge, whereas the nucleus contains a number of positive charges (from 1 to more than 100) equal to the total number of protons. The housing of so much positive charge in the small volume of the nucleus already tells us one important fact about the nuclear realm: the forces that hold those charges together, despite their mutual electrostatic repulsion, must be quite strong. This is the so-called "strong force" or the "nuclear force."

The idea that all matter might be built from elementary subunits has a long history, but progress in understanding the reality of atomic substructure began in the 19th Century with the gradual systematization of chemical knowledge. As more and more pure substances were isolated, laws and regularities of chemical reactions became apparent, thus setting the stage for turning the art of chemistry into the science of atomic physics. The first major theoretical breakthrough came with the proposal of the Periodic Table of Elements by Mendeleev in 1849. While several blanks in the chart remained unfilled and all questions concerning why such periodicity would exist went unanswered, the sequence of elements of increasing mass and the curious periodicity of some important properties were indication that there exists a finite number of physical laws that underlie the huge diversity of chemical phenomena. The importance of the Periodic Table can hardly be overstated, for it brought considerable order to the world of chemistry. Even before an understanding of the structure of the atom had been achieved, the realization that chemical compounds were themselves the result of combinations of a very small number of elements meant that the seemingly endless combinatorial complexity of chemistry was a finite, and therefore decipherable, problem.

The periodicity of the properties of the elements is evident on many grounds (discussed below) and this unambiguous orderliness remains the undisputed bedrock of all knowledge in chemistry. Although the structure of the atom remained unclear for another half-century, the reality of elementary atomic units was widely acknowledged in the mid-1800s in light of the periodicity of the Periodic Table. With the discovery of the negatively charged electron in 1893 and the evidence for a centrally-located, positively-charged nucleus in 1911, the basic conceptual framework for modern atomic theory was completed and, with time, universally accepted.

Once the Periodic Table had been established during the latter half of the 19th Century, the task remained to explain why the periodicities arose at atoms containing certain numbers of electrons, and not others. The most important periodicity was the existence of inert gases at the conclusion of each row of the Periodic Table. That is, when the total number of electrons in an atom was 2, 10, 18, 36, 54 or 86, the atom was found to be chemically unreactive and to show little or no tendency to combine with other atoms to form molecules (the noble gases). In contrast, those atoms containing one more or one less electron than any of the noble gases were highly reactive and readily combined with other atoms. To explain the numbers at which these properties were found – and the other regularities within the Periodic Table, a planetary atomic model was proposed by Niels Bohr in 1913 and was eventually given a formal grounding in the development of quantum mechanics.

The first three decades of the 20th Century, during which the basic theoretical developments in quantum theory were made, constitute the "Golden Age" of atomic physics. This era saw the emergence of a mathematical formulation of the energy states of atoms based on Schrödinger's wave-equation. Several profound controversies concerning the correct *interpretation* of this mathematical formalism arose in the early years of the 20th Century (and have not subsequently disappeared), but the undeniable success of quantum theory was that it allowed for predictions of extreme accuracy regarding the release and absorption of light energy. Particularly during the early years of the 20th Century, most physicists were laboratory experimentalists rather than fulltime theorists, so that a theory that had clear implications for what could be measured in the laboratory was understandably attractive. Theoretical coherency and intellectual beauty may also be highly valued, but above all else a theory must be useful.

The Schrödinger wave equation soon became the basic mathematical tool with which to explain atomic (electron) phenomena. Particularly when many-electron systems are to be considered, various simplifying models must be employed to reduce the complexity of the computations, but there are no competing theories of the atom that are not based on the wave equation of quantum theory. Thus far, the greatest quantitative successes have been concerned with one-electron systems and one-valence-electron systems, but the analytic methods developed for such relatively simple atoms and ions have gradually been generalized, and are now thought to apply to all electron systems of whatever size.

The quantum revolution followed from experimental work by chemists on the light spectra observable when pure elements are heated. Among many such properties, the absorption and release of photons of certain wavelengths was detected. The first essential steps in understanding the regularity of the light spectra of the simplest atom, hydrogen, were taken by Balmer in 1884. That work was later summarized using an *ad hoc* formula devised by Rydberg (1890):

$$\lambda = c/R(1/n^2 - 1/m^2) \qquad (2.1)$$

where lambda, λ is the wavelength of the photons in the light spectrum, c is the speed of light, R is a constant, and n and m are small integers ($n = 1, 2, 3, 4, \ldots; m = n+1, n+2, n+3, \ldots$). The formulae from Balmer and Rydberg are most remarkable in their use of the integers n and m – which indicate that, amidst the plethora of photons emitted by excited atoms, the released light energy is "quantized" in a highly regular manner. Not any amount of energy is possible, but rather only certain distinct wavelengths are found. For atoms with a nuclear charge of more than one, a different Rydberg constant, R, must be used, but the quantized nature of the allowed wavelengths of photons is again found. This theoretical discovery was the beginning of the quantum

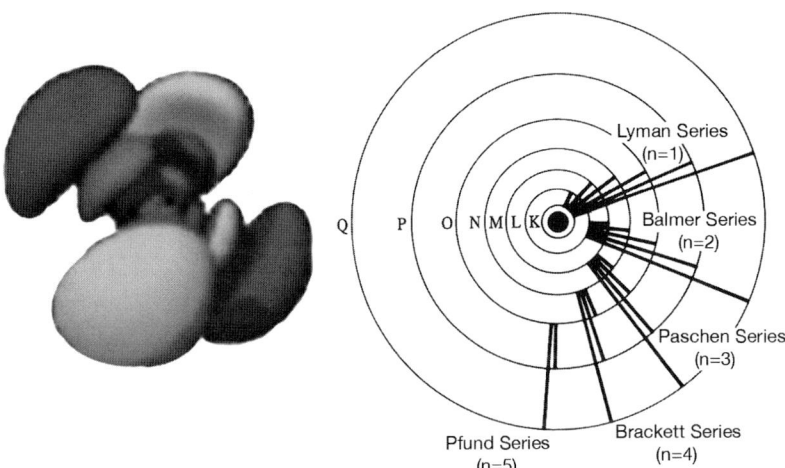

Fig. 2.1. Cartoons of an atom, with a central nucleus and surrounding electron orbitals. The spatial configuration of the $n = 5, l = 2, m = 1$ electron orbitals is shown on the left. The energy shells (K through Q) are shown on the right. The light energy emitted by transitions of electrons from one orbital to another has a quantal regularity neatly summarized by the Rydberg formula (2.1)

If it were just photon energies emitted by electron transitions that can be described by such formulae, then nuclear physics might be as distinct from electron physics as electron physics is from cell biology. On the contrary, however, similarly discrete energy states were found experimentally for the nucleus, and the entire mathematical apparatus devised to explain the phenomena of electron physics was eventually imported into nuclear physics.

2.2 The Nucleus

Developments in chemistry and then atomic physics were rapid throughout the latter half of the 19th Century and the first three decades of the 20th Century. During those years, the basic ideas of atomic and molecular physics were established, and – despite some vociferous objections from the giants of the previous generation of physicists – the quantum revolution was victorious. As valid as classical physics was in its own realm, the phenomena of quantum physics were real and could not be "explained away" on the basis of classical ideas.

Nuclear physics, in contrast, began somewhat later than atomic physics – dating essentially from the discovery of the centrally-located nucleus by Rutherford in 1911. Modern "nuclear structure physics" did not begin, however, until the discovery of the neutron by Chadwick in 1932. Once the stable proton-neutron constituents of the nucleus became known, progress in nuclear theory was also rapid and the "Golden Age" of nuclear physics continued for about two decades until the mid-1950s. During that time the dominant phenomena of the nucleus were discovered and the theoretical quandaries of the nuclear realm were first revealed, pondered and debated. A summary of the major events in the study of nuclear physics is shown in Table 2.1.

Subsequent to 1953, there have been many theoretical and experimental advances, but the main insights and discoveries on nuclear structure occurred mostly during the Golden Age. The experimental and theoretical efforts since the 1950s have been of two main types: One concerns the constituent particles within the nucleus (particle physics) and has proceeded in the direction of higher and higher energies in the hope of fully characterizing the elementary constituents of the atom. The second type has been concerned primarily with nuclei as many-particle systems – and has led to refinements on the liquid-drop model, the alpha-particle model, the shell model, the collective model and their variants. Dealing primarily with low-energy nuclear systems, this field is generally referred to as "nuclear structure theory" proper.

Before discussing the theoretical models employed in nuclear structure theory, the basic empirical facts about which there is complete consensus should be stated. First of all, nuclei as physical systems are *small* in two distinct meanings. All nuclei contain less than 300 constituent nucleons – quite unlike a physical system such as the biological cell containing many trillions of constituent molecules. Nuclei are also small in terms of their spatial

2.2 The Nucleus 15

Table 2.1. The Chronology of the Major Events in Nuclear Physics*

When	What	Who
1896	Discovery of radioactivity	Becquerel
1909	Alpha particle shown to be He nucleus	Rutherford & Royds
1911	Nuclear atomic model	Rutherford
1913	Planetary atomic model	N. Bohr
1919	Artificial nuclear transmutations	Rutherford
1926	Quantum mechanics	Schrödinger et al.
1928	Theory of alpha radioactivity	Gamow, Gurney & Condon
1932	Discovery of neutron	Chadwick
1932	Proton-neutron nuclear model	Heisenberg
1934	Theory of beta radioactivity	Fermi
1935	Meson hypothesis	Yukawa
1936	Compound nucleus model proposed	N. Bohr
1938	Discovery of nuclear fission	Hahn & Strassmann
1939	Liquid-drop model of fission	N. Bohr & Wheeler
1940	Production of first transuranium element	McMillan & Seaborg
1942	First controlled fission reactor	Fermi
1947	Discovery of pi-meson	Powell
1949	Shell model of nuclear structure	Mayer, Jensen, Haxel & Suess
1953	Collective model of nuclear structure	A. Bohr, Mottelson & Rainwater

* modified from Krane (1988)

extent: about $2 \sim 12$ 10^{-15} meters (femtometer or fermi, fm) in diameter. The tiny size of the nucleus means that the nucleus is invisible and lies well beyond the resolving power of even the strongest microscopes. That one fact reveals much about the world of nuclear physics: the basic object of study has never been seen and never will be! Nevertheless, it is important to emphasize that, unlike some of the theoretical concepts of economics or psychology, and unlike some of the more exotic theoretical entities of particle physics, there is no doubt about the physical existence and the physical properties of the atomic nucleus. It is a material object that has mass and spatial extent, stable and quantifiable electric and magnetic properties, and that displays several distinctly quantum mechanical properties, as well. Because of the extremely small size of the nucleus, however, we cannot see it and can learn about it only indirectly in experiments involving huge numbers of similar particles.

The individual nucleus is so small and so widely separated from its neighboring nuclei that if we could enlarge one nucleus enough to place it before us on a table – let's say, about the size of a toy marble – then a nearest-neighbor nucleus would be a half mile away. Even though the nuclei of neighboring atoms are – relative to their own dimensions – far from one another, a typical block of solid matter the size of that marble contains about 10^{15} atoms (or nuclei). Clearly, we are dealing with a microworld that is very different from the kinds of objects that we normally think about, but with the appropriate rescaling we can come to some understanding of what the nucleus is and how it is constructed.

Within nuclei, there are protons and neutrons – that is, the two types of fundamental nuclear constituents, collectively referred to as nucleons. A proton is a stable particle with a positive charge that does not break down into more fundamental pieces – at least over many millions of years. A neutron, on the other hand, is electrostatically neutral and will break into a proton, an electron and a neutrino within 14 minutes when it is emitted on its own from a nucleus. Whether or not neutrons are stable within nuclei remains uncertain. Either they are stable under such conditions and maintain their identities as neutrons or, what is thought to be more likely, they continually transform into protons, through the exchange of mesons, and then retransform into neutrons again. In either case, the mass of any atom is more than 99.9% due to its nucleons. In a word, the study of nuclear physics is, first and foremost, the study of protons and neutrons and their interactions.

A variety of other particles have transient existences within nuclei – most notably mesons, photons and neutrinos – all of which are involved in the interactions among the fermions (protons, neutrons and electrons). As important as these other particles are for understanding the forces holding atoms together, the characterization of atoms does not demand that we have precise knowledge about their numbers. They are the "glue" that holds the principal particles, the fermions, together. In contrast, the fermions themselves determine the nature of individual atoms: their numbers tell us quite explicitly what kind of atom we are dealing with. Specifically, the number of protons tells us the element (Is it an atom of carbon, copper, uranium or some other element?). The number of neutrons tells us the isotope (Is it a stable nucleus with an appropriate balance of protons and neutrons, or is it unstable, with too few or too many neutrons relative to the number of protons?). Together, knowledge of the numbers of protons and neutrons is sufficient for predicting the main properties of any atom (i.e., How many electrons can it hold and will the nucleus itself gain or lose electrostatic charge and transform itself into another element?). Finally, outside of the nucleus, the number of electrons relative to the number of nuclear protons informs us of the ionic state of the atom, and therefore its chemical reactivity. It is important to note that, as real as the other atomic particles are, the numbers of mesons, photons, neutrinos, etc. do not play a role in the initial, first-order description of atoms (Table 2.2). As a consequence, theories of nuclear structure are universally

Table 2.2. What the Numbers of Atomic Particles Tell Us about an Atom

The number of:		determines the:	
	protons		element
	neutrons		isotope
	electrons		ion
	mesons		–
	photons		–
	neutrinos		–

concerned with the relationships between protons and neutrons – mediated, to be sure, by other particles, but not dependent on their precise numbers.

The "particleness" of protons and neutrons is, however, not always evident – and this is a major conceptual difficulty that lacks an easy solution. Being particles that obey the quantum mechanical rules of 1/2-spin fermions, they take on all the weirdness of particle-waves – for which quantum mechanics is notorious. Sometimes a nucleon will exhibit properties that are indicative of a finite, quasi-classical, space-occupying particle and sometimes it appears to be a wave of an inherently "fuzzy" probabilistic nature. This particle-wave duality is a deep philosophical quandary in quantum physics and cannot be resolved by argumentation in the realm of nuclear structure theory, but it is essential to realize that both particle and wave descriptions of nucleons and nuclei are important. For some issues in nuclear physics, the finite size and individual "particleness" of the nucleons are relevant for an understanding of nuclear structure. At other times, a probabilistic view of nuclei and their constituents is appropriate, particularly for understanding nuclear reactions. In contrast, other than the fact that electrons in atoms are always present in integral numbers, the "particleness" of electrons is not normally an issue in atomic physics: if an electron is present, it manifests itself as a diffuse, inherently probabilistic wave and only under contrived laboratory situations does its "particleness" become apparent. Although both realms are essentially quantum mechanical, only in the nuclear realm must we sometimes deal explicitly with the particle qualities of its constituents.

Both forms of the nucleon have definite physical dimensions. Their electrostatic and magnetic radii have been experimentally measured, and are found to be similar to one another. Specifically, the so-called root-mean-square (RMS) electrostatic and magnetic radii of both protons and neutrons are 0.86 fm. As will be discussed in detail in later chapters, nuclei themselves are known to have charge radii ranging between one and six fm. They have an approximately constant density core, outside of which the density gradually falls to zero over a skin region of 2 or 3 fm. These basic facts can be expressed by means of the so-called Fermi density curve shown in Fig. 2.2 and provide a rough idea about nuclear structure.

In approximate terms, the nuclear volume for a large nucleus such as ^{208}Pb containing 208 nucleons can be calculated from the experimentally-known RMS charge radius (5.5 fm): Since the volume of a sphere is $4/3\pi r^3$, that for the lead nucleus is approximately 697 fm^3. Within this nuclear volume, the 208 nucleons (each with a 0.86 fm radius and a 2.66 fm^3 volume) occupy about 554 fm^3. A more careful examination of such numbers will be made in Chap. 6, but this rough estimate of nuclear density indicates that nucleons make up the bulk of the nuclear volume.

The results of the calculation of nuclear density contrast sharply with a similar estimate of the amount of space occupied by electrons in a typical atom. The radius of the lone electron, when considered as a particle, is calculated to be similar to that of the nucleons, i.e., a radius of $\sim$2.8 fm. However,

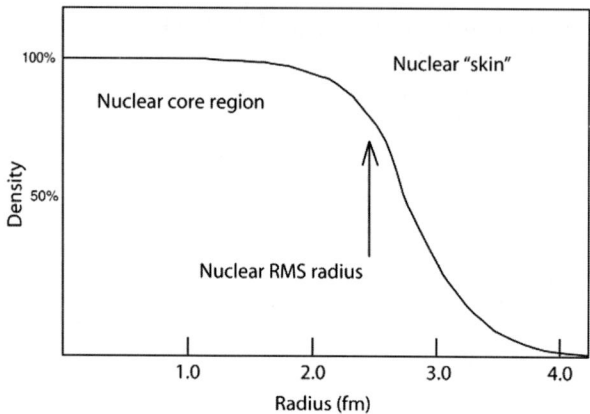

Fig. 2.2. The Fermi curve used to approximate the nuclear texture. The experimental root-mean-square (RMS) charge radius and indications of a constant-density core provide a rough picture of a medium-sized nucleus. Because of the diffuseness of the nuclear surface, the definition of the nuclear "size" is somewhat ambiguous. Neutrons are thought to occupy approximately the same space as protons, so the mass radius follows more-or-less the same distribution as the charge radius

because the electrons are spread over a volume of about one cubic Angstrom, even a relatively electron-dense atom, such as lead, will be only 1/100,000th occupied by the particulate matter of its electrons. If thought of in terms of discrete particles, atoms truly are mostly empty space, but this is not true of the nucleus.

Given that there is a relatively large number of nucleons packed into a relatively small nuclear volume, the next question concerns the spatial relationships of the nucleons to one another within a nucleus, and here our troubles begin. Simply to answer the question whether the nucleus is (i) a diffuse gas of nucleons in rapid, chaotic motion relative to one another, or (ii) a dense liquid of nucleons in slower motion, but interacting with near-neighbors, or (iii) a solid of nucleons locked into definite positions relative to one another, there are not three, but four major hypotheses. They are: (i) the gaseous-phase (shell or independent-particle) model, (ii) the liquid-phase (liquid-drop or collective) model, (iii) the molecule-like, semi-solid-phase (alpha particle or cluster) models, and (iv) several solid-phase lattice models. We have now entered the realm of problems, paradoxes and multiple models for which nuclear physics is notorious.

The topic of nuclear structure is the focus of the chapters that follow, but it is helpful to draw parallels and contrasts with the atomic realm, wherever possible. For that reason, let us begin by examining some of the most basic issues concerning the electron structure of the atom before addressing problems concerning the structure of the nucleus.

2.3 Electron Shells in Atomic Physics

The Schrödinger wave-equation is the basis for the quantum mechanical description of a particle responding to an external force. It can be expressed in the following form:
$$\Psi_{n,l,m} = R_{n,l}(r) \; Y_{m,l}(\phi, \theta) \qquad (2.2)$$
The wave-equation indicates the likelihood (Ψ) of finding an electron at a given distance (R) and orientation (Y) relative to the nucleus located at the center of the atomic system. In a spherical coordinate system, the distance from the nucleus is specified by the variable r, whereas two angles, ϕ and θ, are needed to specify its angular position (Fig. 2.3). At this level, quantum mechanics is nothing more than solid geometry and is not particularly "quantal" since all values of r, ϕ and θ are allowed. As noted in (2.2), however, both the radial and the angular parts of the Schrödinger equation have subscripts that are explicitly "quantal." That is, the non-continuous quantization of electron states is expressed in the radial and angular functions using integer subscripts, n, l and m; all non-integer values are forbidden.

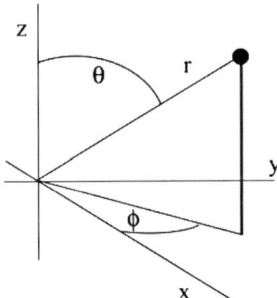

Fig. 2.3. The spherical coordinate system. Conversion between spherical and Cartesian coordinates is: $x = r \sin \theta \cos \phi, y = r \sin \theta \sin \phi, z = r \cos \theta$

The relationships among these subscripts can be derived from the basic principles of quantum mechanics and simply state what possible combinations of quantum numbers, n, l, m, are allowed. They can be summarized as follows:

$$n = 1, 2, 3, \ldots; \quad l = 0, \ldots, n-2, n-1; \quad m = -l, -l+1, \ldots, 0, \ldots, l+1, l$$

For electrons, these energy states can be tabulated, as shown in Table 2.3. Since each electron also has an intrinsic property of spin, s (unrelated to its position within the atomic system), four quantum numbers (n, l, m, s) uniquely identify the electrons in any atomic system.

The detailed 3D geometry of the angular part of the wave equation, denoted by Y, is defined by the so-called spherical Bessel functions and becomes geometrically complex as the subscript values increase (Fig. 2.4). The

Table 2.3. The full set of allowed combinations of quantum values n, l and m for $n < 5$ (after Herzberg, 1937, p. 45). Each combination of nlm corresponds to an energy state in which two electrons with opposite spin ($s = \pm 1$) can exist

n	1	2		3			4			
l	0	0	1	0	1	2	0	1	2	3
m	0	0	−1 0 +1	0	−1 0 +1	−2 −1 0 +1 +2	0	−1 0 +1	−2 −1 0 +1 +2	−3 −2 −1 0 +1 +2 +3

willingness to deal with the mathematical equations describing these strange shapes is perhaps the crucial factor that separates future atomic scientists from the rest of humanity, but the strangeness of the quantum world does *not* lie in the geometry of the Bessel functions, but rather in the integer subscripts. Regardless of the spatial complexity of the electron probability clouds, the important point is that – due to the relationships among the allowed combinations of the subscripts – the energy levels for any given electron are discrete. The simplest energy shells ($l = 0$) are spherically symmetrical, while the shells with larger l and m values become convoluted and disconnected 3D balloons (Fig. 2.4).

The filling of any given energy level is restricted to a maximum of two electrons, as a consequence of the Pauli exclusion principle. In effect, any state, as defined by subscripts n, l and m contains a known number of electrons. Whenever the maximum number of electrons for an l-level has been reached, a "shell" of electrons has been filled, and the next electron must enter the next higher shell. As seen in Table 2.4, the closing of various shells neatly explains the existence of the inert gases.

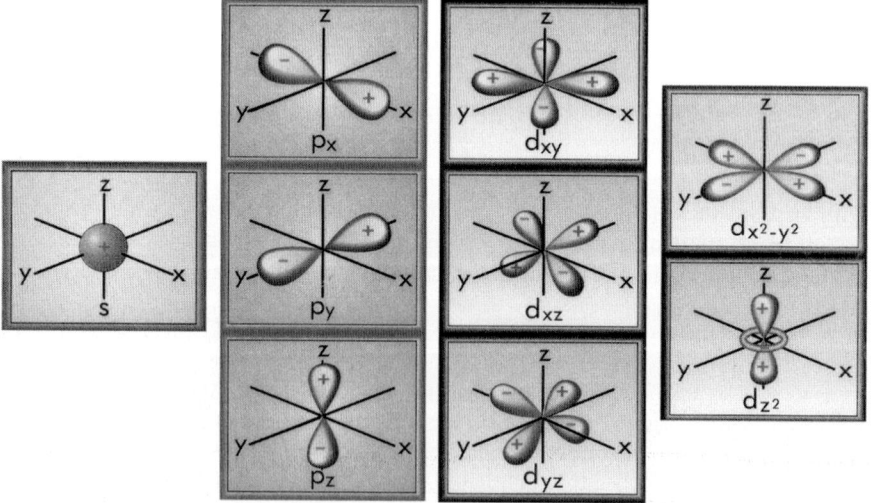

Fig. 2.4. The geometrical configurations of some of the electron orbitals with various n, l and m values

2.3 Electron Shells in Atomic Physics

Table 2.4. The filling of the electron shells

Shell n	Orbital l	States $(n+1)l$	Number of electrons	Total in shell	Grand total (the inert gases)
1	0	1s	2	2	2
2	0,1	2s,1p	2 + 6	8	(2 + 8) = 10
3	0,1	3s,2p	2 + 6	8	(10 + 8) = 18
4	0,1,2	4s,3p,1d	2 + 6 + 10	18	(18 + 18) = 36
5	0,1,2	5s,4p,2d	2 + 6 + 10	18	(36 + 18) = 54
6	0,1,2,3	6s,5p,3d,1f	2 + 6 + 10 + 14	32	(54 + 32) = 86
7	0,1,2,3	7s,6p,4d,2f	2 + 6 + 10 + 14	32	
8	0,1,2,3,4	8s,7p,5d,3f,1g	2 + 6 + 10 + 14 + 18	50	

The sequence and filling of electron states implied by the Schrödinger equation produces subshells that are filled at 2, 8, 18, and 32 electrons. Given the wave equation and its quantal subscripts, electron shell closure can then be deduced to arise at 2, (2 + 8) 10, (2 + 8 + 8) 18, (2 + 8 + 18 + 8) 36, (2 + 8 + 18 + 18 + 8) 54 and (2 + 8 + 18 + 32 + 18 + 8) 86 electrons. The fact that the Schrödinger equation can be used to explain this fundamental periodicity of the Periodic Table (Fig. 2.5) implies that it has captured something of importance about atomic structure. A similar application of the Schrödinger equation to protons and neutrons (discussed below) allows for similar predictions concerning the periodicity of certain nuclear properties.

In summary, use of the Schrödinger equation in atomic physics implies the existence of distinct energy shells. The number of electrons in each shell differs according to the n-value of the shell, but the element that completes each row of the periodic table should have a special stability that is atypical of its horizontal neighbors. Indeed, the elements in all of the columns of the table show remarkable similarities in chemical properties which leave no doubt as to the reality of periodicity (Figs. 2.6 through 2.12). The evidence for shell closure at element-86, Radon, is somewhat weaker than elsewhere, but there is generally good agreement indicating closure at 2, 10, 18, 36, 54 and 86 electrons and not at other numbers. Let us briefly examine these properties individually.

Covalent Radius

When an atom forms covalent bonds with other atoms, the inter-nuclear distance can be experimentally measured by means of crystallography, and the "covalent radius" of each atom calculated. Such values for the first 85 elements are shown in Fig. 2.6. There is a slight trend toward larger radii as we proceed to heavier elements in the periodic table, but more striking than the small increase in average atomic volume with increases in the number of electrons are the five peaks arising at elements 3, 11, 19, 37, and 55. These numbers correspond to elements with one more electron than the closed shells at 2, 10, 18,

Period

	IA	IIA	IIIB	IVB	VB	VIB	VIIB	VIIIB	VIIIB	IB	IIB	IIIA	IVA	VA	VIA	VIIA	Noble gases
1	1 H 1.008																2 He 4.003
2	3 Li 6.941	4 Be 9.012										5 B 10.811	6 C 12.011	7 N 14.007	8 O 15.999	9 F 18.998	10 Ne 20.179
3	11 Na 22.990	12 Mg 24.305										13 Al 26.982	14 Si 28.086	15 P 30.974	16 S 32.064	17 Cl 35.453	18 Ar 39.948
4	19 K 39.098	20 Ca 40.08	21 Sc 44.956	22 Ti 47.90	23 V 50.942	24 Cr 51.996	25 Mn 54.938	26 Fe 55.847	27 Co 58.933	28 Ni 58.70	29 Cu 63.546	30 Zn 65.38	31 Ga 69.72	32 Ge 72.59	33 As 74.922	34 Se 78.96	35 Br 79.904	36 Kr 83.80
5	37 Rb 85.468	38 Sr 87.62	39 Y 88.906	40 Zr 91.22	41 Nb 92.906	42 Mo 95.94	43 Tc (99)	44 Ru 101.07	45 Rh 102.905	46 Pd 106.4	47 Ag 107.868	48 Cd 112.41	49 In 114.82	50 Sn 118.69	51 Sb 121.75	52 Te 127.60	53 I 126.905	54 Xe 131.30
6	55 Cs 132.905	56 Ba 137.33	57 La 138.905	72 Hf 178.49	73 Ta 180.948	74 W 183.85	75 Re 186.2	76 Os 190.2	77 Ir 192.22	78 Pt 195.09	79 Au 196.966	80 Hg 200.59	81 Tl 204.37	82 Pb 207.19	83 Bi 208.2	84 Po (210)	85 At (210)	86 Rn (222)
7	87 Fr (223)	88 Ra (226)	89 Ac (227)	104 Rf (261)	105 Ha (262)	106 (257)	107 (260)											

58 Ce 140.12	59 Pr 140.907	60 Nd 144.24	61 Pm (145)	62 Sm 150.35	63 Eu 151.96	64 Gd 157.25	65 Tb 158.925	66 Dy 162.50	67 Ho 164.930	68 Er 167.26	69 Tm 168.934	70 Yb 173.04	71 Lu 174.96
90 Th (232)	91 Pa (231)	92 U (238)	93 Np (239)	94 Pu (239)	95 Am (240)	96 Cm (242)	97 Bk (245)	98 Cf (246)	99 Es (247)	100 Fm (249)	101 Md (256)	102 No (254)	103 Lr (257)

Fig. 2.5. The periodic table of elements. Each row indicates an energy level ($n = 1 \cdots 7$) and each column (I–VIII) indicates the number of filled electron states within each energy level

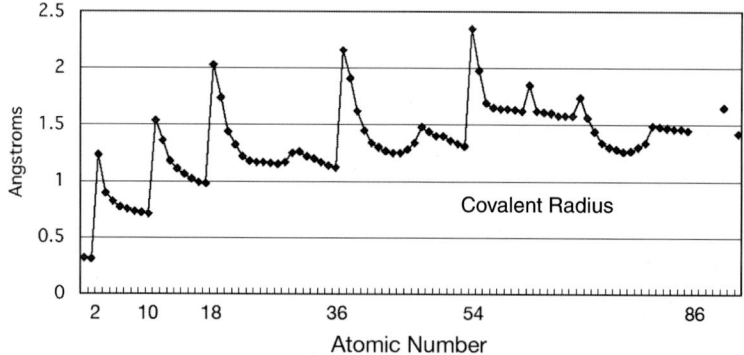

Fig. 2.6. The periodicity of the covalent radii. The x-axis is the number of electrons in the atom (protons in the nucleus) and the y-axis is the radius in Angstroms. The closed shell inert gases have small radii, and the closed-shells-plus-one-electron atoms show clear maxima

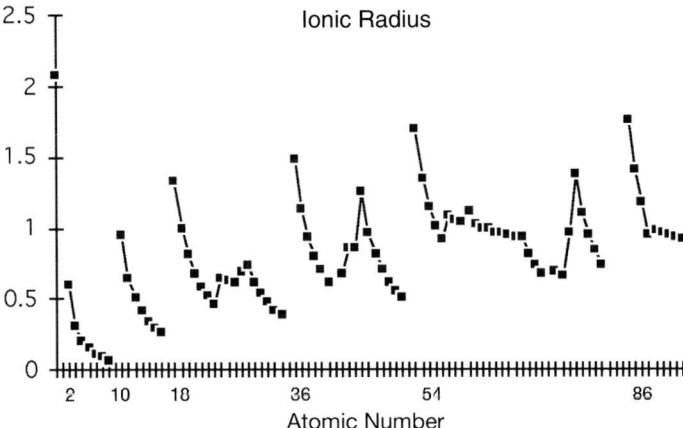

Fig. 2.7. The ionic radius (in Angstroms). For atoms that can exist in several ionic states, the radius of the most positive ion is plotted

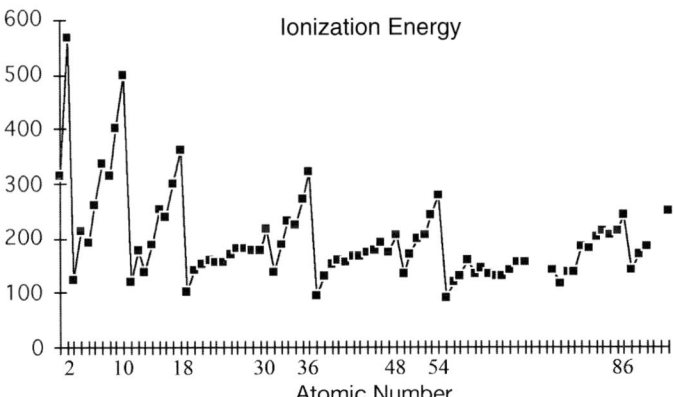

Fig. 2.8. The ionization energies of the elements (kcal/g-mole). The highest ionization energies are found at closed-shells of 2, 10, 18, 36, 54 and 86 electrons. Smaller peaks are seen at $Z = 30$, 48 and 80

36 and 54 electrons. The quantum mechanical explanation is that the closed shells form a core structure, and the one excess electron "orbits" externally. Another way to state this is that, within each period of the periodic table, the first element is largest and there is a gradual decrease in atomic volume across the period. There is some substructure evident within the three larger periods, but the locations of shell closures are in no doubt.

The covalent radii exhibit one of the clearest periodicities of the atomic elements, with shell closure leading to relatively small radii, and a huge jump with the addition of an extra electron. The remarkable effects are at the beginning of each period and this is clear indication that the spatial locations

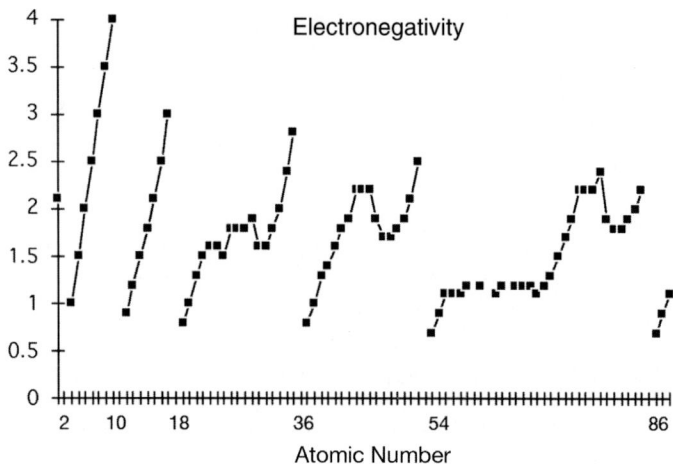

Fig. 2.9. Pauling's electronegativity values (arbitrary units). Each row of the periodic table shows increases from the alkali metals. The inert gases, for which empirical values are not available, would presumably lie at low values near zero in the gaps between the rows of elements

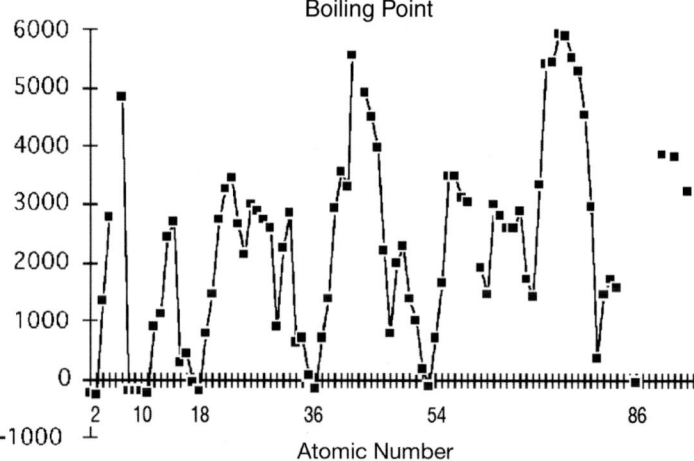

Fig. 2.10. Boiling points of the elements (in degrees Centigrade). The inert gases show the lowest boiling points in each row of the periodic table. Noteworthy substructure is evident at 30, 48, 64, 72 and 80 electrons

of electrons in higher orbitals are external to electrons in lower orbitals – i.e., the valence shell surrounds a filled core structure. There is some substructure seen within the shells, but the only "anomaly" is the rather large radius of Europium ($Z = 63$). Nevertheless, the shell closures at the inert gases are unambiguous.

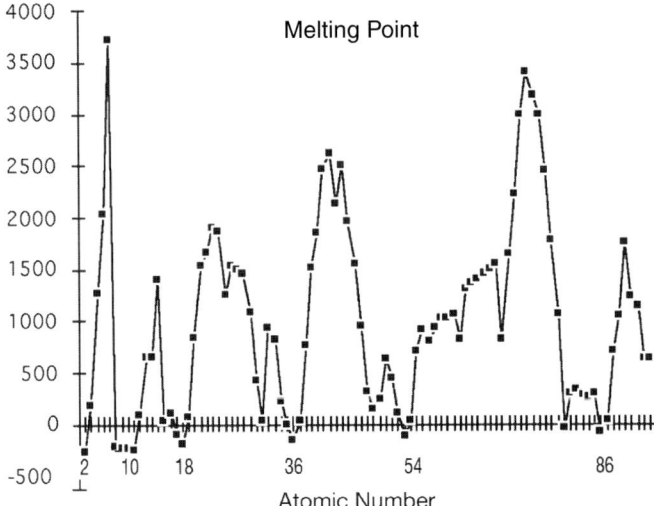

Fig. 2.11. Melting points of the elements (in degrees Centigrade). The inert gases show the lowest melting points in each row. Again, atoms with 30, 48 and 80 electrons show phase transitions at relatively low energies

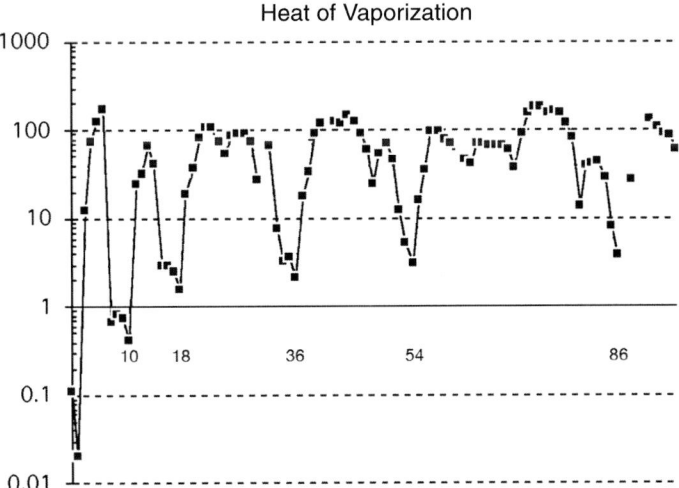

Fig. 2.12. Heat of vaporization (in k-cal/g-atom at boiling point). The lowest values are again seen at the inert gases

Ionic Radius

A related measure of the size of atoms is the ionic radius. Ions are formed by stripping the valence electrons away, so that comparisons can again be made of atomic size when the atoms are crystallized without forming covalent bonds. The periodicity of the periodic chart is again remarkable (Fig. 2.7). Using this measure, the inert gases are not included because they are not ions when they have full valence shells, but there are notable peaks of ionic radius at the alkali (closed shell plus one) elements, followed by a decrease in ion size across each row. Maxima are evident at the locations corresponding to one plus the missing values of: (2), 10, 18, 36, 54 and 86. Some substructure is again evident, suggesting the closure of "subshells" at Zinc ($Z = 30$), Palladium ($Z = 46$) and Mercury ($Z = 80$). But in each period of the periodic table, the closed-shell inert gases fall at positions immediately prior to the maximal ionic radius.

Together, the measures of covalent and ionic radii give clear indication of spatial shells in the build-up of electrons.

First Ionization Energy

In addition to such evidence indicating structural periodicities, chemical evidence concerning the reactivity of the elements strongly supports similar conclusions about shell closures. With regard to the binding energy of the electrons to each atom, several related measures can be examined. The first ionization energy is the energy required to remove one electron from an atom. As seen from Fig. 2.8, removal of an electron from an inert gas requires the most energy in each row of the periodic table. Similarly, removal of an electron requires the least energy for the closed-shell-plus-one-electron alkali metals. Although some subshell structure is again evident [notably the relatively high ionization energies of Zinc ($Z = 30$), Cadmium ($Z = 48$), and Mercury ($Z = 80$)], the nobility of the noble gases is nonetheless pre-eminent in their having the highest ionization energies in each row.

Pauling's Electronegativity Scale

In 1947 Linus Pauling devised a "electronegativity scale" (Fig. 2.9) to summarize the power of attraction for the electrons in a covalent bond. Again, the inert gases are not included in this scale, but the periodicity of the elements within each row and the location of the inert gases is fully evident. The electronegativity finds its lowest value for the alkali elements at the start of each row of the periodic table – where the alkali elements have difficulty hanging on to their lone valence electron – and grows steadily. Again, there is no question concerning the closure of electron shells at all of the inert gases. Substructure in the electronegativity values suggest subshell closure at 30, 48 and 80.

Phase Transitions

Three related properties of the elements concern phase transitions. The first is the boiling point – the temperature at which a liquid-to-gas transition occurs

(Fig. 2.10). The second is the melting point – the solid-to-liquid transition point (Fig. 2.11). And the third is the heat of vaporization – the temperature per unit of mass at which a transition from a solid to a gaseous state occurs (Fig. 2.12). For all three measures, it is evident that the inert gases are happy to separate from their neighbors at relatively low temperatures – always at the lowest temperatures in each row of the periodic table.

Many other chemical properties of the elements are known. Some show periodicities of a similar nature, but most reveal more complex regularities that depend upon variable properties such as bond lengths and bond angles, and are simultaneously influenced by temperature and pressure. Suffice it to say that the most direct measures of atomic size, chemical reactivity and phase transitions – the seven criteria illustrated above – are consistent in indicating shell closures uniquely at the inert gases. In other words, the empirical evidence for shell closure in the case of electrons is strong and unambiguous. Some minor differences in the strength of the shell closures are found, depending on the empirical criteria, and there are suggestions of the existence of a small number of subshells [at 30, (46), 48, (64, 72) and 80], but there is simply no confusion about which numbers indicate the major closed-shells of electrons, corresponding to the inert gases.

2.4 Nucleon Shells in Nuclear Physics

The situation in nuclear physics is more complex. Protons and neutrons are fermions, and have many characteristics in common with those of electrons; most importantly, the energy states of nucleons also can be described using the Schrödinger equation. It follows that the properties of nuclei with many nucleons should show some form of "periodicity" with open and closed shells and subshells – as is experimentally well-established for electrons. Because of differences in the nuclear and atomic realms, the sequence of shells in the two realms differs somewhat: electron shells are filled at: 2, 10, 18, 36, 54, 86 particles, whereas proton *and* neutron shells are separately filled at: 2, 8, 20, 28, 50, 82 and 126. The empirical evidence concerning such shells in the nucleus is diverse, but, as with electron shells, there are both size/shape properties and various measures of nuclear stability/reactivity.

As discussed below, the evidence for the closure of many shells and subshells in the nucleus is rather convincing, but the "periodicity" of nuclear properties comparable to that for atoms is weak. For this reason, the nuclear version of the Periodic Table is normally presented in the non-periodic form of the Segre Plot of the isotopes (Fig. 2.13), where the location of the closure of major shells is merely noted by horizontal and vertical lines.

The criteria most frequently cited in the textbooks for "magic" closure of protons and neutrons are: (i) the number of stable, metastable, or known isotopes (nuclei with the same number of protons, but different numbers of neutrons) and isotones (nuclei with the same number of neutrons, but different

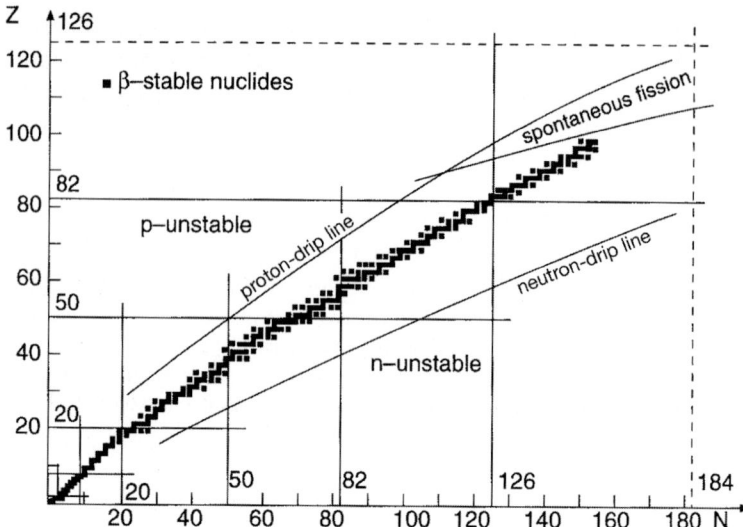

Fig. 2.13. The Segre Plot of nuclear species. Stable nuclei are denoted by black squares. A large number of unstable nuclei lie below the proton-drip line and above the neutron-drip line

numbers of protons); (ii) the natural abundance of nuclei with various numbers of protons or neutrons; (iii) the total nuclear binding energies; (iv) the energy required to remove one neutron (or proton) from a stable nucleus; (v) nuclear quadrupole moments that indicate the spatial distribution of proton charge; and (viii) nuclear radii.

The data and diverse arguments concerning magic shell closures were first collected and systematically studied in the 1930s and 1940s, when a relatively small number of nuclear properties had been measured. Together, they provided support for the idea of certain discontinuities of nuclear properties that an amorphous, structureless liquid-drop account of the nuclear texture could not explain. Today, much more data are available, so that the argument for shell and subshell closure can be re-examined in light of current findings.

Isotopes and Isotones

Figures 2.14 and 2.15 illustrate the empirical evidence for what is often stated to be one of the strongest arguments for shell closure – the numbers of isotopes and isotones for any value of Z or N. Focusing only on the dotted lines showing the *stable* isotopes and isotones, it is seen that there are indeed some noteworthy peaks, but they do *not* reveal an unambiguous pattern of textbook magic numbers. If the shell structure of the nucleus were as clear as that for electrons, we might expect that the special stability that shell closure brings to a nucleus would produce peaks in the number of stable isotopes or

2.4 Nucleon Shells in Nuclear Physics

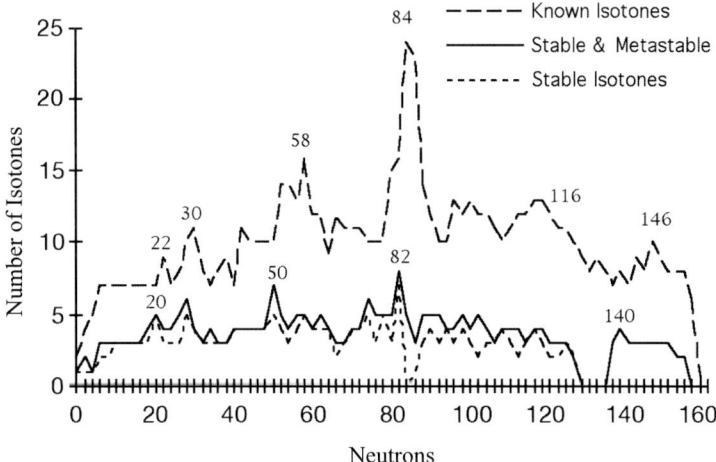

Fig. 2.14. Numbers of stable, metastable and known isotones as a function of the number of neutrons. The stability criterion indicates magic numbers at 20, 28, 50 and 82, but there are many possible candidates for magic numbers using the criterion of known numbers of isotones

isotones. Judging from the *stable* isotones, we find magic numbers at 20, 28, and 82 neutrons, but the peak at 50 is no greater than the peaks at 58, 78, and 80, and there is no indication of magic stability at 2, 8 and 126. Judging from the *stable* isotopes, we get magic numbers at 20, 28 and 50, but other

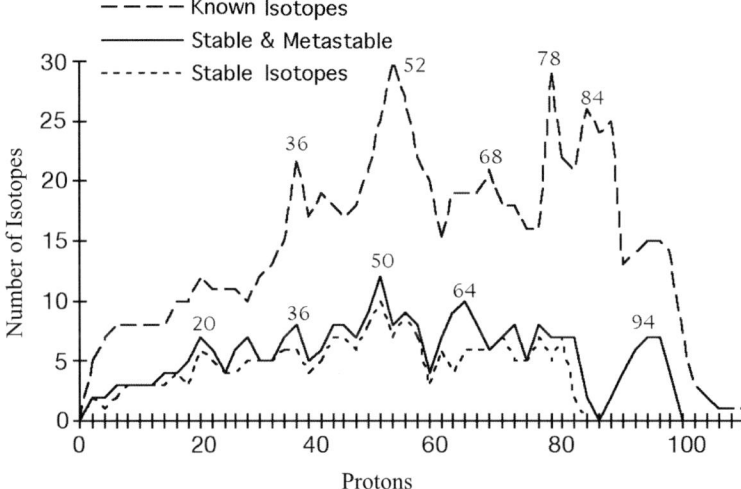

Fig. 2.15. A relatively large number of stable and metastable isotopes with 50 protons are found. Peaks for 20 and 28 are also evident, but other indications of magic stability do not correspond to the well-known magic numbers. Notable anomalies are seen at 36, 64 and 94

small peaks arise at various numbers not normally considered to be magic: 54, 70, 76, and 80.

Since there are many nuclei that are known to be unstable, but that have extremely long half-lives, a somewhat better indication of magical shell closure might therefore be obtained by considering the metastable nuclei with half-lives of, say, one year or more. The stable + metastable curve shows peaks at 20, 28, 50 and 82 neutrons, and another candidate magic number at 140. The stable + metastable curve for protons gives the expected magicness for 20, 28 and 50, but peaks at 36, 64 and 94, and small peaks elsewhere are unexpected.

What about the total number of *known* isotopes or isotones? This measure is the most dubious empirical indication of shell closure because "known nuclei" include both stable nuclei and nuclei that have extremely short half-lives of less than a femtosecond (10^{-12} sec). It is perhaps most amazing that it is even possible to measure these transitory events experimentally, but it is also surprising that the numbers of known isotopes or isotones reveal large differences among the elements, but the peaks *never* include the magic numbers.

Natural Abundance

The natural abundance of elements in the solar system is often mentioned in support of the concept of nuclear magic numbers, but this argument also provides ambiguous evidence. As shown in Fig. 2.16, there is a roughly logarithmic decrease in abundance with increasing nuclear size, but within such

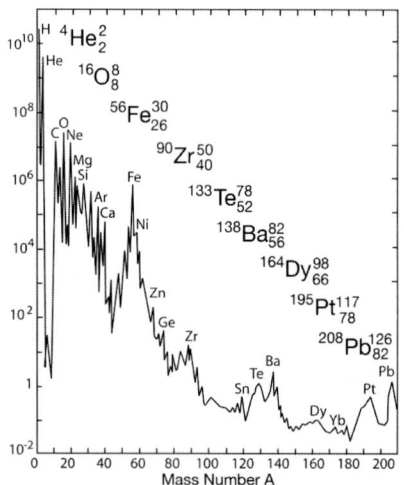

Fig. 2.16. The abundance of the elements in our solar system (normalized to that of Silicon) (after Povh et al., 1996, p. 16). The outstanding peak at Iron ($Z = 26, N = 30$) is not magic. Magic numbers contribute to the abundance of Zirconium ($Z = 40, N = 50 - 54$), Barium ($Z = 56, N = 79 - 82$) and Lead ($Z = 82, N = 120 - 126$), but not those for Tellurium ($Z = 52, N = 74 - 78$) or Platinum ($Z = 78, N = 116 - 118$)

an overall trend there is one outstanding peak at iron with 26 protons and 30 neutrons; these are not magic numbers. Among the other elements that show slight over-abundance, relative to the logarithmic decrease, there are some magic numbers ($N = 50, 82, 126$; $Z = 82$), but many other non-magic numbers ($N = 52, 54, 74, 76, 80, 117, 118$; $Z = 40, 52, 56, 78$). Because the natural abundances are largely determined by the sequence of fusion events leading to the formation of nuclei in the solar interior, simple conclusions are not possible, but it is *not* the case that abundance is strongly influenced by nuclear shell closure.

Binding Energies

The most unambiguous indication of nuclear shell structure comes from the data on total binding energies. By calculating the expected binding energy using the liquid-drop model (without shell corrections) and subtracting it from the experimental value, the deviation from the simple liquid-drop conception can be determined. Relatively large deviations are obtained at $Z = 28, 50$ and 82 and $N = 28, 50, 82$ and 126 (Fig. 2.17). The deviations are indication of slightly higher binding energies for nuclei with these numbers of protons and neutrons, and this supports the idea of relatively tightly-bound, compact closed shells. Note that the shell structure for neutrons is unambiguous, but that for protons is clear only at 28 and 82.

Unfortunately, similar calculations do not show notable deviations for the first three magic numbers, 2, 8 and 20 (Fig. 2.18). There is no dip at 20 neutrons whatsoever, and little regularity below 20 suggestive of neutron shells.

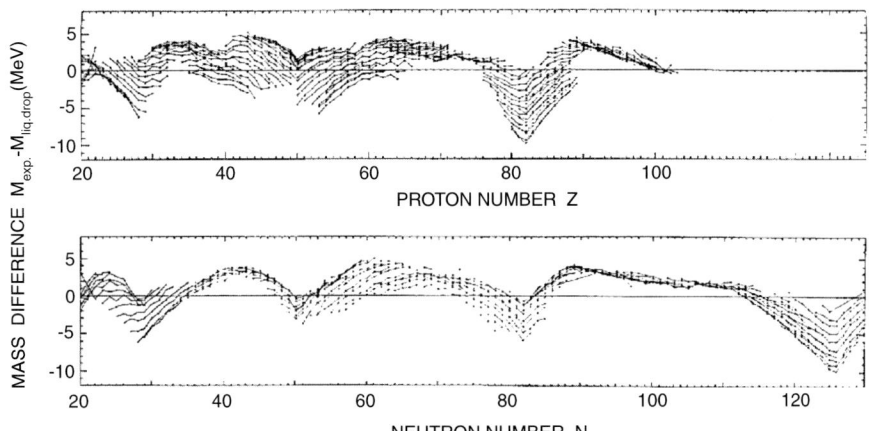

Fig. 2.17. Indications of shell structure from total nuclear binding energies. The graphs show the difference between the theoretical liquid-drop value and the experimental value (Myers and Swiatecki, 1966)

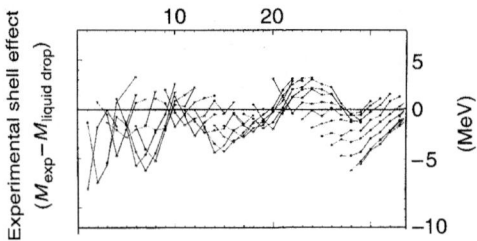

Fig. 2.18. The difference between the liquid-drop calculations and experimental data for the small nuclei plotted against neutron number (Myers and Swiatecki, 1966)

Quadrupole Moments

The changes in the nuclear quadrupole moments across the chart of nuclei are suggestive of a recurring pattern of nuclear structure. Theoretically, a positive quadrupole moment is thought to indicate a prolate (uni-axial or polar) bulge of the positive charges in the nucleus, whereas a negative moment is indicative of an oblate (bi-axial or equatorial) bulge. In contrast, when the quadrupole moment is close to zero, the positive charges of the protons are thought to be evenly distributed over a roughly spherical volume. Since the magic nuclei with closed shells are theoretically spherical, a return of the quadrupole moment from positive or negative values to values near to zero is expected for nuclei with Z near to a magic number.

The experimental data on quadrupole moments (Q) was first collected and displayed in relation to the magic numbers by Townes et al. in 1949, and later updated by Segre in 1965. Segre's figure (Fig. 2.19) is frequently reproduced in support of the argument for quadrupole moment shell structure. Although the variability of Q for the smaller nuclei is difficult to interpret, there is a notable drop in Q at Z or $N = 50$, 82 and 126, and suggestions of shell structure at numbers 8, 16, 28 and 40.

The data plotted in Fig. 2.19 were initially interpreted as supportive of the concept of magic shells, but many more experimental values have been collected over the subsequent half century. It is therefore relevant to plot the nearly-500 known quadrupole moments for stable (or near-stable) nuclei against the numbers of protons and neutrons, as shown in Figs. 2.20 through 2.22. From both Figs. 2.20 and 2.21, there is some indication of shifts from positive to negative values near the magic numbers 50, 82 and 126, but it would be difficult to argue for shape changes uniquely at the magic numbers.

There are, moreover, two well-known problems in drawing conclusions about nuclear shell closure from such data. As seen in both Figs. 2.20 and 2.21, roughly spherical nuclei (as indicated by $Q \sim 0$) are found virtually *everywhere*. It is apparent that something unusual occurs at 82 and 126, but conclusions with regard to the other magic numbers are less certain. The second problem in using quadrupole moment data lies in the fact that the

2.4 Nucleon Shells in Nuclear Physics

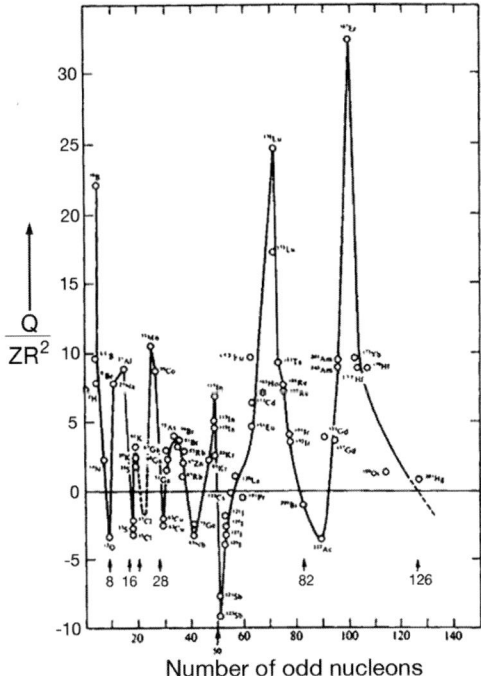

Fig. 2.19. Nuclear quadrupole moments (ca. 1949). The solid line is drawn to guide the eye (after Segre, 1965), but shows a shift from prolate to oblate shapes at the magic numbers

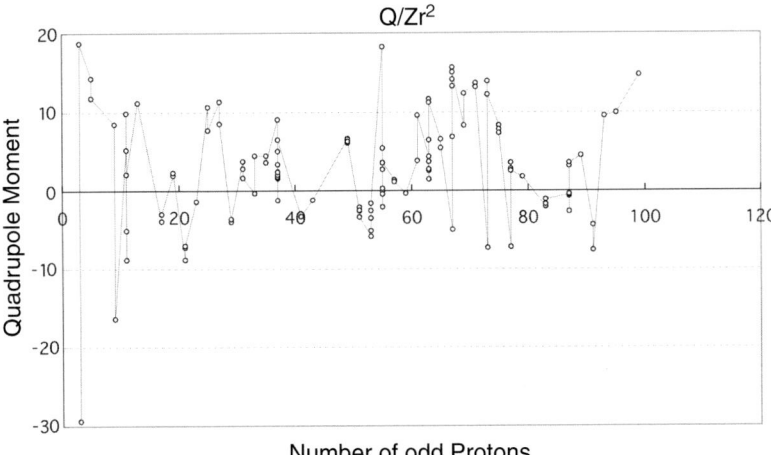

Fig. 2.20. Quadrupole moments of the 129 nuclei with even-N and odd-Z plotted against the number of protons

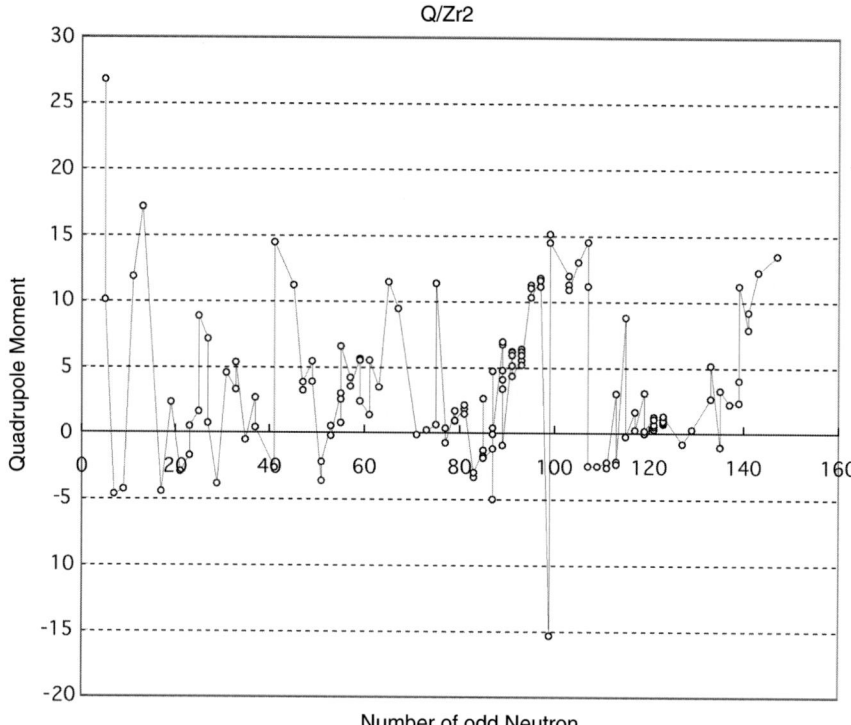

Fig. 2.21. Quadrupole moments plotted against the number of neutrons for 141 nuclei with even-Z and odd-N plotted against the number of neutrons

theoretical prediction is, strictly speaking, valid only for proton numbers. A positive or negative quadrupole moment is predicted only when there is a non-spherical distribution of *proton charge*, so that the influence of neutrons should be minimal. Why then is the quadrupole moment nearly zero for nuclei with magic numbers of *neutrons* ($N = 82$ or 126), when the numbers of protons indicate *unclosed* protons shells (Fig. 2.21)? And why would there be indication of magic shell closure at $Z = 40$, which is not a magic shell (Fig. 2.20)? In contrast to the data available in 1949, the complete data set shown in Fig. 2.22 is far from conclusive concerning shell closure. Clearly, the data indicate some interesting trends in charge distribution that undoubtedly reflect the shape of nuclei, but the reality of "spherical nuclei" uniquely at the magic numbers is uncertain.

Proton and Neutron Separation Energies

The amount of energy required to remove one proton or neutron from a stable nucleus is called the proton or neutron separation energy, and is closely related

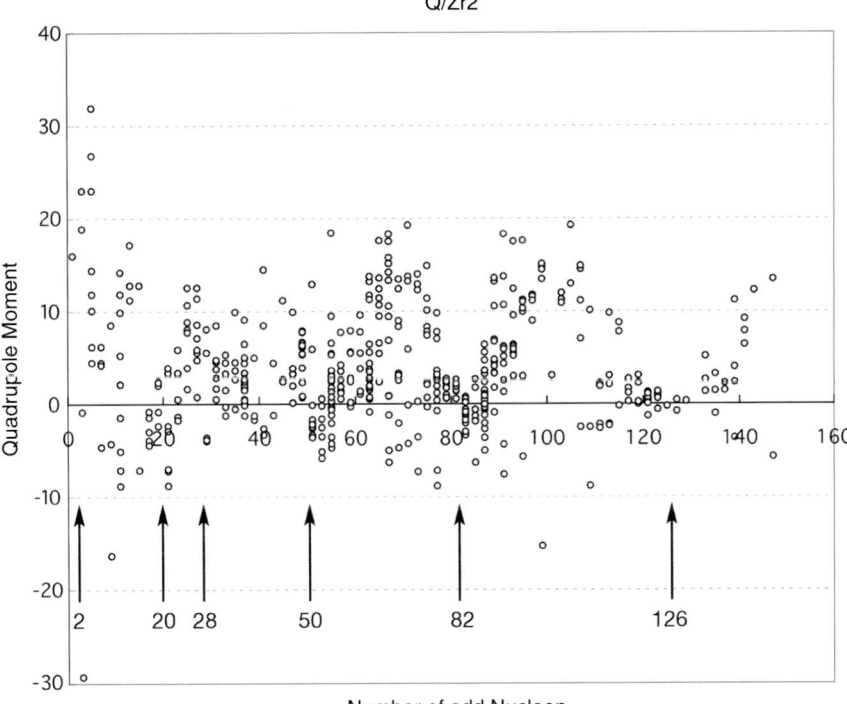

Fig. 2.22. Quadrupole moments plotted in a manner similar to Fig. 2.19 against the number of the odd-nucleon for all 494 nuclei with odd-N, odd-Z or both odd-N and odd-Z

to the idea of ionization energies, as plotted in Fig. 2.8. The neutron separation energy for the even-N nuclei is shown in Fig. 2.23. There are notable jumps at $N = 50$, 82 and 126, some suggestion of a gap at $N = 28$, but no discernible gaps at $N = 2$, 8 or 20. A comparable plot for even-Z nuclei is shown in Fig. 2.24. Magic gaps are suggested at $Z = 50$ and 82, but not elsewhere.

RMS Charge Radii

Precise charge radial values are known for more than 200 isotopes over the full range of the chart of nuclides. If there were spatial "shells" comparable to the changes in the radial measures of the electron shells, they should show up in these measures of nuclear size. As shown in Fig. 2.25, the experimental data on proton and neutron build-up plotted in a manner analogous to that for electrons gives no indication of radial shells. Whether plotted against proton number (Fig. 2.25a) or neutron number (Fig. 2.25b), there are no jumps in radial values following the magic numbers. On the contrary, as expected from the (amorphous, shell-less) liquid-drop conception of the nucleus, there is a

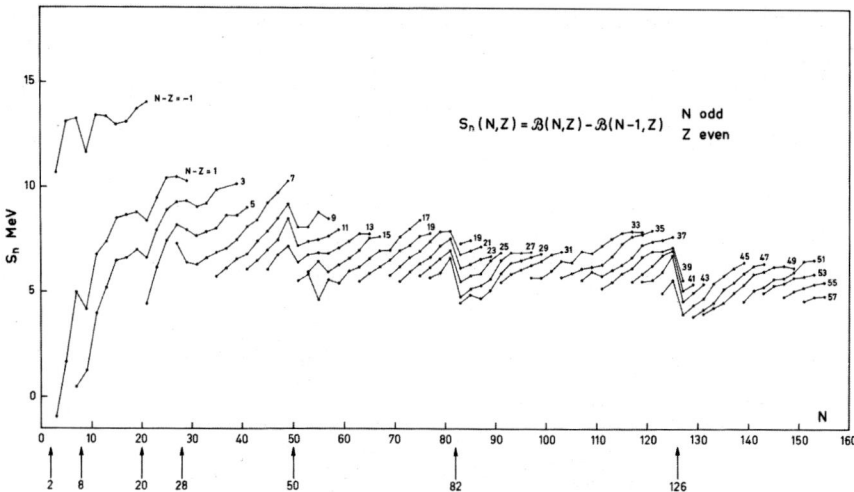

Fig. 2.23. Neutron separation energies for even N-values. Notable gaps are evident at $N = 50$, 82 and 126 (after Bohr & Mottelson, 1968, p. 193)

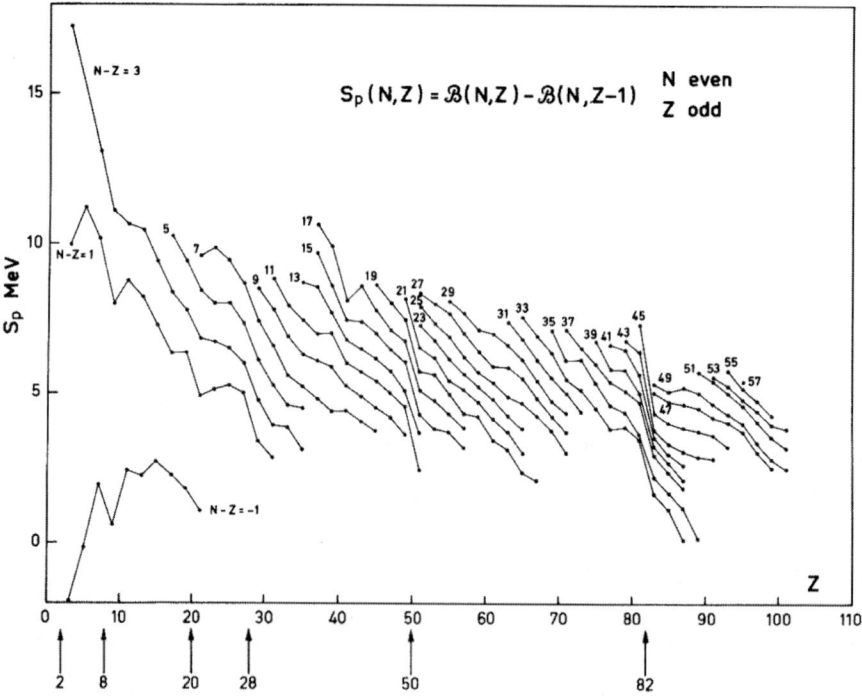

Fig. 2.24. Proton separation energies for even Z-values. A sharp decrease in the binding of the last proton is seen for values larger than 50 and 82 (after Bohr & Mottelson, 1968, p. 194)

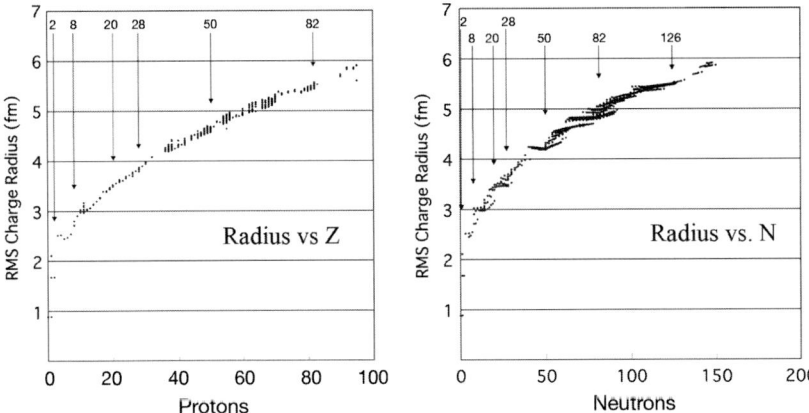

Fig. 2.25. Plots of nuclear charge radii versus proton and neutron numbers. The arrows indicate the magic numbers. Notable jumps in the nuclear radius following the magic numbers are not seen

rather steady increase in the nuclear radius – regardless of the presence or absence of discontinuities in binding energies.

A more sensitive demonstration of spatial shells, however, might be obtained by plotting the difference between the measured value and the expected radius, assuming that each nucleon in a dense nuclear droplet occupies a fixed volume of nuclear space. In other words, how much do nuclear radii deviate from the simple predictions of the liquid-drop model?

To answer this question, the same data on nuclear radii have been plotted in Fig. 2.26. In the plot of size versus proton number, there is a suggestion of a shell in the pattern of radial values for $2 < Z < 9$ – with a gradual decrease across this period. The peak at $Z = 12$ in the next period is more difficult to interpret, since an electron-shell-like explanation would predict a peak at $Z = 9$, followed by a decrease. In any case, there are no notable increases in radial values following the shell-model closures at $Z = 20, 28, 50$ or 82 and several anomalously large radii occurring in the middle of the filling of these shells.

In the plot of this difference versus neutron number (Fig. 2.26b), we see various indications of nuclear substructure, but no sign of shells at the magic numbers. That is, as neutrons are added to a given number of protons, the size of the nucleus (root-mean-square charge radius) increases proportionately less (as witnessed by the various falling diagonal lines in Fig. 2.26 which are chains of isotopes). This is to be expected, since the charge radius is actually a measure of proton size, and neutrons will have influence on the charge radius only indirectly through their influence on the proton distribution. What we do *not* see are any jumps from one closed shell to the filling of a spatially distinct valence shell, as was apparent for atomic radii.

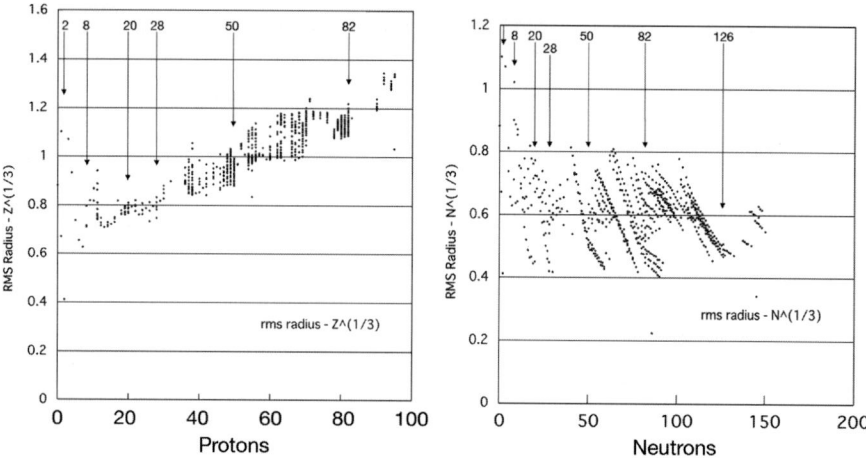

Fig. 2.26. Deviations from liquid-drop model predictions. Plots of the difference between the measured RMS charge radius minus the nuclear charge radius predicted by a simple liquid-drop model vs. the number of protons or neutrons. Radial jumps suggestive of shell structure at the magic numbers (arrows) are weak

Other measures of the spatial and energetic closure of shells have been studied within the framework of the shell model, but the above eight criteria are those most frequently cited in support of the model. It bears repeating that, in contrast to the closure of electron shells, *no single criterion of magic stability produces the textbook magic numbers*, and indeed *no unique set of criteria produces only the magic numbers and not other numbers* (Table 2.5).

2.5 Summary

The purpose of this chapter has been to illustrate both the basic similarities between atomic and nuclear structure, and one of the strongest dissimilarities between these realms. Although the individual particles in both systems are well-described by the Schrödinger wave-equation, experimental data indicating the closure of electron shells are far more robust than the indications of nucleon shells. This does not mean that nuclear applications of the Schrödinger equation are wrong, but there are fundamental differences between the atom and the nucleus that lead to unambiguous shell structure for electrons, but far more subtle shell structure for nucleons. Stated differently, the system of many electrons in atomic physics is relatively well described as the summation of the effects of individual electrons whose positions and energies are dominated by the two-body (nucleus-electron) interaction, whereas it is apparent that other factors, in addition to the attraction of a central potential-well, play important roles in determining the structure of the atomic nucleus.

could account for the gross properties of the nucleus meant that the liquid-phase idea was a reasonably good first approximation.

The reason why nuclear structure theory did not come to a happy and satisfactory conclusion already in the early 1930s is that certain kinds of substructure gradually became apparent. To begin with, some of the smallest nuclei were found to have unusually high binding energies. In other words, they were surprisingly stable relative to nuclei with one or two more or fewer protons or neutrons, and the existence of unusual stability at specific numbers of nucleons did not fit easily into the picture that nuclei are unstructured aggregates of nucleons interacting at random with an ever-changing collection of nearest neighbors. The high binding energy of specifically the 4n-nuclei suggested that there may be certain specific stable configurations – notably, alpha particles – within the nucleus. Clearly, substructure of any kind was not implied by an amorphous liquid-drop model. Eventually, several kinds of substructure became well-established – the first concerning the existence of 4-nucleon clusters within stable nuclei and the second concerning the closure of shells of protons or neutrons at the so-called magic numbers.

The special stability of the small 4n-nuclei led directly to the alpha-particle (or cluster) model, in which it was assumed that the bulk of the nuclear binding energy is due to the binding of nucleons into alpha-particles (two protons and two neutrons) – with much weaker binding of the alphas to each other or to the small number of nucleons not contained within alphas. Together with the finding that many of the largest radioactive nuclei spontaneously emit alpha-particles and later experimental results showing that alphas (and multiples of alphas) are often produced when medium- and large-sized nuclei are bombarded with high-energy particles, it was hypothesized that all nuclei may be essentially clusters of alpha-particles. Over the subsequent decades, much theoretical work has been done on the cluster perspective, and it has been shown that certain small nuclei can indeed be fruitfully thought of as alpha-particles – with or without a few excess nucleons attached. The quantitative successes of the cluster models for specific nuclei have established this approach as one of the three dominant models in nuclear theory. However, despite the fact that the cluster model is one of the oldest in nuclear theory, it has been successfully applied to only a relatively small number of problems – involving principally the small 4n-nuclei, their binding energies and certain of their excited states.

The apparent limitations of the cluster approach aside, it is worth noting that there is no conceptual difficulty in reconciling a fundamentally liquid-phase nuclear interior with the idea that 4-nucleon arrangements might be unusually stable and have at least a transient existence within the liquid interior of nuclei. In other words, nuclei might still be considered to be essentially microscopic liquid-drops containing protons and neutrons tumbling and moving around one another. Occasionally, however, four neighboring nucleons "condense" into a tightly-bound alpha-particle and remain temporarily as a solid cluster within the liquid interior of the nucleus until, eventually,

3

A Brief History of Nuclear Theory

The discovery of a centrally-located nucleus by Rutherford in 1911 marks the beginning of nuclear physics. For two full decades following that discovery, there was speculation about the structure of the nucleus and arguments about the possible relationships of positive and negative charges within the nucleus, but not until the discovery of the neutron by Chadwick in 1932 did ideas concerning nuclear structure appear which can be considered modern. This early phase of nuclear theory is of historical interest (and has been reviewed in detail by Mladjenovic, 1998), but has limited bearing on subsequent nuclear modeling.

Soon after the discovery of the neutron, the first nuclear model was developed based on the apparently short-range, "strong-interaction" among the nucleons, the so-called liquid-drop model. Since then, the idea of a "collective" liquid-drop-like description of the nucleus has proven surprisingly useful in explaining several important nuclear properties. The earliest and still one of the most impressive results, due to Bohr, and Wheeler (1939), was the ability of the liquid-drop model to account for the release of energy during the fission of certain large nuclei. The principal strengths of this model, as developed over subsequent decades, are three-fold: (i) It can explain the general properties of nuclear binding as analogous to the binding of particles in an extremely small droplet. (ii) The model can explain the fact that the size of nuclei is strongly dependent on the number of particles present, as if nuclei were made of impenetrable constituents, each occupying a constant volume. And (iii) the liquid-drop analogy can be used to explain certain vibrational states of nuclei – similar to the vibrations that a liquid-drop can undergo – and the breakup of such a droplet in the act of fission.

These successes of the liquid-drop model were important, but it eventually turned out that the liquid-drop analogy alone did not suffice to account for certain details of nuclear size, binding energy and fission – and theorists have looked elsewhere to explain its quantum mechanical properties. Nevertheless, the fact that so simple a model, built on an analogy with a macroscopic entity, and neutron shell structure at all of the shells/subshells of the Schrödinger equation. In fact, that is not the case, and part of the reason is undoubtedly the strong overlap of the nuclear volume that protons and neutrons occupy. Clearly, nuclear structure is more complex than atomic structure because of the two types of nuclear fermion, even if both are guided fundamentally by the same laws of quantum mechanics.

As illustrated in Table 2.5, 50 and 82 for both protons and neutrons are magic by most criteria, and a neutron shell at 126 is often indicated. In light of the many successes of the shell model, however, it is surprising that the predicted shells at 2, 8, 20 and 28 are sometimes strongly indicated, and other times not at all, while subshell closures at 6, 40, 58 and elsewhere are occasionally suggested. This diversity of evidence for subshell closure in the nucleus shows that the quantum mechanical approach is fundamentally correct in suggesting the shell/subshell *texture* of both proton and neutron build-up, but the magnitude of shell closure effects are oftentimes weak and consequently the "magicness" of these nuclei is mild. Since the experimental evidence indicates somewhat unusual stability at many of the independent-particle model shells *and* subshells, it can be concluded that the theoretical framework that produces the shell/subshell energy steps is going in the right direction, but the mutual interactions of protons and neutrons attenuate the expected effects and there is nothing remotely comparable to an "inert gas" in the nuclear realm.

2.5 Summary

Table 2.5. The complete build-up sequence of nucleons in the shell model

n							j							m							s		Total	
0	1	2	3	4	5	6	1/2	3/2	5/2	7/2	9/2	11/2	13/2	1/2	3/2	5/2	7/2	9/2	11/2	13/2	↑	↓		
2							2							2							1	1	2	a, d, i
	6							4						2	2						2	2	6	d, e, f
							2							2							1	1	8	i
		12							6					2	2	2					3	3	14	h
								4						2	2						2	2	18	
							2							2							1	1	20	a, b, c, d, e, h
			20							8				2	2	2	2				4	4	28	b, c, d, h, i
									6					2	2	2					3	3	34	d
								4						2	2						2	2	38	i
							2							2							1	1	40	b, e, g, i
				30							10			2	2	2	2	2			5	5	50	a, b, c, d, h, i
										8				2	2	2	2				4	4	58	b, f, i
									6					2	2	2					3	3	64	c, i
								4						2	2						2	2	68	e
							2							2							1	1	70	
					42							12		2	2	2	2	2	2		6	6	82	b, d, h, i
											10			2	2	2	2	2			5	5	92	h
										8				2	2	2	2				4	4	100	
									6					2	2	2					3	3	106	
								4						2	2						2	2	110	
							2							2							1	1	112	
						56							14	2	2	2	2	2	2	2	7	7	126	h, i
												12		2	2	2	2	2	2		6	6	138	
											10			2	2	2	2	2			5	5	148	
										8				2	2	2	2				4	4	156	
									6					2	2	2					3	3	162	
								4						2	2						2	2	166	
							2							2							1	1	168	

The numbers in each row correspond to the numbers of protons (neutrons) with the quantum numbers lying directly above it. For example, the first row beginning with 2 has 2 protons (neutrons) with n-value of 0, both of which have $j = 1/2$ and $|m| = 1/2$. Under the s-columns, 1 proton (neutron) has spin up and 1 spin down. In the final column is shown the total numbers of nucleons in each shell/subshell and whether or not there is indication of magicness. The criteria for maagicness are as follows: a, number of stable isotopes; b, number of stable isotones; c, number of metastable (half-life >1 yr) isotopes; d, number of metastable isotones; e, number of known isotopes; f, number of known isotones; g, quadrupole moment; h, neutron separation energy; i, excitation energy of first 2+ state (Cook & Dallacasa, 1987).

What are those other factors? Two are well-known, if complex, in their ultimate effects. The first is the fact that, unlike the build-up of electrons in atomic structure, the build-up of both protons and neutrons occurs simultaneously in the same "nuclear space". This means that the stability and shape of closed proton (neutron) shells is inevitably affected by the stability and shape of neutron (proton) shells. The second and related factor is that the nucleon build-up process occurs in a much more compact space relative to the size of the particles, so that the presence of the other type of nucleon is strongly felt. If protons and neutrons were somehow as sparsely distributed within the nuclear volume as electrons are in the atomic volume, it is conceivable that they would not interact and not influence each other's shell characteristics. In such a case, there might be unambiguous indication of both proton the cluster disintegrates and the individual nucleons dissolve into the nuclear liquid. If the condensation of such clusters occurs frequently enough, then it is a reasonable simplification to assume that the nucleus as a whole might be explained on the basis of the dynamics among such clusters. Furthermore, when any nucleus breaks down into smaller pieces, it would be no surprise if some of the alpha-cluster character remained intact in the fragments of a nuclear reaction. In this way, the liquid-drop and cluster models of nuclear structure can be seen as fundamentally compatible with one another and, indeed, the simultaneous use of the "molecular" cluster models and the liquid-phase liquid-drop model has never been seen as a crisis in nuclear theory. Both models could well be correct and, importantly, numerical work within the framework of either model need not make use of assumptions which are contrary to those of the other model.

This more-or-less satisfactory theoretical situation of the 1930s and early-1940s, was, however, disturbed in the late-1940s and early-1950s. It became known that, in addition to the small 4n-nuclei, certain numbers of protons and neutrons also appeared to give nuclei unusual or "magic" stability. As reviewed in the previous chapter, the stability of magic nuclei is relatively subtle in comparison with the inert gases of chemistry, but the dips and bumps in particularly the binding energy curve were real and demanded an explanation. Eventually, the numbers of protons or neutrons that were identified as magic by Mayer and Jensen in the early days of the shell model were: 2, 8, 20, 50, 82 and 126 – with the numbers 6, 14 and especially 28 sometimes noted as semi-magic. The fact of unusual stability itself presented no real theoretical problem, but the model that was devised to explain the magic numbers (the shell model) was based upon the idea that the nucleus is *not* a densely packed liquid and *not* a semi-solid aggregate of alpha-particles, but is actually a tiny "gas" of protons and neutrons orbiting independently of one another within the nuclear interior. This assumption was conceptually a major revolution and one that – despite many attempts at reconciliation – continues to contradict the conceptual basis of the liquid-drop and alpha-particle models.

As has often been remarked since the early 1950s, the shell model simply should not work! It demands nuclear properties that are at odds with those of the rather successful liquid-drop model and that are fundamentally inconsistent with nucleon-nucleon scattering results. The fact of the matter, however, is that it does work – not by re-explaining the properties that the liquid-drop model can already account for, but by explaining a great many other properties. The essence of the shell model is that nucleons do *not* interact locally with their nearest neighbors, but rather that they interact with the sum force of all nucleons – that is, with a centrally-located "net potential-well" created by the nuclear force contributions of all nucleons acting in concert. In this view, each nucleon is seen as contributing to and moving independently around a central potential-well, with only infrequent interactions with its nearby neighbors in the nuclear volume. The theoretical attractiveness of this idea is that it is similar to the idea of electrons orbiting around a centrally-located positively-charged nucleus – i.e., the concept that is at the heart of the quantum mechanical explanation of atomic structure. Of course, the main *difference* between the nuclear realm and the electron realm lies in the fact that the central potential-well required in the nuclear shell model is *not* associated with a centrally-located material entity. Instead of a more-or-less fixed, massive body to which particles are attracted (the nucleus in atomic theory), in the shell model of the nucleus itself, there is said to be only a ghost-like central "well" that is the time-averaged result of all nucleon effects acting together.

Clearly, the initial premise of the shell model is counter-intuitive, but – like any good science fiction – one unexplained assumption can be tolerated if an interesting story then unfolds. Indeed, it was soon realized that *if* the problems involved in such a nuclear potential-well concept could be ironed out, then it would mean that nuclear structure theory might be put on a firm mathematical basis using the Schrödinger wave-equation techniques that had been developed in atomic physics. The same formalism used at the atomic level could then be reapplied at the nuclear level. For this reason alone, it is no surprise that the shell model was welcomed by many theorists and that the originators of the shell model eventually received Nobel prizes. Although in fact very different in detail, the nuclear shell model provided the conceptual basis for utilizing, at the nuclear level, the successful quantum mechanical approach developed at the atomic (electron) level.

Initially, the shell model left questions concerning nuclear radii, densities and binding energies essentially unanswered (the strengths of the liquid-drop model) and a significant number of theorists argued vehemently against the shell model approach (including Niels Bohr and Victor Weisskopf). Advocates of the model, however, were successful in accounting for the huge volume of experimental data on the total angular momentum (spin) of nuclei and the shell model was the only available means for classifying the many excited states of nuclei. Moreover, it was based on quantum mechanics – a mathematical formalism that brought the promise of exact answers. This was a success that theoretical nuclear physics has not turned away from, despite continuing problems. So, after a stormy beginning, by the mid-1950s the shell model had become the central paradigm of nuclear theory and, with modifications and extensions, it has remained the centerpiece of nuclear physics.

The shell model is based on several assumptions that caused troubles in the early 1950s and indeed still do. But it allowed for a means by which each nucleon could be assigned a unique set of quantum numbers – dependent on its presumed orbit through the nuclear interior. The quantal assignment of each nucleon gave it a unique energy state within the nucleus, and movement of the nucleon to a different energy state corresponded to a change in quantum numbers and therefore an energy transition, as specified in the nuclear version of the Schrödinger equation. Moreover, any nucleus in its ground state could, in principle, be described as the summation of the quantum mechanical states of its constituent nucleons. Implicit to the shell-model approach was therefore a complete and rigorous description of all constituent particles and the total energy state of the nucleus as a whole.

In addition to the theoretical strengths inherent to adopting the quantum mechanical formalism, what in retrospect appears to have been the deciding factor in the general acceptance of the shell model was the fact that this hypothesis concerning a central potential-well (plus an assumption concerning the coupling of the orbital and intrinsic spins of each nucleon) produced the experimentally-known magic numbers where nuclei displayed unusual stability. In effect, nuclear theory was given its own periodic table with shells that could be partially or entirely filled, and with nucleons in valence orbits. By means of the development of the independent-particle model, nuclear physics achieved a quantum mechanical formulation which was conceptually analogous to that in atomic physics and which allowed for a vast range of quantitative predictions, the precision of which was surpassed only by that achieved in electron physics.

In this way, the newest of the nuclear models entered the scene in a strong position to become a unifying theoretical framework for explaining nuclear properties. With no hint of the coming complexities of sub-nuclear (particle) physics, the mid-1950s was a time of unlimited optimism about both practical and theoretical developments in nuclear physics. The previously-established successes and strengths of the liquid-drop and cluster models did not, however, disappear and, as a matter of historical fact, all three models and their modern variants remained as important parts of modern nuclear theory.

The successes of the shell model were impressive and introduced a quantum mechanical formalism that has been in use ever since. Its quantitative strengths were, however, most notable at or near the magic nuclei, while the properties of most other nuclei, particularly the heavy nuclei, did not succumb to similar analytic techniques. To account for such nuclei, Aage Bohr, Mottelson and Rainwater developed the so-called collective model in the mid-1950s. In this model nuclear properties are attributed to the surface motion inherent to a liquid-drop and, by allowing the liquid-drop to assume non-spherical shapes, magnetic and quadrupole moments could also be explained. Despite the added complexity of the collective model, it had the attraction of being a development of the liquid-drop model using a shell-model-like potential-well that could be distorted from a simple spherical shape to more realistic prolate, oblate and even more complex shapes. Predictions of the model included various vibrational and rotational states that have since been experimentally verified. The model was named the "unified model" in celebration of its conceptually complex amalgamation of both liquid-drop and shell model features.

To explain the dynamics of nuclear reactions, further models have also been developed – notably the "optical" model, but since 1948 nuclear structure theory has been elaborated primarily on the basis of two strikingly incompatible ideas – the independent-particle orbiting of nucleons within a nuclear gas and the strong, local nuclear force interactions in a dense nuclear liquid. A

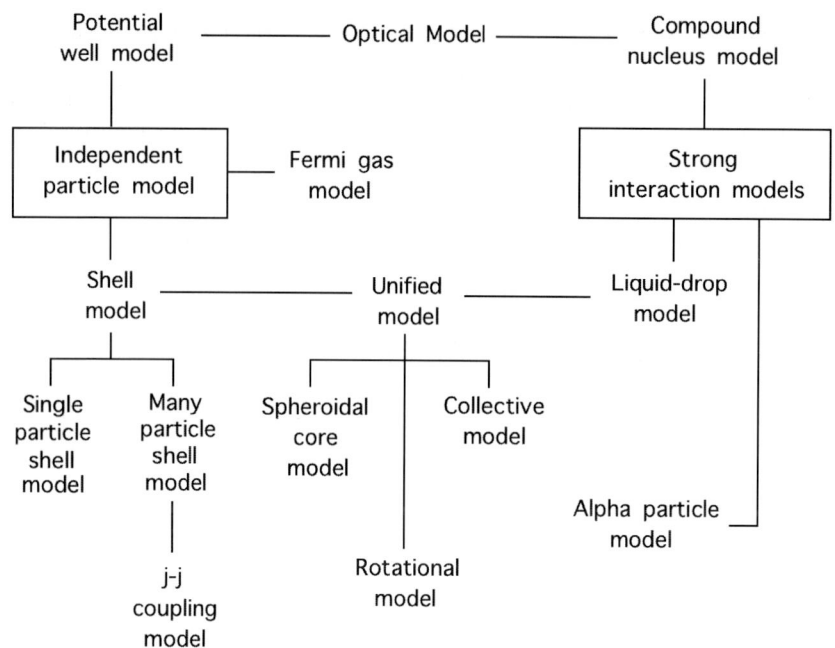

Fig. 3.1. Models of the nucleus (after Moszkowski, 1957)

summary of the models of the nucleus, circa 1957, is shown in Fig. 3.1 and the assumptions underlying the models are summarized in Table 3.1.

There has been much progress in both theoretical and experimental physics since the 1950s, perhaps the most important of which has been the realization that the so-called elementary particles, including protons and neutrons, are themselves made of smaller constituents. For many purposes, thinking of the atom as a system of protons, neutrons and electrons will suffice, but to particle physicists there is a lower level (higher energy) reality that is even more elementary. As a consequence, the majority of one-time nuclear physicists, much of the intellectual enthusiasm and most of the research funding have drifted into the field of particle physics.

Sorting out the issues of the symmetries, energies and structures of the elementary particles themselves has become the central chore of theoretical physics worldwide, but a minority of physicists has focused on the question of what subnuclear particle properties might imply concerning nuclear structure. Although still not developed as a coherent model of nuclear structure, since the 1960s the idea that the constituents of nucleons (quarks or partons) might someday explain nuclear phenomena is widely believed: In principle, a rigorous quark (parton) theory should underlie all of nuclear physics and eventually allow for the deduction of the relatively macroscopic properties of the nucleus on the basis of a more microscopic particle theory. Since the late 1970s, various

Table 3.1. Assumptions of the nuclear models (Moszkowski, 1957)

Model	Type	Assumptions
Independent particle nuclear models (IPM)		Nucleons move nearly independently in a common potential.
Strong interaction models (SIM)		Nucleons are strongly coupled to each other because of their strong and short range interactions.
Liquid drop model	SIM	Nucleus is regarded as a liquid drop with nucleons playing the role of molecules.
Fermi gas model	IPM	Nucleons move approximately independently in the nucleus and their individual wave functions are taken to be plane waves.
Potential-well model	IPM	Nucleus is regarded as a simple real potential-well.
Compound nucleus model	SIM	Whenever an incident nucleon enters the nucleus, it is always absorbed and a compound nucleus is formed. The mode of disintegration of the compound nucleus is independent of the specific way in which it has been formed.
Optical model (cloudy crystal ball model; complex potential-well model)	IPM	A modification of the potential-well model in which the potential is made complex to account for elastic scattering as well as nuclear reactions. The latter effectively remove nucleons from the beam of incident particles.
Alpha particle model	SIM	Alpha particles can be regarded as stable subunits inside the nucleus.
Shell model	IPM	Nucleons move nearly independently in a common static spherical potential which follows the nuclear density distribution.
Single particle shell model	IPM	Same as shell model and: specific properties of odd-A nuclei are due to the last unpaired nucleon.
Many particle shell model	IPM	Same as shell model and: coupling between loosely bound nucleons due to mutual interactions is taken into account.
j-j coupling model	IPM	Same as many particle shell model and: Each nucleon is characterized by a definite value of angular momentum.
Unified model	IPM	Nucleons move nearly independently in a common, slowly changing non-spherical potential. Both excitations of individual nucleons and collective motions involving the nucleus as a whole are considered.
Collective model	IPM	Same as unified model except: only collective motions involving the nucleus as a whole are considered.
Rotational model (strong coupling model)	IPM	Same as unified model and: Nuclear shape remains invariant. Only rotations and particle excitations are assumed to occur.
Spheroidal core model	IPM	Same as unified model and: Deformations of the nucleus into a spheroidal shape results from the polarization of the core, the bulk of nucleons in filled shells, by a few nucleons in unfilled shells.

models of the quark contribution to nuclear physics have been suggested (e.g., Robson, 1978; Bleuler, 1984; Miller, 1984; Petry, 1984), but have not yet had a major influence on the traditional issues of nuclear structure theory. It remains to be seen what quark models might contribute to nuclear physics, but it is clear that the major figures in the developments of 20th Century nuclear structure theory do *not* anticipate that nuclear theory will eventually be rewritten on a quark basis. As Wilkinson (1990) maintains, "quarks are simply not the right language in which to discuss nucleon and nuclear structure in the low-energy regime". Similarly, Elliott (1990) has argued that "the success of the shell-model approach suggests that for low-energy properties ... the inclusion of sub-nuclear degrees of freedom are unimportant".

Finally, it is worth mentioning that, since the discovery of the neutron (1932), there have sporadically appeared solid-phase theories of nuclear structure. These models contain some features of interest, but they also have generally had little impact upon research efforts. Perhaps a mere reflection of the general dissatisfaction with the disunity of nuclear theory or perhaps indication of the attractions of the computational simplicity of solids, the possibility of a nuclear solid has surfaced again and again over the past half-century. Early on, both Niels Bohr and John Wheeler entertained the idea that the nucleus might be an "elastic solid", and Wigner (1979) has commented that a nuclear solid is a reasonable first approximation for the heavy nuclei. Hofstadter (1967) was favorably impressed by an early lattice model (Smith, 1954) that was used to explain the oscillatory character of inelastic scattering results. Despite such comments, the solid-phase approach was, for decades, the *only* phase-state of nuclear matter that was not considered to be a serious contender for explaining the properties of nuclei. Surprisingly, the 1980s brought a huge increase in high-energy heavy-ion scattering data that could not be explained using the shell, liquid-drop or cluster models and it was here that the solid-phase models were shown to have applications. In an attempt to simulate the complex data produced by high-energy nuclear collisions (so-called "heavy-ion" research), a variety of lattice models were developed and exploited – together with new developments in computer technology. Despite lacking a rigorous theoretical foundation to justify the lattice texture, these models have been used with unrivalled success for quantitative explanation of the so-called multifragmentation data produced by high-energy nuclear reactions.

Various attempts at exploring the solid-like characteristics of nuclei have been made over the past 50 years, but the lattice models have generally been advocated, not as comprehensive models of nuclear structure – comparable to and in competition with the liquid-drop and shell models, but rather as computational techniques, suitable for application to a small class of phenomena. Bauer (1988) is representative in arguing for the usefulness of a lattice model approach in predicting the results of high-energy multifragmentation data, but quite explicitly denies the possible significance of the lattice as a "nuclear model":

"Even though nuclei are not lattices and one should be very careful with applying such a concept to nuclear physics, we think we have shown the usefulness of [lattice] percolation ideas in nuclear fragmentation" (Bauer, 1988)

Bauer and others who have used lattices for computational purposes have argued only that the lattice models embody enough of the statistical properties of nuclei that they have a realm of computational utility. A stronger argument will be developed in Chaps. 9 and 10, but during the heyday of active nuclear structure theory (1932–1960), virtually every possibility except a solid of protons and neutrons was thoroughly explored.*

---

*A major factor that has discouraged the use of nuclear lattice models can be traced back to the philosophical debate concerning the interpretation of the uncertainty principle. Although Bohr, Einstein and countless others have argued this issue to a stalemate, Bohr's so-called Copenhagen interpretation has often been used to make ballpark estimates of physical quantities. It must be reiterated that there is no general consensus on the ultimate meaning of Heisenberg's principle – whether it signifies simply an absolute limit in experimental precision or an inherent imprecision in physical reality, but it is certain that the latter interpretation has led to mistaken conclusions. Neither view has been logically or empirically proven, but it is known that one of the earliest *misuses* of the uncertainty principle goes back to the early 1930s when Bohr – convinced of the inherent indeterminacy of nature itself – had the audacity to discourage Rutherford (the discoverer of the nucleus!) from pursuing questions of nuclear structure (Steuwer, 1985). He maintained that in order to theorize about structures less than 10^{-14} m in diameter, the indeterminist *interpretation* of the uncertainty principle would indicate that such an object would have a correspondingly large potential energy – and "common sense" in the 1930s made such energies seem unlikely. Bohr thus propounded on his philosophy of "complementarity" – and argued that the nucleus must be treated as a collective whole, impossible to analyze on the basis of its independent particles. As important as Bohr's many contributions to atomic and nuclear physics have been, his declaration that progress in reductionist science ends at the nuclear level has proven quite wrong. Not only is the independent-particle description of nuclei now the established basis of modern nuclear theory, but the next level of subnucleonic microreality, is known to have spatial structure – with the quarks, as physical objects, now thought to have finite radii of $0.2 \sim 0.4$ fm. No one argues that the uncertainty principle itself is wrong – it is verifiably true that there are limits to simultaneous experimental determination of various physical properties, but use of the Copenhagen *interpretation* of the uncertainty principle *together with* assumptions about the potential energy of elementary particles has clearly led to incorrect conclusions concerning physical phenomena. Similar applications of uncertainty arguments have been made in the textbooks with regard to nuclear structure, and this has been at the root of the longstanding neglect of nuclear lattices. Only by ignoring the textbook warnings that specification of nucleon positions to $\sim \Delta 2.0$ fm is energetically impossible has it been possible to show that lattice simulations of multifragmentation phenomena give results that explain experimental findings – regardless of *philosophical* arguments.

Until the present time, nuclear structure theory has focused primarily on variants of the shell model – with its implications concerning a central nuclear potential-well – and on variants of the liquid-drop model – with its implications concerning a strong, local nuclear force. The cluster models have remained a minority view with several notable successes, while the lattice models have been seen as a curiosity with only circumscribed computational merits.

Much work has subsequently been done on all of these models – gas, liquid, cluster and lattice – and most nuclear properties and nuclear phenomena can be explained in terms of one model or the other. In general, if the properties of one of the nucleons in a many-particle nucleus dominate, then the independent-particle model is most successfully applied. If the nucleus can be considered as consisting of 4n alphas or 4n alphas plus an additional proton or neutron, then the cluster models sometimes prove useful. If, however, the collective properties of all nucleons or of many nucleons in a valence shell predominate, then the strong-interaction, collective models are the best starting point. Finally, for theoretical study of heavy-ion reactions, in which the interactions of many nucleons must be computed, the lattice models have proven useful.

Despite their growing sophistication and even the use of the word "unified" to describe a version of the collective model, the answer as to whether nuclear structure physics is a coherent whole depends critically on whether or not these different models have in fact been brought together within a single conceptual framework. Are they based upon complementary or contradictory assumptions? Is theoretical nuclear physics a self-consistent body of knowledge or can experimental results be explained only with mutually incompatible models? Are the computational successes of conventional nuclear theory indicative of a fundamental understanding of the nucleus or do multiple models indicate the presence of remaining problems? These questions will be raised in more concrete terms in Chaps. 5 through 8.

So, just how many models are in use in nuclear theory today? There is no authoritative answer to such a question, but Greiner and Maruhn provide a reasonable lower limit of 31 in their textbook, *Nuclear Models* (1996). Rather than summarize the nuclear models from a historical perspective, they prefer the conceptual division into collective (=liquid-drop-like) and microscopic (=independent-particle) models, with mixed models lying somewhere in-between (Fig. 3.2).

To this list of 31 models must be added those models not discussed by Greiner and Maruhn – among which three distinct variants of the cluster model are of interest, and the lattice models (among which three variants can be distinguished) – giving a total of 37. Noteworthy is the fact that, taken in pairs, many of these models are consistent with one another. Sometimes, the independence of protons and neutrons is the focus of the model; other times, the distinction between the nuclear core and its surface is emphasized. In yet other models, axial symmetries or deformations are the central issue.

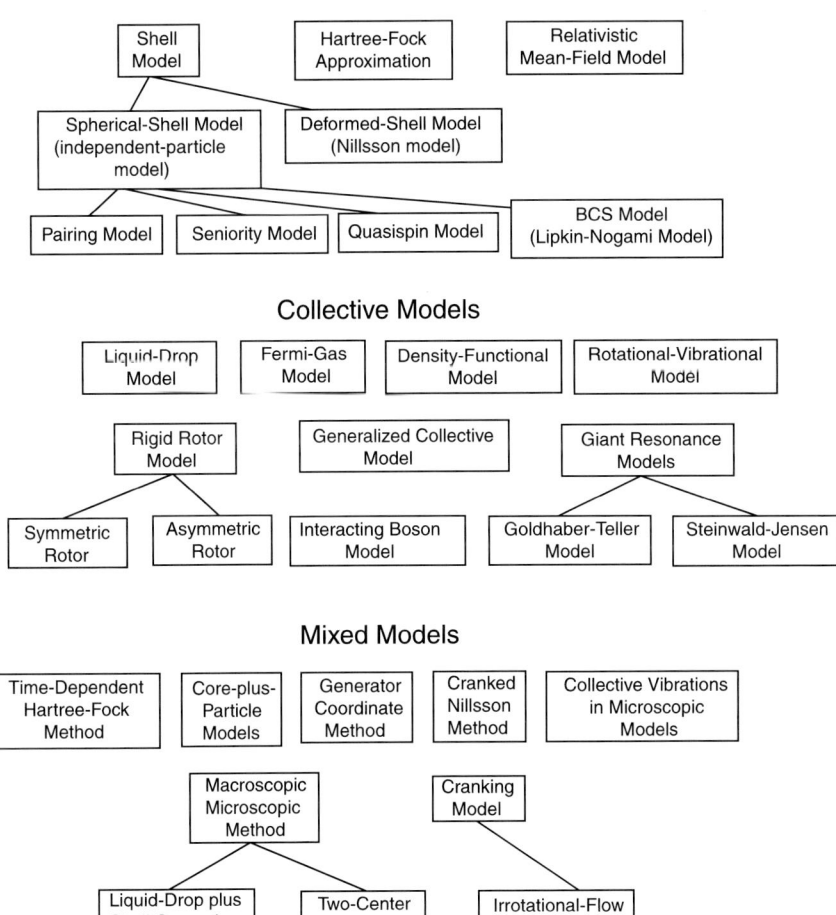

Fig. 3.2. A summary of the three general classes of nuclear models (after Greiner & Maruhn, 1996)

There is in fact no reason to demand that any specific model deal with all nuclear properties at once. Nonetheless, the unsettled nature of nuclear theory is evident in the fact that the assumptions behind these models are often directly incompatible with one another.

It would be inaccurate to leave the impression that nuclear theory came to a halt in the 1950s. On the one hand, it is true that the majority of theoretical developments occurred prior to 1960 and subsequent work on the shell, liquid-drop and alpha-cluster models has focused on the fine-tuning of the ideas originally proposed in the 1930s, 1940s and early 1950s. On the other

Table 3.2. A summary of the four main classes of nuclear model and the types of experimental facts that they can explain

Theory	Supporting Empirical Findings
Weak Nuclear Force Models	
Shell model (1949)	Magic nuclei.
Independent-particle model (1950s)	Systematics of nucleon quantum numbers.
Fermi gas model (1950s)	Nuclear angular momenta and parities.
	Nuclear magnetic moments (Schmidt lines).
Strong Nuclear Force Models	
Liquid-drop model (1930s)	Constant nuclear density.
Compound-nucleus model (1930s)	Saturation of nuclear force.
Collective model (1950s)	Nuclear surface tension effects.
Droplet model (1960s)	Dependence of nuclear radius on A.
Lattice models (1980s)	Phase-transitions in heavy-ion experiments.
Nucleon Clustering Models	
Alpha cluster model (1930s)	Alpha particle radiation.
Spheron model (1960s)	Unusual stability/abundance of 4n nuclei.
2D Ising model (1975)	4n clusters in nuclear fragmentation.
Interacting boson model (1980s)	Nucleon pairing effects.
Very Strong Nuclear Force Models	
Quark models (1980s)	Known properties of the nuclear force.

hand, a great deal of quantitative fine-tuning has been accomplished. It is important to realize that traditional nuclear structure theory (including those developments that led to the harnessing of nuclear power and the invention of nuclear weapons) was accomplished exclusively by pencil-and-paper theorists – prior to the invention of the computer! As more and more theorists got their hands on computers, it became possible to develop vastly more complex models that allowed the fitting of theoretical models to huge volumes of experimental data with extreme precision. The complexity and sophistication of those models are impressive, but the large number of model parameters that a computer can easily handle makes it, if anything, more difficult to determine whether any given model has provided an explanation of the empirical data or, contrarily, has simply been massaged into agreement with data by the proliferation of adjustable parameters. However that issue may be decided, it is today the case that the current models of nuclear structure theory have proven flexible enough that virtually any experimental finding can now be explained within one model or another (Table 3.2).

A chronological classification of the models in nuclear structure theory is shown in Fig. 3.3, where the relative isolation of the models assuming strong nuclear force effects and those assuming a weaker central binding force is apparent. It is unlikely that any classification will find the unanimous support of all theorists, but the undeniable fact is that nuclear structure theory is

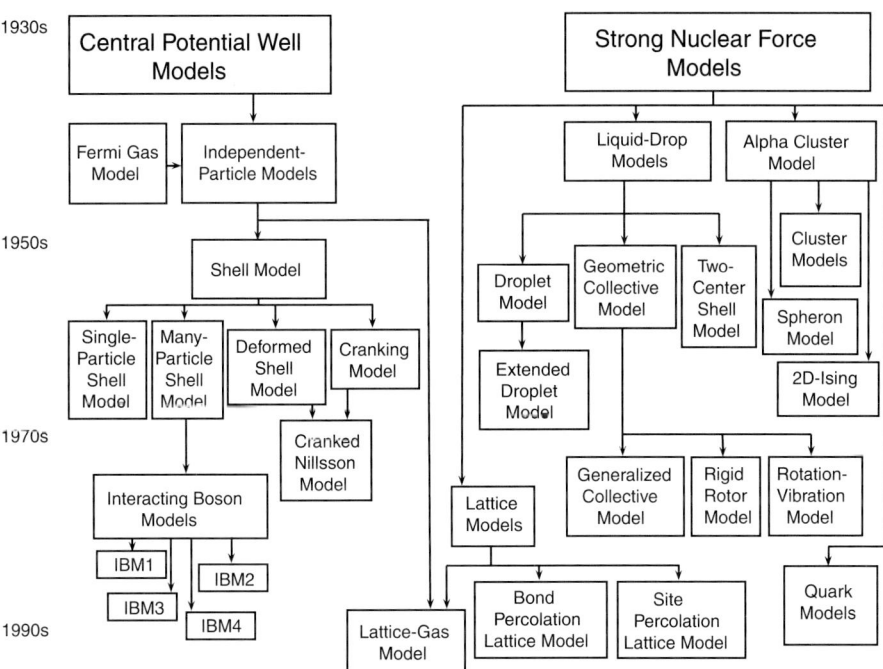

Fig. 3.3. A chronology of the evolution of nuclear models. Note that the roots of the lattice-gas model lie in both the independent-particle model and the strong nuclear force models (Cook, Hayashi & Yoshida. 1999)

unusual among the modern sciences (including quantum chemistry) in employing many different models, embodying many different starting assumptions, to explain essentially one and the same object. This is simultaneously the lure and the bane of nuclear structure physics.

4

Nuclear Models

The words, "theory", "hypothesis" and "model", are often used loosely to mean the same thing: an explanatory framework within which certain facts seem to fit together. To most scientists, however, "theory" is the ultimate goal and implies a small set of universal laws from which firm conclusions and clear predictions can be drawn. "Hypotheses" are more-or-less *ad hoc* ideas that normally can be expressed in mathematical form and that might eventually become formal theories, but hypotheses are developed for the purpose of explaining a limited range of phenomena, with no pretense to being fundamental laws of nature. In contrast to both theories and hypotheses, "models" are little more than rough analogies, and are employed before hypotheses and theories are possible.

For better or worse, nuclear structure physics has seen the use of many models – analogies with macroscopic objects which seem to capture some aspects of nuclear phenomena, but which do not have sufficient power or thoroughness that the analogies can be taken entirely seriously. Because models are not altogether rigorous, their seemingly inevitable employment has been explained as follows: Material systems such as nuclei are too complex and contain too many constituents to be handled precisely with formal "bottom-up" theories, but they are too small and idiosyncratic to be handled with rigorous statistical methods that normally require large numbers to justify stochastic assumptions. Containing less than 300 constituents, nuclei fall in a no-man's land between exact theory and reliable statistics. Here, models can be useful (but be careful!).

The conventional models can be classified into three basic groups corresponding to the implied phase-state of nuclear matter – liquid, gas and semi-solid (cluster) (Fig. 4.1). A fourth type is based on quarks – and is implicitly gaseous, liquid or solid, depending on related assumptions. Each type of model has merits and each has been advocated as a coherent view of nuclear reality, but most theorists consider the models to be, at best, temporary solutions – partial truths that can explain only a few aspects of the structure and dynamics of the nucleus – that await the development of formal theory.

Fig. 4.1. Four models for the $^{114}Sn^{64}$ nucleus (visualized using the NVS program on CD). (**a**) shows the gaseous-phase independent-particle model, where individual nucleons are attracted to the nuclear core by a central potential-well. (**b**) shows the liquid-phase liquid-drop model, where nucleons bind to nearest-neighbors. (**c**) shows the semi-solid cluster model, where the importance of nucleons (*depicted as diffuse probability clouds*) is de-emphasized, relative to that of the configuration of alpha particles (*depicted as solid spheres*). (**d**) shows the quark model, where again nucleons are de-emphasized, while the constituent quarks within the nucleons are emphasized

4.1 The Collective Models

The earliest model of nuclear structure was a collective model based on a liquid-drop analogy – first proposed by George Gamow in 1929 prior to the discovery of the neutron, and later championed by Niels Bohr (1936). Gamow is only infrequently credited (Johnson, 1992) with proposing this most-resilient of models, probably for two reasons. The first is that Bohr effectively used the liquid-drop conception as a means of propounding on his philosophical idea known as complementarity. He was convinced that the uncertainty relations raised insurmountable barriers to knowing physical reality beyond certain limits and, subsequent to his successes in atomic physics, he devoted much of his energy to arguing this philosophical case. In a world tour in 1935, Bohr advocated the complete rejection of the independent-particle model of the nucleus and argued for a "collective" model – not simply because the liquid-drop analogy was more useful, but because the isolation and characterization of independent-particle features implied a precision greater than that allowed by his understanding of the uncertainty principle. Moreover, not only had the liquid-drop model already proved its worth as a theoretical tool with the elaboration of Weizsäcker's semi-empirical mass formula, it was the application of the liquid-drop conception to the startling phenomena of fission by Bohr and Wheeler in 1938 that commanded the attention of nuclear physicists.

Subsequent events indicated that Bohr had overstated the case in arguing that further details of nuclear structure would remain obscure behind the uncertainty veil, but the combination of his outspoken philosophical advocacy and the actual utility of the liquid-drop model meant that this model of the atomic nucleus drew strong support in the first few years after the discovery of the neutron in 1932. The essence of the model was that, quantum physics or not, nucleons are likely to interact in ways that are roughly analogous to other known physical systems. While Bohr and others were loathe to speculate about the properties of individual nucleons, the collective behavior of many-particle systems could be discussed without colliding with the uncertainty principle by assuming that each nucleon was localized to a relatively large volume within the nucleus. It should be recalled that, already by the mid-1930s, physicists had been submerged in the ideas of the quantum world for more than three decades. The difficulties of quantum physics were entirely familiar to them from the known complexities of the electron as a particle-wave and the ongoing philosophical controversies in atomic physics. So, when it came to constructing a model of the nucleus, the difficulties of quantum physics did not initially lead theorists into speculation about time-reversal or non-local forces or multiple overlapping universes. The first task was to account for some of the bulk properties of nuclei – and that was where the liquid-drop analogy was strong.

The early successes of the model were impressive. The only features of the nuclear force that were known with some certainty in the 1930s – its saturation and thus the constant binding energy per particle – were consistent with the nearest-neighbor interactions implied by the droplet conception. Those features appeared to present problems for a gaseous model. Most important, however, was the fact that, by the late 1930s, fission had become the central concern of experimental, theoretical and applied nuclear physics. Fission of a large nucleus could be usefully conceptualized as the break-up of a liquid-drop under the disruptive influence of excessive positive charge within the nucleus. For these reasons, while independent-particle ideas continued to evolve throughout the 1940s, the dominant paradigm for all practical issues was the liquid-drop.

Some of the strengths of the liquid-drop model will be examined below, but it is appropriate to note here the general character of this model. From the start, the liquid-drop model was a *collective* model: it set out to explain the gross properties of medium-sized and large collections of nucleons – not the smallest nuclei for which the ideas of a nuclear "core" and a nuclear "surface" were of questionable relevance. By the 1950s, a considerable volume of experimental data on binding energies and radii had been accumulated, and the first attempts to systematize such data were made within the context of the liquid-drop model. The model *was* not and, to this day, *is* not, a "first-principles" theory of the nucleus based strictly on detailed knowledge of the nucleon and the nuclear force itself. Whether or not this is a fatal flaw in the model depends on how it is to be employed. For the description of the general features of nuclei, the liquid-drop model has remained without rival, but there are few firm conclusions that can be drawn specifically about the nuclear force. The liquid-drop analogy describes the nuclear "community," and has very little to say about the nature of individual nucleons.

The Weizsäcker Mass Formula

One attraction of the liquid-drop model was that it lent itself directly to theoretical calculations of the binding energies of nuclei. By the mid-1930s scores of binding energies of nuclei were known, and provided the raw material for theoretical work. Notable was the so-called semi-empirical mass formula developed by Weizsäcker (1935).

In principle, the total binding energy of a droplet of liquid could be calculated by counting the number of molecules and adding up the total number of nearest-neighbor bonds, but a more realistic technique is to do approximate calculations based upon the presumed size of the constituents, the total volume of the droplet and its total surface area. The latter values can be taken as proportional to (i) the number of particles that interact with a maximum number of neighbors (in the nuclear core region) and (ii) the number of particles on the surface of the droplet. Insofar as the overall effect between neighboring particles is attractive, particles on the surface will be bound somewhat less strongly than particles imbedded in the nuclear interior. Moreover, the surface particles will feel a net attraction pulling them toward the center of the nucleus, and this effect will be responsible for maintaining an approximately spherical shape of the droplet due to surface tension-like effects. The core and surface properties were thus taken as the two main terms that determine nuclear binding in the Weiszäcker formula. A third term with a basis in classical physics was the Coulomb repulsion acting among all pairs of protons (Fig. 4.2).

The volume term is proportional to the number of nucleons present, A, and the surface term is proportional to the surface area of a sphere containing A nucleons, $A^{2/3}$. Both are a simple consequence of the approximately spherical geometry of the liquid-drop (Hasse & Myers, 1988). The Coulomb term is proportional to the square of the number of protons present, $Z(Z-1) \sim Z^2$, since the electrostatic repulsion among all pairs of protons must be included.

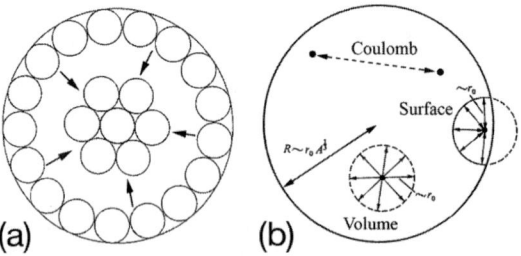

Fig. 4.2. A summary of the basic components of the liquid-drop model. (a) depicts the volume and surface components. (b) shows the three main terms in the formula for binding energies are the volume term, the surface term and the Coulomb term. The parameter r_0 is used in the calculation of nuclear size (after Das and Ferbel, 1994)

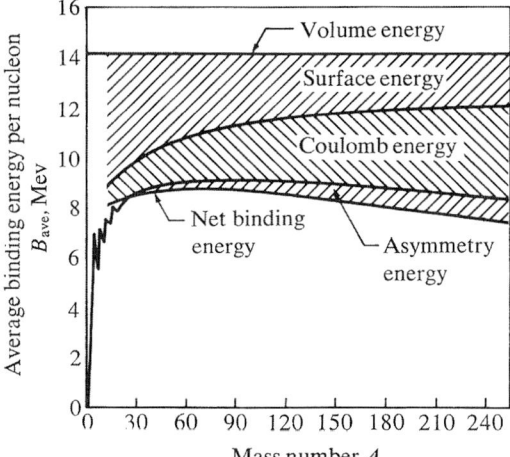

Fig. 4.3. The contributions of the four major terms in the semi-empirical mass formula. The pairing energy term makes corrections too small to include in the figure. The experimental values lie near to the curve labeled "Net binding energy"

The effects of these three terms are clearly seen in Fig. 4.3, where the successive addition of each term, volume, surface and Coulomb, gradually brings the theoretical binding energy curve down toward the experimental (net binding energy) curve. As shown in the figure, the surface term becomes slightly *less* important as the size of nuclei increases because there is proportionately less surface area as the liquid-drop increases in size. In contrast, the Coulomb term becomes more important because each additional proton feels the repulsion of an ever-increasing number of protons. Together, the volume, surface and Coulomb terms produce the general trend of nuclear binding, but the liquid-drop analogy with only these three terms overestimates the binding energies of the largest nuclei. For this reason, two further terms were included: a "symmetry" term that reflects the relative numbers of protons and neutrons, and a "pairing" term that reflects the tendency of like-nucleons to couple pair-wise for additional nuclear stability.

The symmetry term is proportional to the number of neutrons relative to protons, $(A - 2Z)^2/A$. The necessity of this term for estimating the total binding energy is clearly indicated from the data, but its physical justification does not lie in the liquid-drop analogy. Finally, the pairing term is also a data-driven correction to the formula, since it was found that nuclei with an even-number of protons or neutrons have slightly higher binding energies than nuclei with odd-numbers.

Under the assumption that nuclei are approximately spherical, the basic Weizsäcker formula can be stated as:

$$\text{BE}(Z, N) = k_1(\text{Vol})A + k_2(\text{Surf})A^{2/3} + k_3(\text{Coul})Z(Z-1)$$
$$+ k_4(\text{Sym})(A - 2Z)^2/A + k_5(\text{Pair})/A^{1/2} \quad (4.1)$$

Since the total number of protons and neutrons are known for any nucleus whose binding energy has been measured, all that was required was to deduce the values of the five constants, $k_1 \sim k_5$, that would give a best fit to the experimental data. One set of such parameters (Yang & Hamilton, 1996, p. 392) is: $k_1 = 15.8, k_2 = -18.3, k_3 = -0.72; k_4 = -23.2, k_5 = -11.2$.

The Weizsäcker formula is often referred to as the *semi-empirical* mass formula for two reasons. Most obviously, the formula is "semi-empirical" because the values for the constants ($k_1 \sim k_5$) are determined from a best-fit of the formula to the empirical data, i.e., it is not a calculation based on first-principles concerning nucleon-nucleon interactions. Given a large enough database of binding energies for a variety of nuclei, it is a straight-forward computational matter of determining what values will, on average, best reproduce the experimental data. It is noteworthy, however, that this formula is "semi-empirical" for a second reason, as well. The pairing and symmetry terms are introduced – not because the liquid-drop analogy suggests the reality of such terms, but because empirical findings indicate increased binding when the numbers of protons and neutrons are even and when there is an appropriate balance between the number of protons and neutrons. They are truly empirical corrections to the physical model and remain foreign to the liquid-drop analogy.

Because the Weizsäcker binding-energy formula is not based on a fundamental theory of the nuclear force, it does not reveal quantitative features of the nuclear force, but it is nonetheless an impressive summary of nuclear binding. As illustrated in Fig. 4.3, the general shape of the binding energy curve is well reproduced using only five adjustable parameters whose physical meanings are clear. A more detailed study of the results of (4.1) indicated that the largest errors in the five-parameter droplet model lie primarily at the so-called magic numbers (Fig. 4.4). For this reason, more terms were later appended to the five terms that were in Weizsäcker's initial formulation in light of indications of nuclear shell structure:

$$\text{BE}(Z, N) = k_1 A + k_2 A^{2/3} + k_3 Z(Z-1) + k_4 (A - 2Z)^2/A$$
$$+ k_5/A^{1/2} + \text{Shell Corrections} \quad (4.2)$$

The shell corrections are necessarily complex – and vary by shell and by isospin, thus giving 13 additional parameters that are adjusted to minimize the difference between experimental and theoretical binding energies.

The volume term has a numerical value of about 16 MeV, and tells us that the sum of all near-neighbor nucleon interactions must be close to that value. Similarly, the surface term has a value of -19 MeV and corresponds to the decreases in binding energy per particle on the surface of a tightly-packed droplet. The Coulomb term is a simple expression of the repulsion among

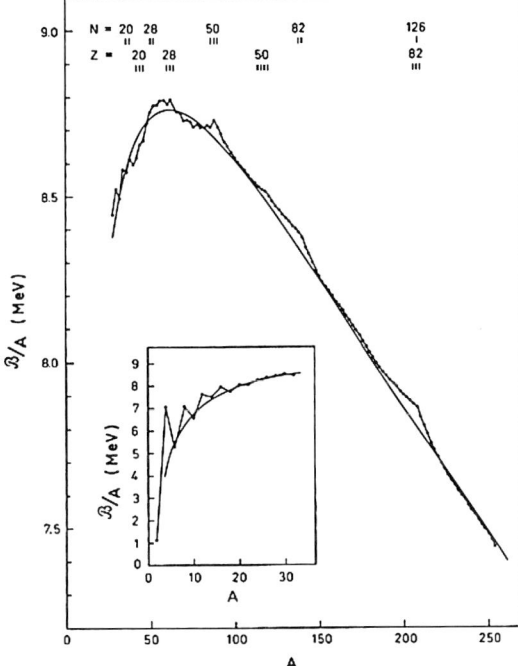

Fig. 4.4. Deviations from the theoretical (*liquid-drop*) binding energy curve (*solid line*) indicated the need for corrections near the magic numbers (adapted from Myers, 1977)

the protons and can be calculated precisely. A quantitative understanding of the symmetry and pairing terms is less obvious, but the necessity of those corrections reveals aspects of the nuclear force that the simple liquid-drop analogy does not.

As illustrated in Fig. 4.5, it is clear not only that there is a most stable ratio of protons-to-neutrons for any given number of nucleons, but the binding

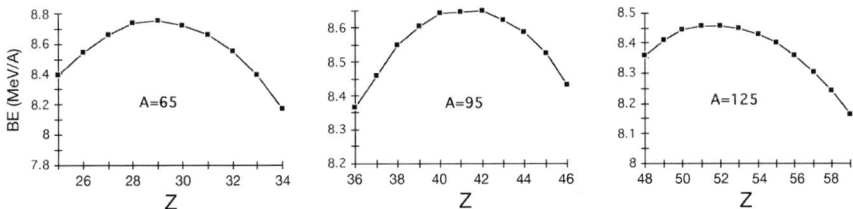

Fig. 4.5. Illustration of the symmetry term. The binding energies (BE) of nuclei with the same $A = 65, A = 95$ and $A = 125$ indicate that for any A, there is a combination of Z and N that gives a maximal binding energy. The smoothness of the parabolas is indication of the collective nature of nuclear binding energies

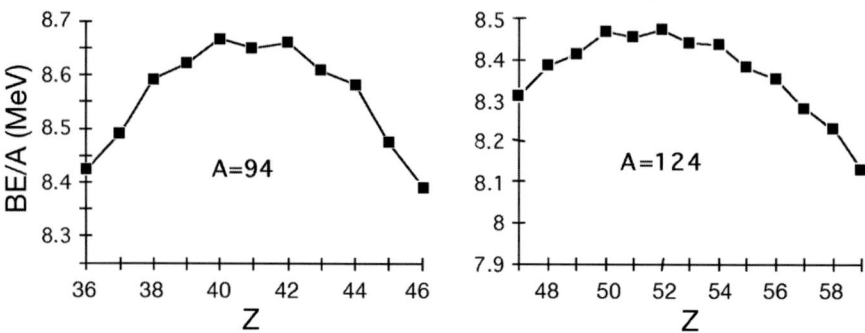

Fig. 4.6. The pairing effect. Binding energies of nuclei with even-A show distinct zig-zag patterns. Whether plotted against Z or N, the alternating pattern indicates that even-Z and even-N nuclei have slightly greater binding energies than their odd-Z and odd-N neighbors

energies of all nuclei with the same number of nucleons form a remarkably precise parabolic function. This is found for any isobar of nuclei with an odd-number of nucleons.

The importance of the pairing term in the mass formula can be seen in Fig. 4.6, where the binding energies of typical even-A nuclei are plotted. Here the jaggedness of the parabolas shows that an even number of protons has slightly higher binding than an odd number. A similar effect is found for neutrons, indicating that there is a "pairing" of the last two same-isospin nucleons. The magnitude of the pairing effect changes slowly over the periodic table (Fig. 4.7), and it is this trend that is modeled with the pairing parameter.

Finally, shell corrections allow for further improvements – and ultimately an extremely good fit between the binding energy formula and experimental data (Fig. 4.8). With a mean error of 0.655 MeV for binding energies ranging from 120 to 1800 MeV, clearly the model captures the basics of nuclear binding. As seen in the figure, however, the fit for the smaller nuclei is less good, and for $Z < 6$ the formula cannot be applied. In its modern form (Möller and Nix, 1995), the formula explains nuclear binding energies across the entire periodic chart (more than 2000 isotopes) with a mean error of 0.06%.

The Nuclear Radius

A second and equally fundamental property of any nucleus is its spatial extent. Implicit to the liquid-drop model is the prediction that the radius of a nucleus will depend on the number of protons and neutrons it contains. Since nuclei are three-dimensional objects, this implies that the radius will be proportional to $A^{1/3}$:

$$\text{RMS}(A) = r_0 \cdot A^{1/3} \tag{4.3}$$

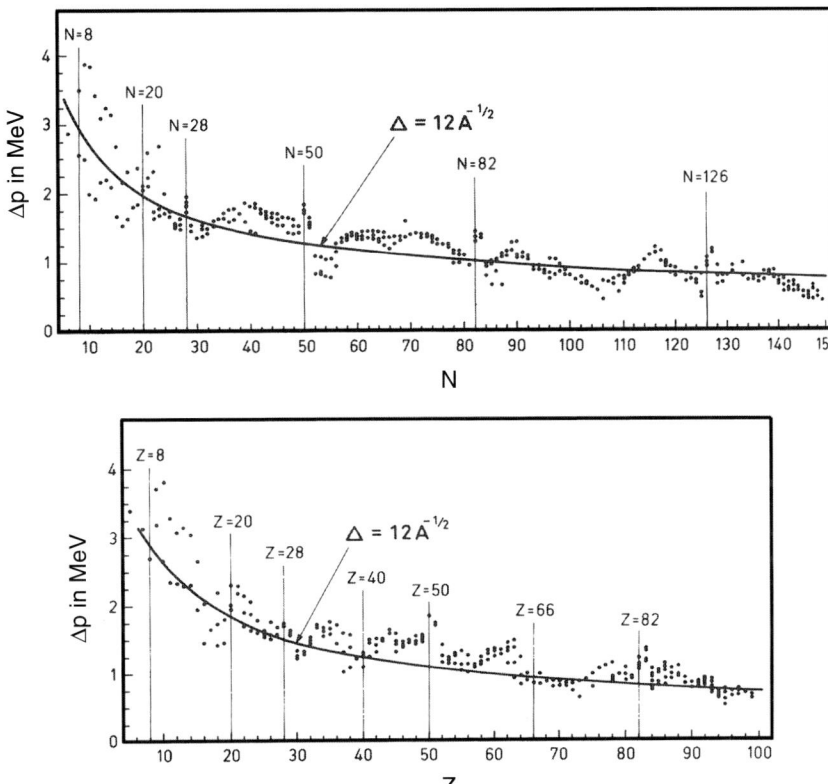

Fig. 4.7. The empirical pairing effect for protons and neutrons (Bohr & Mottelson, 1969, p. 170)

where the best fit with the experimental data is obtained with $r_0 = 0.853$ fm for the 617 known radial values. The near-linear dependence of the radius on the number of nucleons is clearly seen in a plot of $A^{1/3}$ versus R (Fig. 4.9). A similar plot of atomic radii shows a slight trend toward larger radii when more

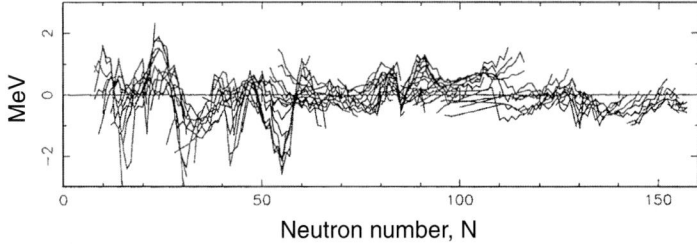

Fig. 4.8. The difference between experimental and theoretical binding energies for 1654 isotopes (Myers and Swiatecki, 1966)

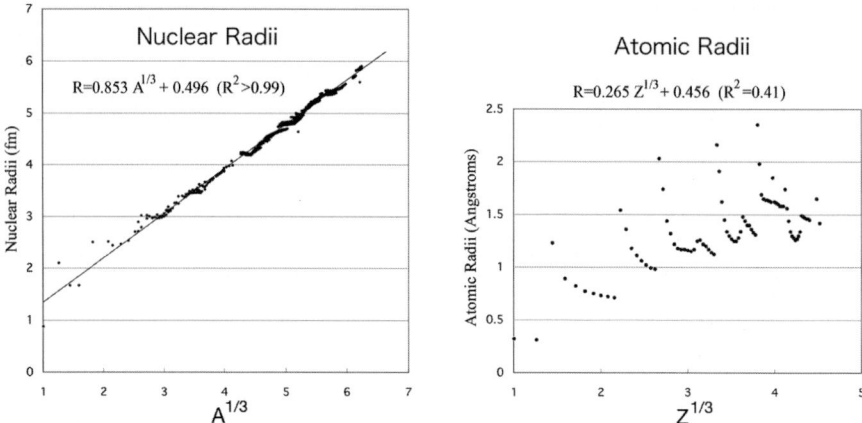

Fig. 4.9. The dependence of the nuclear radius (*left*) and atomic radius (*right*) on the number of nucleons (A) or electrons (Z) present

electrons are present ($R^2 = 0.4$), but there is striking substructure within each shell. The fact that such substructure is *not* seen in nuclei is strong indication that nuclei have a constant-density, liquid-drop-like texture.

In detail, the data on nuclear size show some notable deviations, especially among the lightest nuclei, but the overall linear correlation ($R^2 > 0.99$) suggests that each nucleon occupies a constant volume, as predicted by the liquid-drop model, and that nucleons are *not* overlapping probability clouds comparable to electrons.

The liquid-drop model has been a success in explaining the general trend of both nuclear binding energies and size (Myers, 1977; Hasse & Myers, 1988) , but shows its limitations when dealing with the smallest nuclei ($Z < 20$), as well as nuclei with protons or neutrons near to the magic numbers. These and other nuclear properties have indicated the need for other models, in addition to the liquid-drop model.

Collective Motion

Modern developments of the liquid-drop model have been done primarily within the context of the so-called collective or "unified" model, championed by Aage Bohr and Ben Mottelson (1969). The collective model has focused on nuclear phenomena, where all or most nucleons move in concert. Representative of such effects are the giant dipole resonance and the giant quadrupole resonance (Fig. 4.10). The importance of collective vibrations, oscillations and rotations lies in the fact that the majority of higher-energy states can be understood only as the statistical effects of many nucleons in motion together. In this respect, the collective model is a direct descendant of the liquid-drop model.

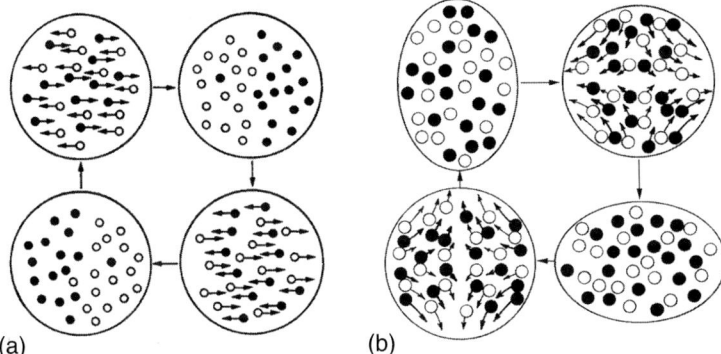

Fig. 4.10. Collective nuclear phenomena known as the giant dipole resonance (a) and the giant quadrupole resonance (b). Protons are the solid circles, neutrons are the open circles (from Bortignon et al., 1998, pp. 3, 27)

4.2 The Cluster Models

The cluster models are based on the assumption that nuclei can be usefully thought of, not only in terms of protons and neutrons, but also as aggregates of small "clusters" of nucleons, the most important of which being the alpha-particle (two protons and two neutrons). The alpha-particle model has some face validity because of the fact that the binding energy per nucleon for the smallest 4n nuclei ($A \leq 40$) is higher than those of neighboring nuclei – indicating unusual stability whenever nuclei have a 4n-multiple of nucleons (Fig. 4.11).

If the simplest geometrical structures for the small 4n nuclei are assumed (Fig. 4.12), then it is found that the binding energies of these small nuclei is equal to the sum of the binding energy of the alpha-particles themselves (~28 MeV) plus a small contribution from the bonds between alpha particles (Table 4.1). The trends apparent in Fig. 4.11 and Table 4.1 were the initial supporting evidence for the alpha-particle model and remain a strong indication that the cluster perspective on the nucleus may have some validity. Moreover, historically, a major impetus to studying alpha-particle configurations was the fact that alpha-particles are spontaneously released from many of the largest radioactive nuclei. All theoretical speculation about nuclear substructure aside, we *know* empirically that alphas are emitted from nuclei – and must therefore have at least transient existence on the nuclear surface or in the nuclear interior.

It is noteworthy that the average binding force between alpha particles (B_{alpha}/Bond) remains fairly constant at about 2.5 MeV per bond. Since this "bond" is the net effect between nearest-neighbor alpha-particles, its meaning in terms of the nuclear force (acting between nucleons) is not in fact clear. But, if the assumption is made that clusters assemble themselves into geometrical

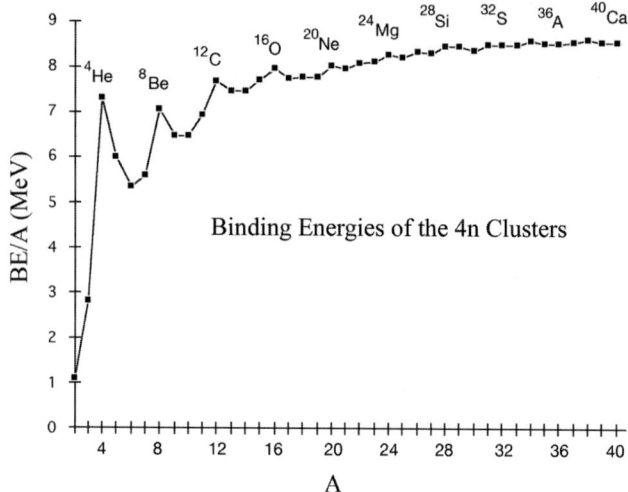

Fig. 4.11. The binding energy per nucleon (BE/A) for the smaller stable nuclei. There is a peak for each of the alpha-cluster nuclei, indicating a special stability of such nuclei

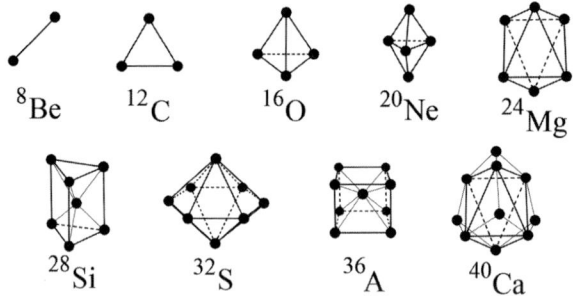

Fig. 4.12. Possible molecular structures of the simpler 4n-nuclei (after Brink et al., 1970)

states of equilibrium, then it is reasonable to suppose that the intercluster bond would have a constant value.

Deciding on what is the "optimal" molecular configuration for a given number of clusters is, however, not obvious for any but the smallest 4n-nuclei and different configurations will often have different numbers of intercluster bonds. The approximate constancy of B_{alpha}/Bond in Table 4.1 is therefore somewhat fortuitous and modern versions of the cluster models have been developed such that arbitrary decisions concerning the configuration of alphas can be entirely avoided. Specifically, by assuming local binding effects among either nucleons or among alphas themselves, and letting a complex many-body system settle into various quasi-stable states using computer simulation techniques, there is no need to select among possible alpha configurations

4.2 The Cluster Models

Table 4.1. The binding of the small 4n nuclei

Nucleus	Number of alphas	Number of Bonds between Alphas	B_{alpha} (MeV)	B_{alpha}/Bond (MeV)
^{4}He$_2$	1	0	0.00	
^{8}Be$_4$ (unstable)	2	1	−0.10	−0.10
^{12}C$_6$	3	3	7.26	2.42
^{16}O$_8$	4	6	14.40	2.40
^{20}Ne$_{10}$	5	8	19.15	2.39
^{24}Mg$_{12}$	6	12	28.46	2.37
^{28}Si$_{14}$	7	15	38.44	2.56
^{32}S$^*_{16}$	8	18	45.38	2.52
^{36}Ar$^+_{18}$	9	20	52.02	2.60
^{40}Ca$^{**}_{20}$	10	24	59.05	2.46

*hexagonal bipyramid, + body-centered cube, ** intersecting tetrahedron and octahedron (after Goldhammer, 1963)

(see, for example, Fig. 4.13). The simulations themselves provide candidate structures.

The alpha-cluster model has been developed over a period of more than six decades, and binding energies and vibrational energies of certain of the small 4n-nuclei have been worked out in impressive detail. While still a minority view, there are those who argue that the alpha cluster model is as good a nuclear model as there is (Hodgson, 1982), and indeed that unification can occur only within the cluster conception (Wildermuth & Tang, 1977; Wuosmaa et al., 1995).

Despite various successes of the cluster models, however, any model that applies to only a small percentage of the known nuclei cannot be considered as a general theory. The fact that all of the larger stable nuclei ($A > 40$) have

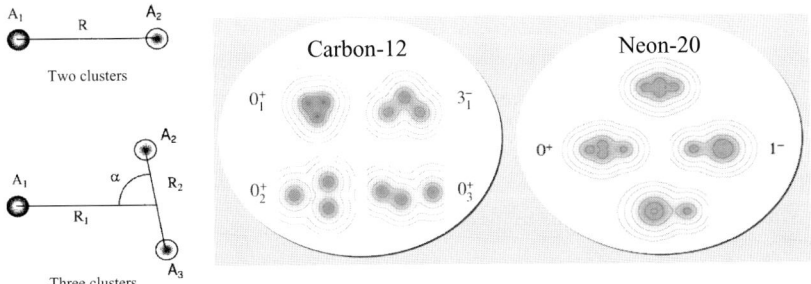

Fig. 4.13. Examples of clustering in the modern alpha-cluster model (Neff and Feldmeier, 2003). Instead of assuming fixed geometries, calculations on small numbers of clusters (e.g., A_1, A_2, A_3, ...) can be done by allowing distances (e.g., R_1 and R_2) and angles (α) among the clusters to vary

unequal numbers of protons and neutrons means that a simple summation of alpha particles will no longer work and, in fact, most theorists have ventured into the framework of the collective model or the shell model in search of a general theory of nuclear structure – except when the small 4n nuclei must be dealt with. There have been two unconventional exceptions to this trend away from the cluster models, however – the models developed by Malcolm MacGregor and Linus Pauling.

Other Cluster Models

MacGregor's (1976) cluster model was an ambitious and internally self-consistent attempt to explain all nuclei in terms of the build-up of a small set of 2-, 3- and 4-nucleon clusters. The basic assumptions behind the model are that: (i) all nuclei can be considered as aggregates of alpha-particles, tritons, helions, and deuterons, and that (ii) all such clusters are arranged into so-called two-dimensional Ising layers. The second assumption is what is particularly unusual about MacGregor's work and allows for a systematic (if somewhat non-intuitive) build-up procedure for the clusters. The 2D Ising layer is a convenient idea for use in a cluster model for it demands specification of the number of clusters within each layer, but the actual "molecular" configuration within each layer is left unspecified. This avoids the problems of deciding which among several conceivable geometries must be chosen as the correct structure for a given number of clusters. In other words, the 2D Ising model contains more structure than a liquid, but less structure than a full specification of the location and binding of all clusters.

In defending the idea that nuclei consist of nucleon clusters, MacGregor cited a large body of evidence indicating the existence of alpha clusters in the nuclear interior. Clustering can be explained only by assuming a strong, *local* force acting between nucleons, evidence for which comes from relatively high-energy experiments that were the focus of much research in the 1970s. The first is the so-called multi-alpha knockout experiments, in which a target nucleus is bombarded with projectiles of sufficient energy that the target is shattered by the collision. When the masses of the nuclear fragments are measured, an abundance of alpha particles or their multiples is consistently found. Cluster fragmentation occurs regardless of the projectile used for bombardment and has been found with high-energy negative pions and high-energy protons, as well as low-energy pions, low-energy kaons, and positive pions.

MacGregor (1976) commented, as follows:

> "The results on alpha-particle knockout ... offer a challenge to one of the basic tenets of the conventional shell model. In the standard shell model formulation, all of the nucleons in a nucleus (and not just the unpaired nucleons) are assumed to occupy single-particle shell model orbital states, just as in atomic physics, and these single-particle orbitals, which are prevented by the Pauli exclusion principle from

interacting with one another, do not contain internal 'alpha-particle' correlations."

From the perspective of the independent-particle model, there is no reason to expect alpha particles to pre-exist in nuclei and to be a major break-up product following nuclear collisions. If nucleons are truly independent, then their numbers in fragmentation experiments should be a simple function of the collision energy.

> "However, Lind et al. bombarded ^{40}Ca with 220 MeV pions and found that large cross-sections were obtained for the production of gamma-radiations from final-state nuclei, which correspond to the knocking out of 1, 2, 3, 4, or 5 alpha-particles (or equivalent nucleons) from the ^{40}Ca nucleus.... In somewhat similar experiments, Jackson et al. bombarded ^{58}Ni and ^{60}Ni nuclei with both pions and protons and found 'significantly larger cross-sections' for the removal of 'integral numbers of alpha particles', and Chang et al. bombarded ^{56}Fe and ^{58}Ni nuclei with 100 MeV protons and obtained large cross-sections for the removal of from one to three alpha-particles." (MacGregor, 1976)

When relatively large nuclei are used as the projectiles in so-called heavy-ion reactions, the collision between the two nuclei results in a large number of decay products, but the projectile will often transfer an alpha particle to the target nucleus, or vice versa. The number of alpha particles within medium and large nuclei, and the mechanisms of their formation and break-up are simply not known, but the fact that alpha particles are frequent products in a variety of nuclear reactions is *prima facie* evidence that they normally exist in such nuclei.

MacGregor (1976) also pointed out that there is a huge penalty paid by, particularly, the large nuclei for having a core region with equal numbers of protons and neutrons because that implies an abundance of proton charge in the nuclear interior. The penalty is roughly 300 MeV for ^{208}Pb, but electron-scattering experiments indicate that the nuclear charge is rather evenly distributed throughout the nuclear core region with a *decreasing* charge density in the skin. Electrostatic considerations would suggest, however, that a large nucleus would have a binding energy gain of several hundred MeV if it were simply to allow its protons to separate from one another by drifting toward the nuclear surface. The fact that this does *not* occur and that there is no significant proton "hole" at the nuclear center is indication of some kind of mechanism for binding protons (possibly as alpha clusters) within the nuclear core.

Based on the known properties (binding energies, spins, parities, magnetic moments, quadrupole moments and RMS radii) of the 2-, 3- and 4-nucleon clusters and assuming a particular quasi-2D build-up procedure, MacGregor showed that the nuclear properties of all known nuclei could be well explained as summations of the properties of the smallest nuclei. There is in fact some arbitrariness in what should be summed (for example, should ^{6}Li be considered the summation of ^{4}He and ^{2}H, or the summation of ^{3}He and ^{3}H?) and the number of possible permutations for large nuclei progressively grows. Unlike the distinct limitations of the standard alpha-particle model, however, the 2D-Ising cluster model does indeed apply to all nuclei (including odd-Z and odd-N nuclei, and $N > Z$ nuclei). If this model had been advocated in the 1930s or 1940s, it might well have become a major force in nuclear structure theory, but, perhaps as a matter of historical accident, it was developed *after* the liquid-drop and shell models and has been able to account primarily for nuclear features which the other models *already* successfully accounted for.

Pauling's Spheron Model

The spheron model advocated by Linus Pauling in the 1960s and 1970s was devised explicitly to produce magic shells using nucleon clusters, and it also found support in the evidence indicating nucleon clustering in the nuclear interior. As discussed below, the structural geometry of Pauling's model has some common-sense, intuitive appeal, but shows strengths and weaknesses distinct from those of MacGregor's model.

The basic idea behind the spheron model is that nucleons aggregate ("hybridize") into *spherical* 2-, 3- and 4-nucleon clusters ("spherons"), which can then be arranged geometrically into more-or-less close-packed symmetrical structures that correspond to the magic numbers (Fig. 4.14). The molecular build-up procedure is similar to the standard alpha-particle model, but the spheron model is unusual in its emphasis on magic shells and in the fact that an attempt was made to formalize the rules for how nucleon clusters aggregate. The cluster build-up procedure was then applied to all nuclei (Pauling 1965, 1976; Pauling & Robinson, 1975).

Fig. 4.14. Spheron model structures giving approximately spherical nuclear shapes. (a) shows an icosahedral spheron structure containing 12 spherons on the surface and one at the center. (b) shows a 14-spheron structure obtained by using spheres of two different diameters (used to explain the magic number 28). (c) shows a 22-spheron structure used to explain the partially-magic character of number 40 (from Pauling, 1965)

As is evident in Fig. 4.14, the model attempted to reconcile the idea of magic stability with the idea of a short-range force acting between spherons, but it turned out that a straight-forward build-up procedure using spherical 4-nucleon alpha-clusters (all of which with the same diameter) will not produce spherically symmetrical nuclei at most of the known magic numbers. A central alpha-particle on its own corresponds to the magic number 2 (^{4}He), and a tetrahedron of spherical alphas gives a doubly magic nucleus for ^{16}O, but building upon either the central alpha particle or a centrally-lying tetrahedron of alphas does not produce magic numbers at 20 or 28 (or indeed any of the other magic numbers). An icosahedron built around a central tetrahedron predicts incorrectly a magic number at 26 (1 central and 12 peripheral alphas) or 24 (if the central alpha particle is removed). Adding an alpha-particle to each of the four faces of the ^{16}O tetrahedron of alphas predicts incorrectly a magic number of 16.

These difficulties forced Pauling to use two adjustable parameters: (i) a variable number (2 ∼ 4) of nucleons in each cluster, and (ii) a variable cluster diameter to allow for different kinds of cluster packing. In effect, these two parameters did permit the construction of the desired spherically-symmetrical structures at certain of the magic numbers, but they also introduced a host of problems. Specifically, the arbitrariness of cluster filling and cluster dimensions meant the inevitable occurrence of similarly symmetrical structures at non-magic numbers.

The theoretical dilemma here arises from the fact that the magic numbers are not as unique and unambiguous as those indicating the closure of electron shells (see Chap. 2 for a discussion of the electron and nucleon shells). The magicness of closed shells is far less certain in the nuclear than in atomic realm, and *any* model that is designed to explain uniquely *some* of the magic numbers is bound to fail in explaining others, unless the model reproduces the entire quantum number systematics (shells and subshells) of the independent-particle model (**2**; 6, **8**; 14, 18, **20**; **28**, 34, 38, 40; **50**, 58, ...).

The attraction of the spheron model is the implied unification of the cluster and shell concepts; given the starting assumptions that allow for the creation of spherons by hybridizing a small number of nucleons into a cluster, the model leads to a rather common-sense molecular build-up of nuclei and has an internal logic that is hard to deny. It is relevant to note that Linus Pauling was the inventor and driving force behind the concept of the hybridization of electron orbitals, that has been successfully used to explain many aspects of molecular structure. In essence, the nuclear spheron model employed an identical technique for assembling 2-, 3- or 4-nucleon "hybridized" spherons with diameters adjusted to allow for the construction of magic nuclei. If electron orbital hybridization works so nicely for molecular structures, how could it not work equally well for nucleons at the nuclear level? In fact, however, despite two decades of advocacy by Pauling, nuclear theorists have not elaborated on the idea of nucleon spherons, and Pauling's model has not entered mainstream nuclear theory.

4.3 The Independent-Particle Models

Mainstream nuclear theory *is* the shell model.

Although the "compound nucleus" model advocated by Niels Bohr became dominant in the 1930s with several quantitative successes involving nuclear binding energies and fission, by the late 1940s, the discontinuities in binding energies and the changes in nuclear properties associated with certain numbers of protons and neutrons indicated nuclear substructure that was not explicable on the basis of the collective properties of nucleons. Data suggestive of shells of nucleons similar to the well-understood electron shells of atomic structure encouraged Mayer, Jensen and colleagues (Mayer & Jensen, 1955) to devise a model that emphasized the properties of the independent particles within the nuclear collective. While related ideas had been entertained in the 1930s (e.g., Wigner, 1937), a coherent description of the quantum mechanics of the nucleons was not devised until the late 1940s with the introduction of a spin-orbit coupling force. That "independent-particle" description of nucleon states also explained the nuclear shells and consequently became the dominant theoretical paradigm in nuclear structure physics.

Conceptually, the independent-particle (or shell) model is built on the idea of a Fermi gas: nucleons are assumed to be point particles that are free to orbit within the nucleus due to the net attractive force of a potential-well. The model assumes that the nuclear force acting between nucleons produces a net potential-well that pulls all nucleons toward the center of the nucleus, and not directly toward other individual nucleons. Nucleons under the influence of that potential-well can exist in distinct quantal energy states, the magnitude and occupancy of which determines the detailed nucleon build-up procedure. The problem facing the developers of the shell model was to find a well of a depth and shape that would reproduce the experimental discontinuities in nuclear binding energies. The starting point was the formalism for standing-waves in a rectangular box that had been used to explain atomic states (Fig. 4.15a). Discontinuities are predicted, but a more realistic potential-well was the harmonic oscillator, shown in Fig. 4.15b. Again, the discontinuities in binding energies arose at the wrong numbers, but an explanation of the so-called magic numbers could be arrived at by postulating a spin-orbit interaction for each nucleon.

The harmonic oscillator alone predicted shell closures at proton and neutron numbers: 2, 8, 20, 40, 70, 112, ...; whereas the known magic numbers were: 2, 8, 20, 28, 50, 82, 126, ... In order to rectify this slight mismatch, the energy steps implied by the harmonic oscillator were modified by means of one further assumption, i.e., so-called spin-orbit coupling, the idea that the intrinsic spin angular momentum of each nucleon is coupled to its orbital angular momentum. The coupling theoretically gives each nucleon a total angular momentum, j, that can be compared to experimental values. In fact, nuclear application of the spin-orbit coupling model known from atomic theory caused several problems (e.g., Bertsch, 1972; Bertsch et al., 1980). The strength of the coupling is an order of magnitude too weak to allow significant

4.3 The Independent-Particle Models

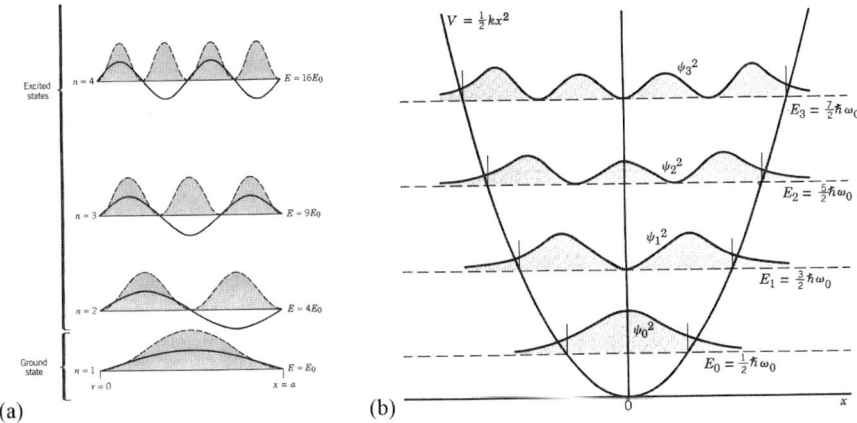

Fig. 4.15. A wave in a box (**a**) and the harmonic oscillator (**b**) both show integral numbers of wave-states (Krane, 1988, pp. 21, 25)

shell splitting, and the proposed sequence of the splitting ($l + s$ before $l - s$) was the exact opposite of the spin-orbit coupling for electrons.

Despite these difficulties, it was found that, if it is assumed that most nucleon j-values cancel out in pairs and only unpaired nucleons contribute to the nuclear spin, then the overall pattern of nuclear J-values corresponded well with experimental data on nuclear spins. Both the concept of a nuclear potential-well and the spin-orbit effect were problematical, but the model was ultimately well-received because it produced a new series of shell and subshell closures with the following occupation numbers: **2**, 6, **8**, 14, 18, **20**, 28, 34, 38, **40**, 50, 58, 64, 68, **70**, 82, 92, 100, 106, 110, **112**, 126, ... – among which were all of the magic numbers. By making appropriate assumptions about the depth and width of the harmonic oscillator and about the spin-orbit coupling, the correct (experimental) sequence of j-subshells and the appearance of gaps at various magic shells were produced (Fig. 4.16).

As discussed in Chap. 2, the experimental support for magically-closed shells uniquely at the magic numbers is rather mixed, but the implications concerning the sequence of j-subshells found overwhelming experimental support. Although there were known cases of "configuration-mixing" (where energy levels were so close that the j-sequence was temporarily reversed or mixed) and other cases of "intruder states" (where a seemingly distant energy level intruded into the shell model sequence), the vast majority of spins and parities of nuclei agreed completely with the shell model predictions. Say what one might about "magic" closure, the j-subshell texture implied by the shell model found abundant support from the experimental data.

The j-subshell texture and the explanation of the magic numbers were the two factors that most impressed theorists in the early 1950s, but another factor that ultimately weighed in favor of the shell model was the prediction of nuclear magnetic moments (Fig. 4.17). Since the magnetic moments of the

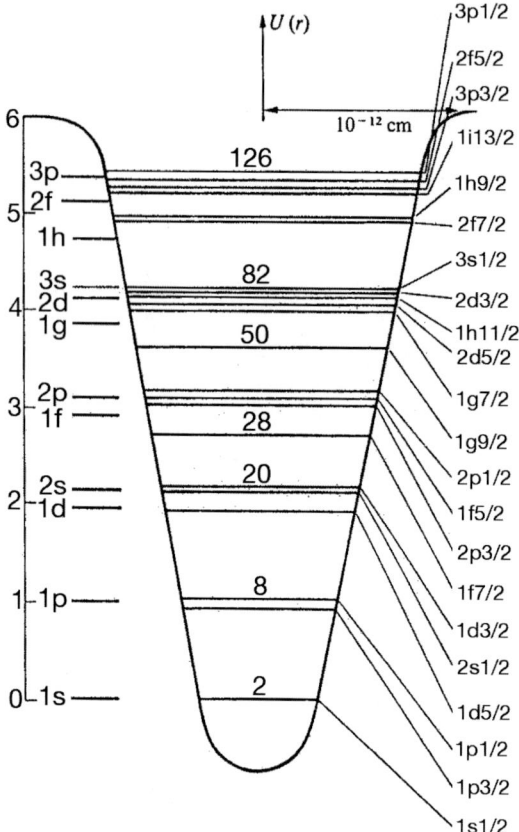

Fig. 4.16. The shell model predictions of the harmonic oscillator when spin-orbit coupling is assumed (after Blin-Stoyle, 1959). The sequence of nucleon states is close to the experimentally-known sequence

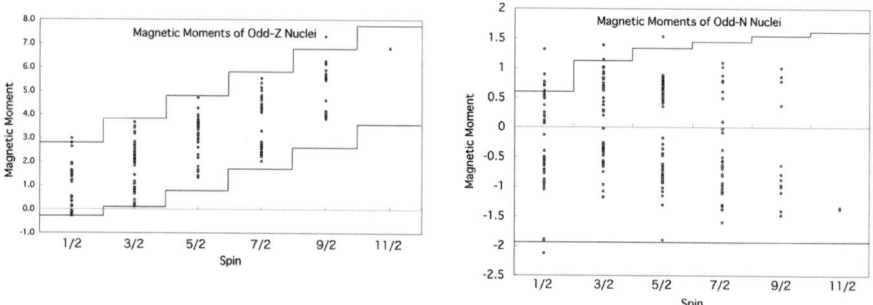

Fig. 4.17. The magnetic moments of odd-Z and odd-N nuclei. The thick lines are the so-called Schmidt lines that predict the upper and lower bounds of the magnetic moments. The dots show the experimental data for all of the 235 measured nuclei with odd-N or odd-Z

isolated proton and neutron were known, and the influence of the nucleon's orbit on its magnetic properties easily calculated, the upper and lower bounds of the magnetic moment of any nucleus with an unpaired proton or neutron could be determined (the so-called Schmidt lines). Comparison with experimental data showed two main effects. On the one hand, only a very small number of nuclei had magnetic moments that are precisely as predicted by the single-particle shell model – i.e., lying on the Schmidt lines. But, on the other hand, and quite remarkably, nearly all values fall between the predicted upper and lower bounds. In other words, as judged from their magnetic properties, most nuclei do not exist in pristine shell model states, but they do lie in states that are bounded by the shell model limits.

The successes of the shell model have been notable, but already by the mid-1950s the internal contradictions inherent to the shell model approach were well-known:

> "Few models in physics have had such a persistently violent and chequered history as the nuclear shell model. Striking evidence in its favour has often been followed by equally strong evidence against it, and vice versa." (Elliott & Lane, 1957)

Controversy has continued, but the combination of predictions concerning closed shells, numerous spin/parity states and the upper and lower limits of magnetic states provided ample empirical support for the otherwise equivocal theoretical foundations of the shell model – and the shell model and its variations have been in use ever since. As noted above, the particular strength of the shell model was in explaining nuclei that contain one proton or one neutron more or less than a magic shell. To account for the nuclear properties of nuclei away from closed shells, however, it has been found necessary to distort the nuclear potential-well from a spherical shape to various ellipsoidal shapes.

Manipulation of the Potential-Well

The vast majority of nuclei fall between the magic numbers, where nuclear properties are not dominated by the properties of one unpaired nucleon, lying outside of a spherical closed shell (Fig. 4.18). In principle, any nucleus might be accounted for by assuming an inert magic core, and some number of valence nucleons outside the core. In practice, however, the presence of even a few valence nucleons implies a huge number of nucleon-nucleon and nucleon-core interactions that the single-particle shell model cannot handle.

To account for the properties of nuclei that are not near to closed shells, the shape of the potential-well can be distorted (Nilsson, 1969). Manipulation of the potential-well in effect changes the sequence of j-shells from the spherical shell model sequence to a new sequence that depends on: (i) the magnitude of the distortion, and (ii) the number of axes involved. Fig. 4.19 shows the simplest distortions of the harmonic oscillator using one parameter, ε. It is

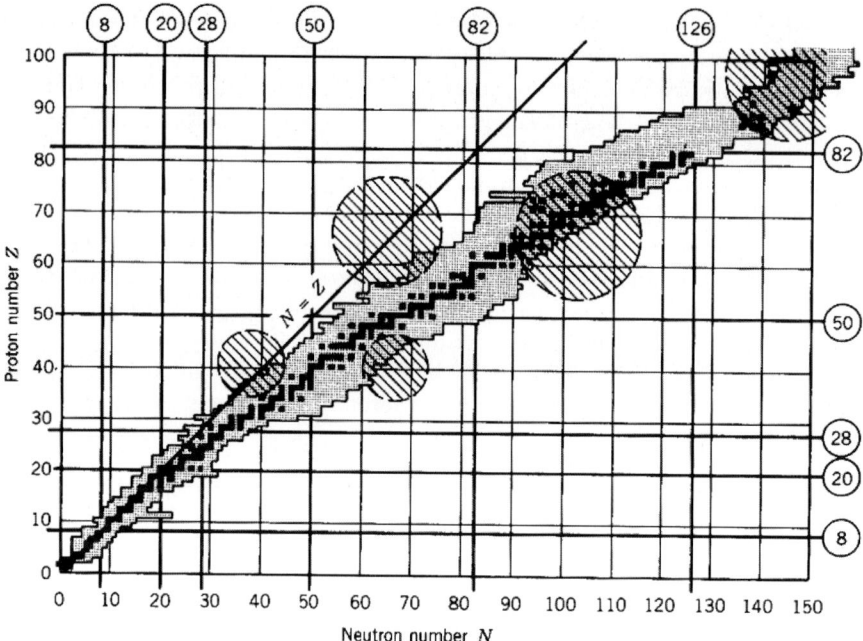

Fig. 4.18. The regions (*cross-hatched*) where nuclei are far from closed shells and where permanent deformations of the nuclear potential might be expected (from Krane, 1988, p. 148)

seen that, even for small distortions ($\varepsilon \sim \pm 0.2$), the sequence of j-subshells is drastically changed.

Distortions in all three dimensions lead to a variety of prolate or oblate shapes, each of which will show different sequences of energy states (Fig. 4.20). The complexity and the permutations of nucleon states increases rapidly with more realistic 3D distortions of the potential-well. This can be seen in a magnification of shell model states near to $Z = 82$ when all three axes are separately manipulated (Fig. 4.21).

The complexity of nuclear states in Fig. 4.21 illustrates both the strength and weakness of the central potential-well approach to nuclear structure. On the one hand, deformations of the well allow for a *post hoc* explanation of virtually any experimentally-detected sequence of energy states that differs from the predictions of the spherical shell model, but, on the other hand, the extreme flexibility of the allowed manipulations makes its evaluation difficult. Do the manipulations of the potential-well actually correspond to the spatial deformations in real nuclei? No one knows for certain:

> "It may well be that we shall always be able to account for the phenomena of nuclear structure physics by appropriate elaborations such as the introduction of more-and-more complicated forms of configuration

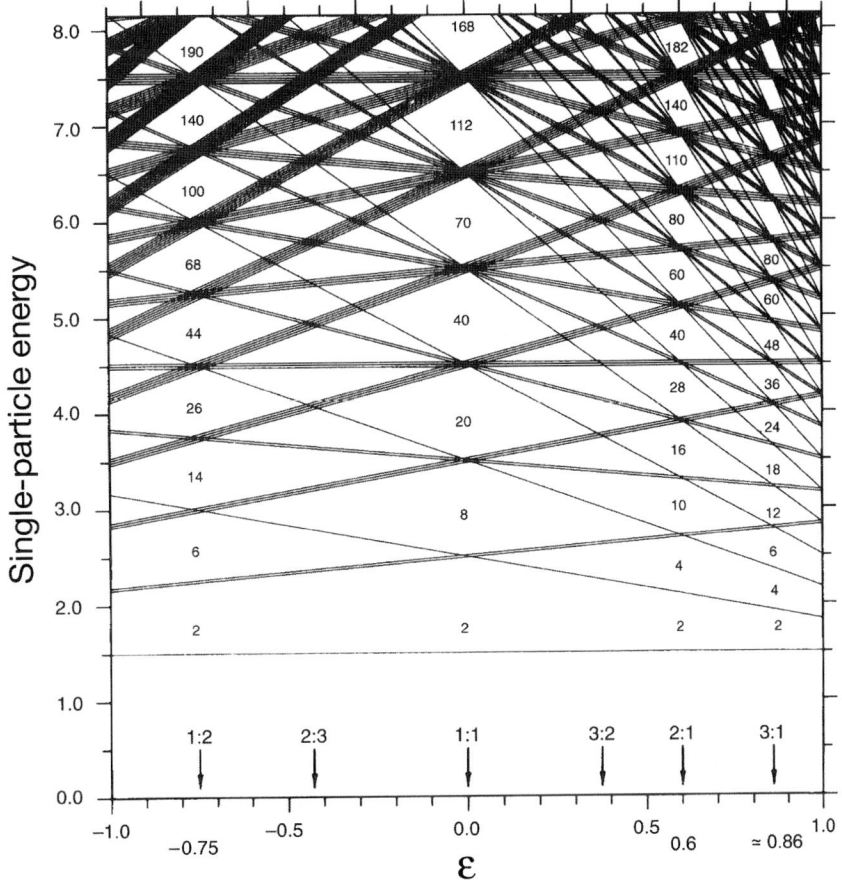

Fig. 4.19. Distortions of the spherical potential-well lead to the separation of single-particle states that differ from the spherical harmonic oscillator (from Nilsson and Ragnarsson, 1995, p. 116)

mixing or by using effective matrix elements that depart more-and-more from what we expect on the basis of free-space interactions and so on. We should not welcome such rococo extravagance but it may be necessary and, furthermore, be a true description of what is going on." (Wilkinson, 1990, p. 286c)

4.4 Other Models

Whether or not the independent-particle models can be fairly accused of being rococo is debatable, but it is certainly the case that the extreme flexibility of the conventional, shell-model approach to nuclear structure has led other

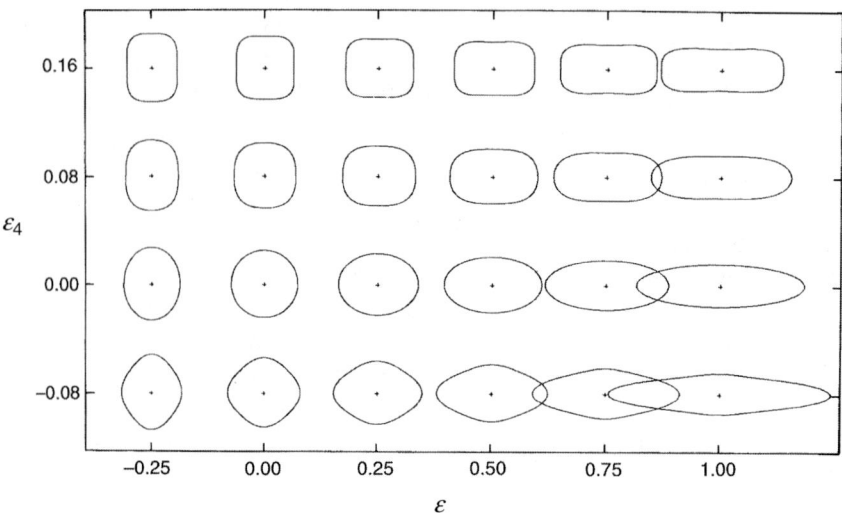

Fig. 4.20. Distortions along two dimensions (ε and ε_4) produce a variety of nuclear shapes and a corresponding variety of j-subshell admixtures (from Nilsson & Ragnarsson, 1995, p. 125)

theorists to develop models that offer the promise of similar predictive power with firmer constraints on the underlying nuclear force. Specifically, instead of a potential-well that can always be manipulated into agreement with experimental findings, an attractive alternative is to postulate a small number of two- or three-body nuclear force effects, and then build nuclei on that basis. This approach is not unlike that of the cluster models, but insists on a realistic nucleon-nucleon interaction as the mechanism underlying all of nuclear structure.

Exemplary of this type of "bottom-up" model is the quark-based model of Robson (1978). Starting with conventional (ca. mid-1970s) quark ideas about the internal structure of nucleons, he argued that "the interaction between nucleons at $R = 2$ fm will be entirely dominated by the exchange of quarks" (p. 391), implying that the approximation of a time-averaged nuclear potential-well need not be considered. As shown in Fig. 4.22, a quark-based explanation of the nuclear force necessarily makes the two-nucleon interaction geometrically more complex, but this is counter-balanced by an increased simplicity of nuclear binding effects in many-nucleon nuclei, insofar as all of the parameters of the nuclear force can be specified at the 2-nucleon level.

Unlike most subsequent quark-based accounts of nuclear structure, Robson proposed concrete quark/nucleon structures for nuclei up to $A = 24$, and maintained that their binding energies, radial measures, rotational, vibrational and cluster states could be explained on such a basis. More recent quark models have examined the binding energies and radii of $A = 3$ and $A = 4$ nuclei on the basis of the distribution of valence quarks (Goldman et al.,

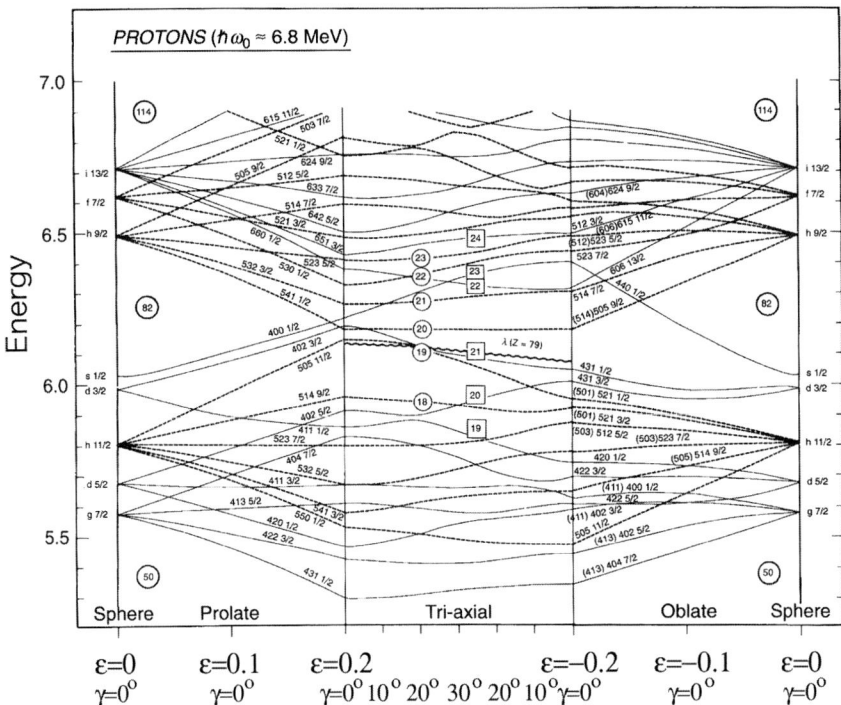

Fig. 4.21. Details of the changes in shell gaps near $Z = 82$. The spherical potentials on the far left and far right give energy gaps at the traditional magic numbers of the shell model. Deformation along one axis produces prolate nuclear shapes, along two axes gives oblate shapes, and along all three axes independently gives so-called triaxial shapes (from Nilsson & Ragnarsson, 1995, p. 131)

1988; Goldman, 1991; Maltman et al., 1994; Benesh et al., 2003). Goldman has concluded that quark delocalization leads to a lattice-like "egg crate" potential that, in principle, can be extended to all nuclei. Similarly, Musulmanbekov (2003) has advocated a "strongly correlated quark" model for light nuclei

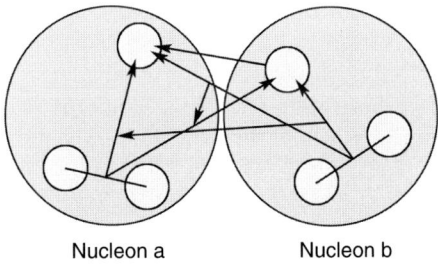

Fig. 4.22. Coordinates for the six-quark system (after Robson, 1978)

($A < 17$) based on the relationship between current and constituent quarks; that model also implies a close-packed lattice of nucleons.

Lattice Models

Until the development and pervasive use of digital computers in the 1980s, the structural simplicity of lattices for objects containing less than 300 particles did not mean that they had computational advantages over liquid- or gaseous-phase models. On the contrary, all models necessitated the use of simplifying assumptions that would allow for numerical calculations. For the liquid-drop and Fermi-gas models, the well-understood stochastic phenomena of liquids and gases could be employed, even though there remained difficulties inherent to the relatively small numbers of nucleons in nuclei. When high-speed computing became available, however, it was possible to construct individual nuclei and to make exact calculations on particle interactions. In the case of the nucleus, since neither the underlying force nor the configuration of the nucleons was known for any given nucleus, reliable results could be obtained only by systematically varying nuclear force parameters and by randomly changing the nucleon configuration. By repeating the calculations many times and collecting statistics, comparisons between the simulations ("experimental theory") and experimental data became possible. Various techniques for such computer simulations were rapidly developed and, here, lattice models had certain inherent computational advantages over liquid and gas nuclear models because of the regularity of lattice structures.

Application of the lattice simulations has provided insights into the breakup of nuclei in high-energy nuclear reactions involving many nucleons – so-called multifragmentation studies (e.g., Bauer et al., 1985; Campi, 1986; Richert & Wagner, 2001). Two methods that have proven useful are the so-called bond percolation and site percolation lattice techniques. In either case, a lattice containing as many coordinate sites as nucleons for a specific nucleus is constructed and then fragmented, following a set of rules for the disintegration of the lattice. In bond percolation, the bonds between nearest-neighbors in the lattice are broken (with some probability $0.0 < p < 1.0$), whereas in site percolation the lattice sites themselves are depopulated with some probability.

A typical simulation can be summarized as shown in Fig. 4.23. For simplicity of calculation, simple-cubic packing is most often employed; an scp lattice is randomly filled with particles (or, equivalently, fully filled but with bonds between nearest-neighbors formed at random) (Fig. 4.23a). The simulation proceeds by boring a whole through the target nucleus of given size (R) using a projectile of given size (r) (Fig. 4.23b), and tabulating the number of fragments of various sizes (A_f) that remain, following a prescription for calculating when neighboring nucleons are bound into a stable state or unbound (Fig. 4.23c–e).

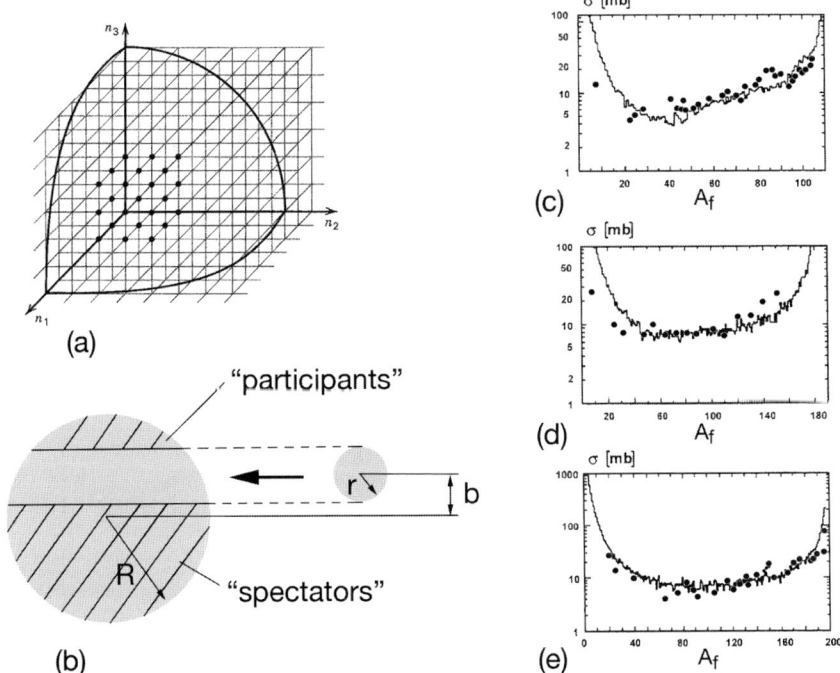

Fig. 4.23. The basic lattice structure (**a**), and dynamics (**b**) of simulations of heavy-ion collisions. Fragmentation results (**c**–**e**) of a lattice simulation, showing experimental data (*dots*) and simulation data, are from Bauer (1988)

Lattice simulations normally examine the break-up products following the collision of projectiles and targets at various high-energies, where many nucleons are involved in the reaction. For example, Bauer (1985) has reported good agreement between lattice results and experimental data obtained for collisions between protons and silver nuclei at 11.5 GeV (Fig. 4.23c), protons and tantalum nuclei at 5.7 GeV (Fig. 4.23d) and protons and gold nuclei at 11.5 GeV (Fig. 4.23e). Because individual events can be tracked in such simulations, detailed comparisons between theory and experiment can also be made for effects such as the maximum fragment size (Fig. 4.24a), the range of intermediate fragments (Fig. 4.24b) and the incidence of alpha products (Fig. 4.24c).

In order to compare lattice simulations with the rich variety of high-energy experimental data that became available in the 1980s and 1990s, the adjustment of certain parameters of the lattice is required. Typically, this entails the choice of rules concerning what constitutes an intact or broken bond within the lattice, and therefore what constitutes a nucleon cluster that has survived the collision. The effects of various parameters have been extensively studied (e.g., Richert & Wagner, 2001; DasGupta et al., 1995–2003) and conclusions

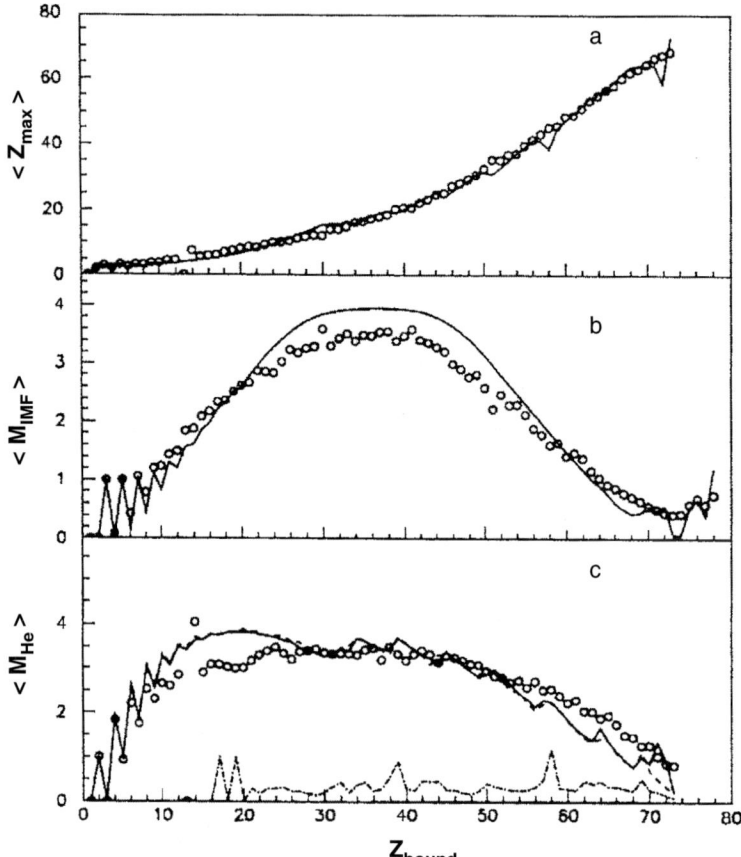

Fig. 4.24. A comparison of high-energy multifragmentation data and lattice simulation results (Elattari et al., 1995). (**a**) shows the frequency of maximum fragment sizes. (**b**) shows the incidence of intermediate fragments. (**c**) shows the incidence of ^{4}He fragments

drawn concerning phase transitions in nuclear matter. Interestingly, the lattice configuration itself plays a significant role. Whenever direct comparisons have been made between lattice types, close-packed lattices always reproduce the experimental data more accurately than simple-cubic lattices (e.g., Canuto & Chitre, 1974; D'yakonov & Merlin, 1988; Chao & Chung, 1991; Santiago & Chung, 1993) (Fig. 4.25).

Unlike the liquid-drop and shell model approaches – that simply cannot deal with the complexity of the many-body effects involved – the lattice simulations are inherently many-body interactions that do not suffer from the combinatorial explosion encountered when all possible particle interactions must be computed. The lattice simulations reproduce experimental findings well, but the successes of these models in accounting for the

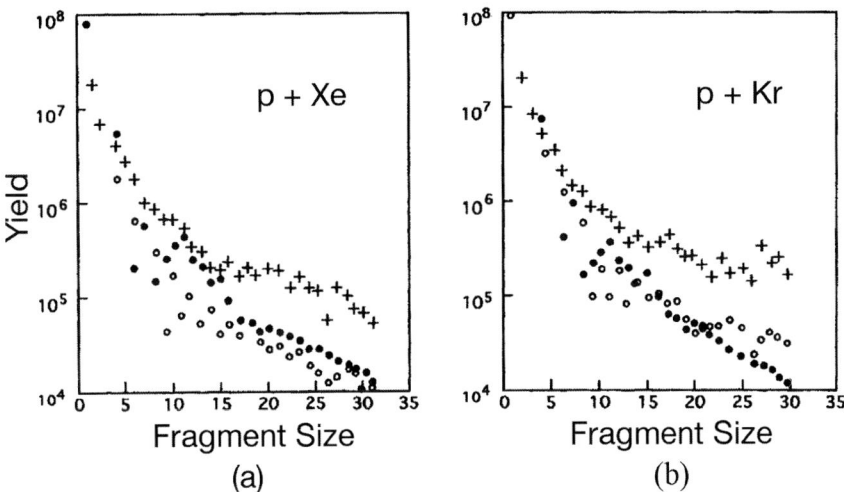

Fig. 4.25. A comparison of lattice simulations of heavy-ion experiments. Crosses are theoretical results obtained using a simple-cubic packing (scp) lattice and solid circles using a face-centered-cubic (fcc) lattice in simulations; experimental data are shown with open circles (Chao & Chung, 1991). The fcc lattice reproduces the data more reliably due to the higher packing fraction and larger number of near-neighbors

multifragmentation data have remained puzzling. On the one hand, a "bottom-up" explanation of phenomena is always to be preferred – where the starting assumptions concerning the nuclear force are explicit and expressed in such a way that underlying physical principles are evident. The lattice models have generally been the antithesis of such good methodology: they begin with assumptions that are apparently unrealistic and, instead of working to justify the approach, they proceed by adjusting parameters to reproduce experimental data. What is surprising, therefore, is the level of success in simulating the empirical data. In a review of the various approaches used to explain heavy-ion reaction data, Moretto and Wozniak (1993) noted that:

> "Remarkably this [percolation lattice] theory predicts many features of the experimentally observed mass distributions and fragment multiplicities" (p. 382).

And, with apparent reluctance, they concluded their review with the comment that:

> "Percolation [lattice] models can describe many features associated with multiplicities and mass fluctuations with perplexing accuracy, despite their dearth of nuclear physics content" (p. 450).

The successes of the lattice models can be partially explained by the fact that a lattice has certain statistical properties in common with any many-body system, and the spatial regularities of a lattice model provide computational advantages that are not found in other models. While not being grounded in a rigorous theory of the nuclear force, the physical assumptions intrinsic to a lattice of nucleons are clearly more realistic than those underlying a Fermi gas of nucleons. Justification of a specific lattice configuration may be more difficult, but the assumption of a short-range nuclear force and constant nuclear density are essentially restatement of the assumptions underlying the strong-interaction models. Stated contrarily, the utility of the lattice models – even if confined solely to the high-energy multifragmentation studies of the 1990s – argues against a central potential-well model of the nucleus, insofar as the lattice results are dependent entirely on nearest-neighbor nucleon-nucleon interactions.

4.5 Summary

Starting with a nuclear phase that is either gaseous, liquid, cluster or solid, the diverse models of nuclear structure theory have been able to account for a wide variety of experimental data, but it remains a paradox that these very different models are describing the same physical reality. Aspects of each model have led to verifiable predictions and explain some portion of the experimental data, but "unification" of nuclear theory within any one model has not been achieved. It may be tempting to accept the "many-models" theoretical *status quo* and ask philosophical questions concerning how one-and-the-same physical object can be described in so many different ways, but a more critical, and ultimately more constructive approach is to inquire what kinds of experimental findings cannot be readily explained by any model. This question will be addressed in the following four chapters.

Part II

Long-Standing Problems

Having outlined the principal models employed in nuclear structure theory and the history of their development, let us turn to the question of whether or not any genuine contradictions among the models remain today. It is of course a matter of historical record that the nuclear models were built upon contrary assumptions concerning the nuclear force and diametrically opposed ideas about the internal texture of the nucleus. But "violent disagreements" are a thing of the past, and nuclear structure theory is better characterized by its tolerant diversity of opinion than a sense that a crisis of theory demands that firm decisions be made. So, is it fair to say that nuclear theory is a finished chapter in the history of science or do problems remain?

Here in Part II, I rely entirely on the published literature to show the history of debate on several core issues. Chapter 5 reviews the problem of the mean-free-path; Chapter 6 tackles nuclear size and shape topics; Chapter 7 discusses issues concerning the nuclear force and super-heavy nuclei; and Chapter 8 reviews the facts concerning asymmetrical fission. In all four chapters, the literature from the late 1930s through until the 21st Century demonstrates the reality of unresolved issues.

5

The Mean Free Path of Nucleons in Nuclei

A question raised early in the development of the shell model concerns the average distance over which a nucleon inside of a nucleus travels before it collides and interacts with another nucleon, the so-called "mean free path" (MFP, often referred to as λ). The length of the MFP is a value of theoretical importance because the independent-particle ($\sim$shell) model requires the undisturbed orbiting of each nucleon in a unique energy state within the nucleus before it experiences nucleon-nucleon collisions and detectable changes in its energy. Shell theorists have therefore argued that, in order to have a fixed momentum, each nucleon must travel in a fixed orbit over a distance that is relatively long (in terms of nuclear dimensions) before it interacts with other nucleons.

As a consequence of such general considerations, one of the initial doubts raised at the time when the shell model was first devised (1948–1949) was whether or not nucleon "orbiting" was in fact a reasonable starting point. After all, for more than a decade the liquid-drop model had proven fairly successful in describing nuclear features based on a dramatically different idea – a nuclear texture that explicitly does not allow the free movement of nucleons within a dense liquid. The apparent contradictions between a weak-interaction (gaseous) model and a strong-interaction (liquid) model can be approached in the context of various other issues, but the mean-free-path is perhaps the most direct expression of the independence of nucleons in nuclei. The liquid-drop model requires the MFP to be short – approximately equal to the average interparticle distance in the liquid-drop interior – while the independent-particle model requires the MFP to be long – at least equal to the distance of several nucleon orbits in order to establish the nucleon in its unique state of angular momentum.

It is worth noting that many textbook discussions of the MFP dismiss the entire issue as being resolved by the exclusion principle. This argument will be examined below, but, whatever the quantitative implications of the exclusion principle, it should be understood that the MFP was at one time thought to be a major obstacle to acceptance of the shell model. How can independent nucleon orbiting – that is, a long MFP – be justified in a substance as dense as nuclei?

In an early discussion of the apparent contradiction regarding the MFP in the shell and liquid-drop models, Blatt and Weisskopf (1952) stated uncategorically:

> "The effective mean free path of a nucleon in nuclear matter must be somewhat larger than the nuclear dimensions in order to allow the use of the independent-particle model. This requirement seems to be in contradiction to the assumptions made in the theory... We are facing here one of the fundamental problems of nuclear structure physics which has yet to be solved." (p. 778)

And 34 years later Siemens and Jensen (1987) expressed a similar concern:

> "The riddle of the mean-free-path is one of the most striking features of nuclear physics." (p. 54)

And yet, this "most striking feature" receives curiously little discussion in most modern textbooks. So, let us focus some attention both on the basic physics involved and on the historical record of how the MFP has been handled.

In Chap. 2, we saw that the size of nucleons relative to nuclei raises some questions concerning the likelihood that protons and neutrons can actually orbit within the nucleus. Here we will ignore the issue of the size of the individual nucleons and discuss the question of the distance over which they can be expected to enjoy free movement within the nuclear interior. Shell model theorists have not in fact provided estimates of how many revolutions through the nucleus will suffice to give the nucleon a well-defined orbit (and therefore a unique energy state), but have often stated the need for "several" orbits or a mean free path that is "long" relative to the diameter of the nucleus. So how long is "long"?

For the large nuclei (with charge radii of 5–6 fm), an MFP of several revolutions implies unimpeded movement of several tens of fermi. One circular revolution at the average distance, r, from the nuclear center would be $2\pi r$, or 31–37 fm for a large nucleus such as lead, and "several" revolutions would imply a free, unimpeded motion of about 100 fm. Even a linear "orbit" across the nuclear volume would on average be $4r$, or 20–24 fm, and several such orbits would need 60–72 fm of collision-free movement. Within the context of nuclear dimensions, even 20 fm is a very long distance for some 200 particles (each with a diameter of 1.7 fm) to move freely within a 10–12 fm diameter sphere. On the face of it, unimpeded movement over such a distance seems unlikely.

The qualitative picture is roughly that depicted in Fig. 5.1: nucleons have electrostatic RMS radii of 0.86 fm, but cannot be on average more than 2.0 fm apart, if a core density of 0.17 nucleons/fm^3 is to be maintained. It is not, therefore, obvious where the orbiting nucleons can go in their intranuclear

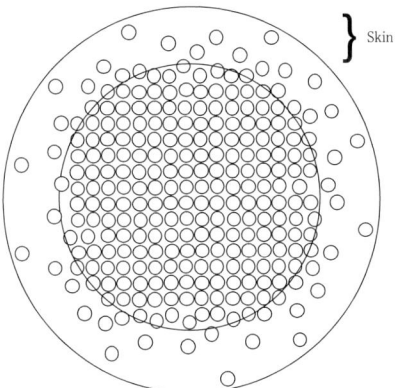

Fig. 5.1. The mean-free-path problem in two dimensions. The constant density core radius of ^{208}Pb is approximately 5.5 fm, outside of which the density of the "skin" region falls to zero over 2–3 fm. How far can a nucleon travel before colliding with another nucleon?

travels. The qualitative picture is more-or-less what one would expect from the liquid-drop model of nuclear structure, but there are, in fact, experimental data and theoretical calculations that address this question quantitatively and are worth examining in some detail. Since nucleons within nuclei are known to have kinetic energies around 20 MeV, the most relevant experimental data are for nucleons that have been injected into nuclei at about this energy (in the range between 10 and 30 MeV).

In Segre's text, *Experimental Nuclear Physics* (1965), Morrison tabulated typical MFP values (ranging from 0.1 to 4.0 fm for low and high energy protons) and noted that the calculated value of 0.1 fm for very low energy (1 MeV) nucleons:

> "is so small that the picture evidently fails, but the conclusion seems confirmed. As soon as the particle crosses the surface of the nuclear sphere, it will interact strongly with the nuclear matter." (p. 30)

A range of similarly small values can be found in a variety of sources from the late 1940s to the mid 1990s (Table 5.1).

Clearly, the energy of the incoming particle and the assumptions underlying the calculations lead to different results. Most values lie between 3 and 6 fm, but there is remarkably little consensus of opinion and no discernible trend in results over five decades of theoretical concern about the movement of intranuclear nucleons. Nevertheless, all such calculations – based on various experimental data and theoretical calculations – are, from the shell model perspective, surprisingly short (an average MFP value between 3 and 4 fm) and considerably less than the 100 fm figure that would unquestionably give each nucleon an independent orbit. The liquid-drop model demands precisely such low MFP values and, for this reason, findings on the MFP were initially

Table 5.1. Some Published Estimates of the Mean Free Path

Incident Nucleon Energy (MeV)	Mean Free Path (in fermi) Reference														
	a 1948	b 1954	c 1958	d 1965	e 1966	f 1967	g 1969	h 1969	i 1974	j 1980	k 1981	l 1983	m 1983	n 1985	p 1987	q 1996
100	5.5	.	5.03	4.0	.	.	.	.	.	.	.	.	5.0	.	.	3.9
90	.	.	.	.	.	.	5.2	.	.	.	2–4	.	.	2.6	.	.
60	.	.	.	.	.	.	.	1.6	.	.	.	.	.	.	.	.
50	.	4.34	.	.	.	.	.	.	2.2	3–5	.	2.5	6.0	.	.	.
40	.	4.09	2.66	.	.	5–8	.	.	2.3	.	.	2.3	.	.	0.4–4.0	.
30	.	3.83	.	.	.	.	.	.	.	.	3–5	.	.	.	.	.
20	.	3.54	4.47	.	.	.	.	.	.	.	.	.	.	.	.	.
10	.	3.23	5.19	0.4	3.23	.	.	.	.	.	.	.	.	.	.	.
1	.	.	.	0.1	.	.	.	.	.	.	.	.	.	.	.	.

a: Goldberger, 1948; b: Feschbach, Porter & Weisskopf, 1954; c: Glassgold & Kellogg, 1958; d: Morrison, 1965; e: Enge, 1966; f: Fricke et al., 1967; g: Bohr & Mottelson, 1969; h: Fulmer et al., 1969; i: vanOers et al., 1974; j: Schiffer, 1980; k: lower values refer to MFP in the nuclear core, higher values in the nuclear skin, DeVries & DiGiacomo, 1981; l: Dymarz & Kohmura, 1983; m: Negele, 1983; n: Aichelin & Stocker, 1985; p: Siemens & Jensen, 1987; q: Caillon & Labarsouque, 1996.

thought to support the strong-interaction liquid-drop model and to pose serious problems for the future of the orbiting-nucleon shell model. (In a search of the literature, I have found only one quantitative argument for a long [16.7 fm] MFP [Gadioli et al., 1976]. The parameters of the "exciton model" together with the long MFP were adjusted to deduce cross-sections in reactions involving 40–80 MeV incident protons, but the long MFP argument was abandoned in subsequent work by the same authors [Gadioli et al., 1981]. Moreover, Blann (1975) also worked on the exciton model and made MFP calculations for 15 and 60 MeV protons in Po^{210}. In the nuclear core [$R < 6.5$ fm], the MFP was 6 and 3 fm, respectively.)

It should be noted that, prior to the early arguments for the compound nucleus presented by Niels Bohr (ca. 1936), the possibility that nucleons move around within the nucleus as freely as electrons do within the atom was a common conception of nuclear structure. In promulgating the compound nucleus model, Bohr argued explicitly that the independent particle conception was fundamentally wrong and that the nucleus was a collective ensemble of particles whose individual properties were, in principle, indecipherable. Moreover, the first real achievements of nuclear theory (concerned with nuclear binding and fission energies) were based on the short-MFP compound nucleus idea; the identity of individual nucleons played no role. A decade or so later, the shell model – despite its implication of a long MFP – also proved to be theoretically useful and, contrary to Bohr's philosophical arguments, it became known that sometimes the properties of individual nucleons (e.g., spins and magnetic moments) play a major role in determining the properties of the nucleus as a whole. Until today, it has not been possible to draw any simple conclusions concerning the relative importance of independent-particle and collective properties, and indeed these contrasting perspectives have been at the heart of nuclear debates for decades.

Most discussions of the shell model during its first years in the spotlight included at least some mention of this potential difficulty. For example, Hamilton (1960) stated the problem as follows:

> "For the shell model states to be reasonably well-defined, it is essential that a nucleon be able to make several circuits of its orbit before it is knocked out of the orbit. Now we have a paradox, as follows: The nucleons inside a nucleus have kinetic energies of 20–30 MeV. It is well-known that if a nucleon whose kinetic energy is 20 or 30 MeV is shot into a nucleus it will be stopped in a very short mean free path, of the order of one fermi in length. Indeed, Niels Bohr used this fact as the basis of his compound-nucleus model by assuming that the incoming nucleon almost immediately shares all its kinetic energy with all the nucleons in the nucleus. It is therefore not easy to see how the shell-model states can exist ..."

Several years later in a textbook on nuclear theory, Smith (1965) made similar remarks:

> "The nuclear density is very large compared to the electron cloud density and therefore it is difficult to understand how a nucleon can have a well-defined nuclear orbit. One would expect the mean free path of a nucleon between successive nucleon-nucleon collisions inside the nucleus to be small in comparison with the nuclear diameter." (p. 664)

And Reid (1972) acknowledged the problem without any further discussion:

> "We shall see that the shell model depends on assumptions which appear incompatible with those of the liquid drop model. The reason for these two models, based on apparently contradictory assumptions, each having its areas of useful application, has for long been a central problem in nuclear physics." (p. 85)

Hodgson (1975) raised the same issue:

> "It was [sic] puzzling that the compound nucleus model requires the nucleons to interact strongly, while the shell model requires that they interact so weakly that they can follow relatively undisturbed orbits in the nucleus."

Also worried about the theoretical foundations of the independent-particle model, Pearson (1986) stated the problem as follows:

"Now although [the wave-function for the shell] model is easily soluble, there does not seem to be any reason why it should constitute a valid description of the nucleus, for it does not appear to be possible that the motion of one nucleon should be independent of the detailed motion of the other nucleons, particularly in view of the fact that nucleons can interact only with their nearest neighbors. This last observation is a consequence of the simple fact that all nuclei have more or less the same density.

"... there is nothing inside the nucleus that plays the same role there as does the charge of the nucleus inside the atom, and it would seem impossible for the nucleon-nucleon forces to smooth themselves out into a common field if each nucleon interacts only with its nearest neighbors. Another way to look at the paradox is to say that one would expect the strong short-range force to lead to a rapid sharing of energy and momentum between colliding nucleons." (p. 55)

More recently, Povh et al. (1995) stated the problem this way:

"The fact that nucleons actually move freely inside the nucleus is not at all obvious" (p. 223)!

And Hodgson et al. (1997) were straight-forward in stating the problem:

"The successes of the collective model which assumes the nucleons to be strongly interacting and of the shell model which assumes that they move independently in an overall potential raise the question of the relationship between the two models. What are the nucleons actually doing? It is not possible for them to be both strongly interacting and not interacting at all. We know the cross-section for the interaction of two free nucleons, and this gives a mean free path that is far too short to be compatible with independent motion inside the nucleus. We can accept that different models should reflect different aspects of the nucleus, but they should be consistent with one another." (p. 315)

The paradox remains, so that, over the decades, many suggestions have been made to resolve this issue. In a celebration of Niels Bohr's contributions to nuclear physics, Mottelson (1985) presented a list of only *seven* (!) crucial discoveries during the two golden decades of nuclear theory (mid-1930s to mid-1950s). Among these was the "discovery" of a long MFP in 1952, but the argument has a decidedly problematical ending, since Bohr's contribution was in the establishment of the short-MFP compound nucleus model. On the one hand, Mottelson noted that

"[Bohr] recognized that the assumed single particle motion, copied from atomic physics, was being falsified and he suggested in its stead an idealization which focused on the many-body features and the strong coupling of all the different degrees of freedom of the nuclear system – the compound nucleus." (p. 11)

Moreover,

> "[t]he core of Bohr's thinking is the recognition that the densely packed nuclear system being studied in the neutron reactions forces one to place the collective, many-body features of the nuclear dynamics at the center of attention... [He drew] attention to the far-reaching consequences for the course of a nuclear reaction of the assumption of a short mean free path for the nucleons." (pp. 13–14)

Nevertheless, despite the importance of the compound nucleus model to nuclear theory, its underlying assumptions were, according to Mottelson's account, found to be unnecessary:

> "... the assumption that the mean free path was short compared to nuclear dimensions, believed to be a cornerstone of the compound nucleus, was shown to be wrong! But still the compound nucleus has survived and continues to be the basis for interpreting a large part of the data on nuclear reactions." (p. 20)

As shown below, however, such contrary ideas cannot be easily assembled into a coherent argument.

Weisskopf was one of the early workers to squarely face the problem of the MFP (and propose a possible solution, discussed below). In the textbook by Blatt and Weisskopf (1953), they noted:

> "The 'mean free path' λ of an entering nucleon in nuclear matter is very much smaller than the nuclear radius if the incident energy ε is not too high ($\varepsilon < 50\,\mathrm{MeV}$). It can be estimated to be roughly $\lambda \sim 0.4\,\mathrm{fm}$ if the entering nucleon has a kinetic energy up to about $\varepsilon = 20\,\mathrm{MeV}$." (p. 340)

Later, when discussing the shell (independent-particle) model vs. the liquid-drop (compound nucleus) model, they returned to this issue:

> "The existence of orbits in the nucleus with well-defined quantum numbers is possible only if the nucleon is able to complete several 'revolutions' in this orbit before being perturbed by its neighbors. ... Hence, the effective mean free path of a nucleon in nuclear matter must be somewhat larger than the nuclear dimensions in order to allow the use of the independent-particle model. This requirement seems to be in contradiction to the assumptions made in the theory of the compound nucleus." (pp. 777–8)

A possible resolution of this problem was in fact suggested by Weisskopf himself, but the basic dilemma – resolved or unresolved – is crucial to a proper understanding of nuclear structure theory. In a word, is there a contradiction between the fundamental assumptions of the liquid-drop model and those of the shell model, or not? For good reason, the potential contradiction between the realistic "strong-interaction" liquid-drop model and the (unrealistic) "weak-interaction" independent-particle model dominated early discussions concerning the nuclear models. Despite the greater sophistication of each of the models in their modern forms, their conceptual bases have remained much as before, and the same issue of the relative movement of the nucleons has bedeviled nuclear structure theory for many years.

Weisskopf's solution to this paradox was to invoke the Pauli exclusion principle and, for the moment, let us assume that it avoids an MFP crisis in the shell model. Nevertheless, it should be clear that the MFP was widely thought to be a major problem in nuclear theory at the time of the development of the shell model. If nucleons do not have independent orbits, then, the many successes of the shell model notwithstanding, the theoretical foundations on which it is based would have to be reformulated.

As Siemens and Jensen (1987) remarked:

> "The modern era of nuclear physics began with the surprising revelation that...the nucleons can for the most part be considered to be moving in a single, smoothly varying force field." (p. 56)

But is this "surprise" an empirical fact or a theoretical wish?

Wheeler (1979) has also attempted to draw favorable conclusions from what he called "The Great Accident of Nuclear Physics" by declaring that the MFP is neither fish nor fowl, pondering the hopeful position:

> "that the mean free path of particles in the nucleus is neither extremely short compared with nuclear dimensions (as assumed in the liquid drop picture) nor extremely long (as assumed in the earliest days of nuclear physics), but of an intermediate value. When we took this fact into account we found that we could understand how a nucleus could at the same time show independent particle properties and yet behave in many ways as if it were a liquid drop." (p. 267)

What is not clear in such a compromise position is what an "intermediate value" for the MFP means. Assuming an MFP that is neither short nor long, do strong interactions take place between nearest-neighbors, or not? Is the nucleus held together by local nucleon interactions, or is it the net nuclear potential well that allows for the stability of multinucleon nuclei? Even in the paradoxical world of quantum physics, both views cannot be simultaneously correct: whatever the MFP value we decide on, it will have major impact on our calculation of nuclear force effects (or, conversely, whatever assumptions are made about the nuclear force, they will have numerical implications regarding the MFP). Clearly, this "accident" requires some forensic work.

5.1 Avoiding the Issue

Given that there is at least an apparent paradox with regard to the strong interaction between neighboring nucleons in the liquid-drop model and the free orbiting of nucleons in the shell model, how do textbooks deal with the question of the nucleon's mean free path? Certainly, if the exclusion principle successfully avoids the necessity of making very different assumptions about the MFP in the liquid-drop and shell models, then this can be considered as one of the *resolved* paradoxes of nuclear physics and deserves full explication to illustrate how theoretical problems can be dealt with. Or, if it is yet an unresolved problem or one where only tentative answers are available, then there is all the more reason to go into detail to stimulate minds and provoke further debate.

In fact, the most common "resolution" of the MFP problem in nuclear textbooks is a resounding silence. While of course no text on nuclear structure theory can avoid discussion of the different nuclear models, most simply skip over the entire issue of a numerical estimate of the MFP of nucleons in nuclei. Despite the fact that the independence of nucleon movement is perhaps the single, most direct, quantitative expression of the internal texture of the nucleus, no mention at all of the MFP issue is made in textbooks by Kaplan (*Nuclear Physics*, 1955), Evans (*The Atomic Nucleus*, 1955), Mayer and Jensen (*Elementary Theory of Nuclear Shell Structure*, 1955), Eisenbud and Wigner (*Nuclear Structure*, 1958), de-Shalit and Talmi, (*Nuclear Shell Theory*, 1963), Rowe (*Nuclear Collective Motion*, 1970), Bertsch (*The Practitioner's Shell Model*, 1972), Irvine (*Nuclear Structure Theory*, 1972), Preston and Bhaduri (*Structure of the Nucleus*, 1975), Lawson (*Theory of Nuclear Shell Structure*, 1980), Valentin (*Subatomic Physics: Nuclei and Particles*, 1981), Eisenberg and Greiner (*Nuclear Theory*, 1987), Burge (*Atomic Nuclei and Their Particles*, 1988), Dacre (*Nuclear Physics*, 1990), Burcham and Jobes (*Nuclear and Particle Physics*, 1994), Walecka (*Theoretical Nuclear and Subnuclear Physics*, 1995), Nilsson and Ragnarsson (*Shapes and Shells in Nuclear Structure*, 1995), Greiner and Maruhn (*Nuclear Models*, 1996) and Cottingham and Greenwood (*An Introduction to Nuclear Physics*, 2001).

This collective neglect of what once *was* a problem and, according to those researchers investigating the MFP today, still *remains* a problem in nuclear structure theory is understandable only in the sense that there simply is no satisfactory solution of the MFP problem. To bring the issue up is to invite trouble.

In Enge's (1966) text, the MFP was mentioned, as follows:

"The question of whether a nucleon in the nucleus has a sufficiently long mean free path, so that one is justified in assuming a fairly independent motion, will be postponed. Here, we will assume that this is the case..." (p. 143).

The only subsequent reference to the MFP was an estimate of 3.23 fm (p. 411) – a value typical of MFP estimates, and one that is notably small.

Others have simply declared that the MFP is "large" or "very large" or even "infinite," and there is consequently no problem. In their authoritative textbook, *Nuclear Structure Theory*, Nobel Prize-winners, Bohr and Mottelson (1969), argued that:

> "A fundamental characteristic of any many-body system is the mean free path for collisions between constituent particles. A wide variety of evidence testifies to the fact that, in the nucleus, this mean free path is large compared to the distance between the nucleons and even, under many circumstances, is longer than the dimensions of the nucleus." (p. 139)

The importance of the MFP issue is evident from its prominent place in *Nuclear Structure Theory*: After 137 pages introducing the ideas of quantum mechanics, this comment appeared within the first two pages concerned with nuclear structure theory. Indeed, the mean free path is a "fundamental characteristic of any many-body system"! Questions about *which* evidence, *which* circumstances and *which* dimensions are important (and will be examined below), but it is clear from these remarks introducing the topic of nuclear structure that Bohr and Mottelson believe that the MFP paradox is no longer a significant problem in nuclear theory.

Similarly, Antonov, Hodgson and Petrov (1993) began their book, *Nucleon Correlations in Nuclei*, with brief discussion of the issue of the validity of the independent-particle model. On the first page of text, they noted the underlying assumption of the model:

> "It is assumed that the nucleons [in the Fermi-gas model] are moving freely in volume Ω." (p. 3)

Every model has starting assumptions, and Antonov and colleagues are to be commended for making theirs explicit at the outset. But in the very next sentence they made a questionable assertion about an empirical issue:

> "The mean free path of a nucleon in the nuclear medium is comparable with the size of the nucleus" (p. 3).

For pursuing the Fermi gas model, the assumption of a lengthy mean free path for nucleons would be convenient, but neither experimental nor theoretical work supports the view that the mean free path is as large as the nuclear diameter, much less large enough to allow nucleon orbiting. Whatever successes the Fermi gas model may have, empirical findings on the MFP of nucleons must be discussed on their own merits. The MFP of low-energy nucleons is generally considered to be only 3–4 fm (Table 5.1), and that would mean that the typical nucleon making a typical orbit across the nuclear diameter of a typical large nucleus ($\sim$10–12 fm) would interact with three or four other nucleons – each time changing its angular momentum and disturbing its independent orbit. That is not what one could characterize as "moving freely." Over the course of several revolutions inside the nucleus, something on the order of 50 collisions would be expected – *if nucleons wander freely like particles in a gas.* It should be noted that this difficulty with the issue of the MFP does not mean that the Fermi gas model is categorically wrong, but the evidence that the MFP is short remains an empirical problem for the gas model, not theoretical support.

Marmier and Sheldon (1969) acknowledged that there is a quantitative issue at hand and calculated the MFP to be 3.226 fm. Guided by the idea that a large MFP is at the heart of the shell model, they then paradoxically remarked that:

"Clearly, the mean free path is by no means minute compared with nuclear dimensions." (p. 1090)

Their own calculations gave an MFP that is about 1/3rd the diameter of a large nucleus. Nevertheless, clinging to the notion of the free orbiting of nucleons, they suggest that their value for the MFP is somehow large...... "by no means minute."

McCarthy (1968) raised the problem of the MFP in the opening chapter of *Introduction to Nuclear Theory*. There, he noted that:

"The mean free path of a nucleon in such a [nuclear] gas is very long." (p. 11)

and having made this statement on page 11 – without reference to experimental data or theoretical calculations, no further mention was made concerning the MFP.

Jones (1986) also did not hedge his bets and asserted, without qualification and without further comment, that:

"The mean free path of a bound nucleon in the ground state of a nucleus is effectively infinite, and the concept of particle orbits is meaningful." (p. 52)

With confidence like that, who needs experimental data or theoretical calculations?

5.2 The Persisting Problem of the MFP

Such bold assertions notwithstanding, all published attempts at determining the magnitude of the MFP of nucleons either bound within a nucleus or injected into nuclei from outside reveal that the MFP is rather short. Already in 1948, Goldberger calculated that the MFP of 100 MeV protons in ^{208}Pb is 4.30 fm when the exclusion principle is *not* included in the calculations, and 5.52 fm when it *is* included. The latter figure is approximately equal to the radius of the lead nucleus, but even this relatively large MFP value is calculated for 100 MeV protons – well above the energy range for bound nucleons. When the more relevant issue of low energy projectiles is considered, however, the MFP becomes embarrassingly small, and this fact and its periodic rediscovery has led to occasional discussions in the physics literature. Some representative examples will be discussed below, and it is worth noting that the actual numerical results are often quite interesting. Unfortunately, whatever the outcome of the MFP calculations, the independence of nucleon movement has become such an indispensable part of the shell theorist's mental image of the internal texture of the nucleus that no amount of experimental data or theoretical clarity can change that picture. Efforts made to protect this notion – despite obvious contradictions – inevitably lead to problems.

The calculations presented by Bohr and Mottelson (1969) are particularly revealing of the difficulties theorists face when trying to explain nuclear structure in light of MFP calculations. At the beginning of their discussion of the independent-particle model (p. 138), they provided a standard nuclear density value of 0.17 nucleons/fm^3 and a standard formula for calculating the nuclear RMS radius: $R = r_0 A^{1/3}$, where, they note, the best fit to the data suggests a value for r_0 of 1.1 fm.

On the next page, Bohr and Mottelson addressed the question of the MFP of nucleons in the independent-particle model. They began with the discussion quoted above, and then went into details:

"A fundamental characteristic of any many-body system is the mean free path for collisions between constituent particles. A wide variety of evidence testifies to the fact that, in the nucleus, this mean free path is large compared to the distance between the nucleons and even, under many circumstances, is longer than the dimensions of the nucleus.

"A very direct way to explore [this issue] is provided by scattering experiments involving incident protons and neutrons... Fig. 2.3, p. 165 shows typical examples. [These findings] establish the fact that the MFP is at least comparable with the nuclear radius...

"The relatively long MFP of the nucleon implies that the interactions primarily contribute a smoothly varying average potential in which the particles move independently..." (p. 139).

So, what is the numerical result and how valid is it? Let us not quibble about the assertion that "comparable to the nuclear radius" should be considered "long", but the examples of the Bohr and Mottelson text are worth examining to see how conclusions were reached about the independent movement of nucleons.

In the relevant section (p. 165), entitled "Estimate of mean free path from neutron total cross sections," the figure that illustrates the MFP value, λ, includes data on ^{63}Cu, ^{112}Cd and ^{208}Pb. For detailed discussion, they chose the results for ^{208}Pb when the incident neutron energy is 90 MeV. It is questionable that the MFP of a 90 MeV nucleon (that is, 3–5 times the energy of neutrons bound within nuclei) has any relevance to the bound-nucleon MFP problem, and, what is more, they used a value of r_0 (= 1.4) that is 27% larger than the value used throughout most of their book. Using that value for r_0, they obtained a radial value for ^{208}Pb as:

$$R = 1.4 \times A^{1/3} = 8.3 \text{ fm} \tag{5.1}$$

Experimental radial values are of course available for ^{208}Pb and most other stable nuclei, and there is no real necessity of reverting to an approximate formula, such as (5.1), to discuss the nuclear radius. But, following their train of thought, if this unusually large radial value for the lead nucleus is to be used for calculating the MFP, then certainly we must later compare the calculated MFP value against the same radial value when we discuss the relative size of the MFP and the nucleus. Clearly, what we are interested in, when addressing the issue of the MFP, is the approximate volume of a nucleus within which its nucleons are thought to orbit.

Although a qualitative result concerning the MFP can be obtained using an approximate figure for R, the exact result depends crucially on whether we take r_0 to be 1.1 or 1.4. Negele (1983) noticed this problem and made a similar complaint about the Bohr and Mottelson MFP calculations. He remarked that a relatively small ($\approx$3 fm) or a relatively large ($\approx$5 fm) MFP can be obtained from the *same* experimental data, depending solely on the value assumed for r_0. Clearly, we gain no insight into the underlying physics of the situation if our result is so strongly influenced by our starting assumptions.

Given the experimental data from a nuclear reaction at an energy level three-fold higher than the range of interest and a questionable theoretical value for the lead radius, we get an MFP that is nonetheless consistent with the idea of a rather *short* MFP. Bohr and Mottelson obtained an MFP that is less than *one* nuclear radius. And if we use the *known* RMS radial value for Pb^{208}, we get an MFP value of 3.4 fm. As seen in Table 8.1, this and most other attempts at estimating the MFP indicate that the MFP for bound nucleons is 3–5 fm – less than the radius of lead.

Even an MFP value of 5 fm suggests an abundance of intranuclear nucleon-nucleon interactions and the impossibility of nucleon orbiting, but, like most theorists, Bohr and Mottelson are committed to an independent-particle nuclear texture and later inflated their own estimate still further. Instead of an MFP "comparable to the nuclear radius," the same numerical value was subsequently (p. 189) referred to as "larger than the dimensions of the whole system".

Bohr and Mottelson are two of the great pioneers in 20th Century physics and, despite these critical comments about their textbook, they have made innumerable contributions to nuclear structure theory. Their discussion of the mean free path is, however, noteworthy for the difficulties it generates and leads to the inevitable conclusion that their calculations (and most other textbook treatments of the MFP problem) are, with or without numerical calculations, little more than assertion of a preconceived notion that the independent movement of nucleons simply *must* be allowed if we are to enjoy the benefits of the shell model.

Some 16 years later, Mottelson (1985) was again at pains to show that the MFP is long. To make his point, he cited a letter to the editor dating from 1953 (one of the "seven milestones" of nuclear physics prior to 1954!). There, an MFP of 20 fm had been suggested (Feshbach, Porter and Weisskopf, 1953), but it was obtained using optical model parameters that those *same* authors abandoned less than a year later in a more detailed study where they concluded that the MFP is 3–4 fm (Feshbach, Porter and Weisskopf, 1954; see Table 5.1).

Unfortunately, a great many textbooks and journal articles continue to make reference to the authoritative work of Bohr and Mottelson when it comes to discussion of the MFP. Usually, this takes the form of repeating the qualitative argument that the MFP is greater than the "dimensions" of the nucleus. When a quantitative estimate is needed, the Bohr and Mottelson result of 5.2 fm is sometimes referred to as "yielding $\lambda \sim 6$ fm." Many authors have picked up on the fuzzy jargon used by Bohr and Mottelson. Koonin (1981), for example, stated that:

> "Nucleons are largely independent within the nucleus and interact with each other only through their common time-independent mean field. In other words, the mean free path for nucleons, λ, is many times larger than the nuclear dimensions." (p. 234)

Whatever is meant here by "nuclear dimensions," numerical calculations are not supportive.

The theoretical estimate by Koonin (1981, p. 235) – designed specifically to illustrate how long the MFP is – is based on the following equation:

$$\lambda_p \sim 530 \text{ fm}/(E_p - E_F)^2 \qquad (5.2)$$

The crucial effect lies in the difference between the energy of the free particle, E_p, and that of the so-called Fermi surface, E_F, of nuclear matter. Clearly, the MFP can be made as large as desired in this type of theoretical argument by selecting E_p and E_F values that are close to one another, but if the energy of the particle is a rather modest 10 MeV different from the Fermi surface energy, then the MFP is typically short (relative to any chosen "dimension" of the nucleus). Although the result is highly dependent on the starting assumptions, Koonin nonetheless suggested that

> "the conclusion is inescapable ... that nuclei are ... a nearly collisionless ... gas ... rather than a short-mean-free-path liquid drop." (p. 235)

But this conclusion is more accurately described as a restatement of a notion regarding the presumed nuclear texture within the independent-particle model rather than an "inescapable" deduction of the value for the MFP.

Bohr and Mottelson's phrasing has also been borrowed by Bortignon, Bracco and Broglia (1999) in their otherwise excellent text, *Giant Resonances*:

"the mean free path of a nucleon at the Fermi energy is expected to be of the order of nuclear dimensions" (pp. 8–9)

"There is extensive experimental evidence, which testifies to the fact that a nucleon moving in an orbital close to the Fermi energy has a mean free path which is large compared with the nuclear dimensions." (p. 71)

Other recent examples of the same difficulty in discussing the MFP are to be found in the textbooks by Heyde (*The Nuclear Shell Model*, 1995) and (*Basic Ideas and Concepts in Nuclear Physics*, 1994). Similar to Bohr and Mottelson, Heyde devotes the first 57 pages of *The Nuclear Shell Model* to general considerations of quantum mechanics, and begins the discussion of nuclear structure with the assumption of the independent movement of nu cleons in the shell model. On the second page of discussion on nuclear theory, a paragraph consisting of only one sentence was devoted to the issue of the MFP:

"One of the most unexpected features is still the very large nuclear mean free path in the nuclear medium." (p. 59)

A figure is provided (similar to Fig. 5.4, below), but there is no further mention of the mean-free-path in the remaining 379 pages on the shell model.

In a similar vein, in *Basic Ideas and Concepts in Nuclear Physics* (1994) Heyde began the discussion of nuclear structure with a brief description of the liquid-drop model, but then gave three reasons why it will not suffice to explain the structure of the atomic nucleus. The first is that alpha-clustering is known and the third concerns the nuclear force. The second reason why he considers the high-density, collective conception of the nucleus to be insufficient is that

"the mean free path of a nucleon, traveling within the nucleus, easily becomes as large as the nuclear radius ... and we are working mainly with a weakly interacting Fermi gas." (p. 191)

For clarification, reference is made to a subsequent chapter on the Fermi-gas model (in which no discussion of the MFP is to be found) and a figure (his Fig. 7.3, shown here as Fig. 5.2).

For anyone not committed to the independent-particle model approach to nuclear physics, the information in that figure illustrates how remarkably *short* the MFP is. *All* of the theoretical values lie between 1.6 and 4.8 fm and *all* of the experimental values lie between 2.4 and 6.6 fm. All of those values are less than the diameter of a large nucleus. In what sense can we consider movement of a few fermi equivalent to unimpeded orbiting?

Virtually every coherent theoretical discussion and virtually every experimental finding related to the MFP of bound nucleons or low-energy nucleons interacting with nuclei indicate a short MFP – usually values well less than the nuclear radius, but *always* much less than a distance comparable to several revolutions of a nucleon through the nuclear interior. Textbook authors

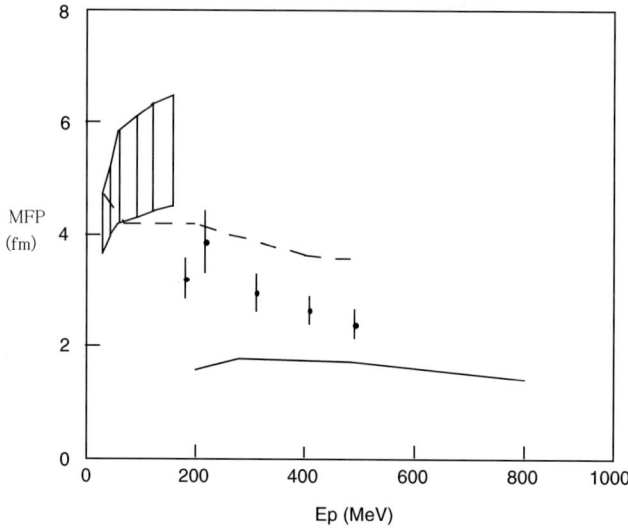

Fig. 5.2. A summary of theoretical and experimental findings on the MFP (from Heyde, 1994, p. 193). The *shaded band* indicates the range of data for ^{40}Ca, ^{90}Zr, and ^{208}Pb. Theoretical values are indicated by the dashed curve (Dirac-Brueckner) and the solid curve (the optical model)

cite the theoretical and experimental work that indicates a short MFP, and then draw the opposite conclusion. So, let us move away from the textbooks and examine some of the technical literature on the MFP.

With characteristic frankness, Schiffer (1980) introduced the problem as follows:

> "One would have thought that by now our knowledge of the nucleon mean-free paths in nuclear matter would be well-established. Yet this is far from being the case. If we look at essentially any experimental data, we find the mean-free-path one expects is of the order of 5–10 fm... On the other hand, if one studies any a priori optical model, the mean free path one obtains after taking into account all Pauli blocking and similar effects just from nuclear density and the free nucleon-nucleon cross-section, is of the order of $2 \sim 2.5$ fm. This discrepancy has not been understood and in fact it has not really been recognized in the literature. Yet it is a serious anomaly..." (p. 348).

The actual discrepancy between experiment and theory may be smaller than indicated by Schiffer, because the so-called "experimental" values rely on the use of adjustable parameters. Nevertheless, it is clear that resolution of the MFP problem at a value anywhere between 2 and 10 fm still leaves the question about the validity of the independent-particle model unanswered.

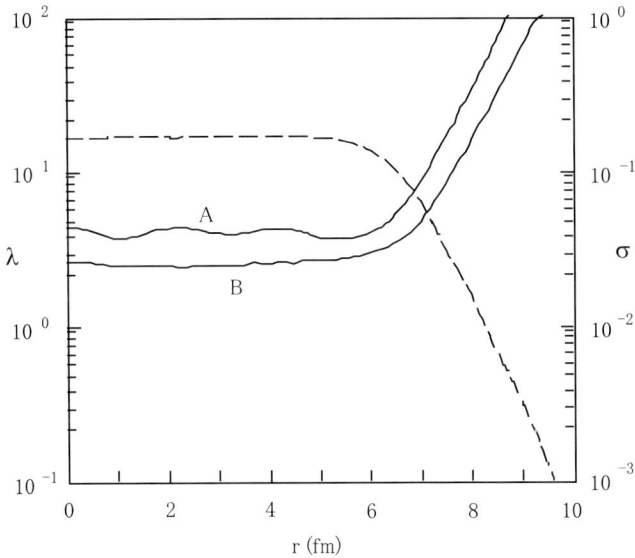

Fig. 5.3. The effect of the nuclear density on the MFP (after DeVries and DiGiacomo, 1981). The two *solid* lines are the MFP results for 30 MeV (A) and 100 MeV (B) protons in ^{208}Pb. Note that the MFP, λ (fermi), is less than 5 fm for all nucleons up to nearly 7 fm from the nuclear center. The nuclear density, σ, (nucleons per cubic fm), is shown with the *broken line*

Some insight into the problem of the MFP was provided by DeVries and DiGiacomo (1981) who calculated that a proton shot into a lead nucleus at 30 MeV (an energy level that is comparable to the energy of bound nucleons) would classically have an MFP of only 0.27 fm. In other words, it would hardly enter the nucleus at all before colliding with other nucleons – precisely as a liquid-phase theory would predict. They noted, however, that

> "a more realistic evaluation has been performed. This approach involves not only realistic densities but effective nucleon-nucleon cross sections which include Pauli-blocking, Fermi motions and real nucleon and Coulomb potential effects. Pauli blocking and Fermi motion can increase the MFP by more than an order of magnitude."

So, is this the kind of calculation that demonstrates a long MFP? Does inclusion of the effects of the Pauli exclusion principle vindicate the independent-particle model? The figure illustrating their results (reproduced as Fig. 5.3) shows an MFP (in the core of ^{208}Pb for protons of 30 and 100 MeV) that is only 2–5 fm, whereas "several revolutions" through a nucleus with a 10 fm diameter still requires an MFP of at least several tens of fermi. In other words, the magnitude of the MFP that can be obtained by consideration of the Pauli principle is still too small by 20- or 30-fold.

In fact, the calculations by DeVries and DiGiacomo reveal an interesting effect. That is, the length of the MFP is dependent on the location of the nucleon relative to the nuclear center. As illustrated in Fig. 5.3, the MFP is smallest within the nuclear core region and increases progressively for nucleons as they enter the low density skin. This is of course to be expected because the high density core should allow far less free movement than the diffuse skin region. They therefore warned that "the concept of a constant (nuclear matter) mean-free-path is rather meaningless" and concluded that "only a local mean-free-path, which is highly dependent on the location of the nucleon in the nucleus, should be considered when finite nuclei are involved."

That is an interesting result. In exactly the same way that a "mean" value for the nuclear density (that implicitly neglects the large difference between the density of the nuclear core and that of the skin) is not very informative, a universally-valid MFP value that applies to all density regions within the nucleus simply does not exist. As DeVries and DiGiacomo maintained, it is meaningless or even misleading to speak of "*the*" MFP value for all nucleons. But, as clearly illustrated in their own figure, a small MFP value is found throughout the nuclear core region – extending about 6 fm from the center of the Pb^{208} nucleus! Gradually larger MFP values are found in the low density skin region, but, surely, for the vast majority of nucleons located in the nuclear core, an MFP value of less than 5 fm is still fundamentally in contradiction with the "independent-particle" assumption underlying the shell model.

Therefore, in reply to the question asked in the title of their paper "Is the concept of a mean free path relevant to nucleons in nuclei?" it can be said that, if we are looking for a *single, universally-valid* MFP value to use in all possible contexts, the answer is "no," but the answer is certainly "yes" if we are interested in examining the theoretical basis of the independent-particle conception of the nucleus. The inevitable conclusion is that most nucleons in most nuclei have short MFPs: nucleons are *not* free to orbit and the short MFP remains an experimental fact that is supported by theoretical calculations and that flatly contradicts the "orbiting nucleon" conception of the shell model.

Negele and Yazaki (1981) and Negele (1983) have also made serious attempts at analyzing the problem of the MFP at nucleon energies of $50 \sim 150$ MeV. As illustrated in Fig. 5.4, three different sets of theoretical calculations produce MFP values roughly at 2, 4 or 6 fm, with the largest theoretical results thought by them to be correct. "Experimental" MFP values (i.e., those deduced from experimental results using the optical model) are found at $4 \sim 6$ fm, so they concluded that the bulk of the discrepancy between theory and experiment had been accounted for.

Within this energy range, there is clearly no prospect of an MFP of several tens of fermi from such experimental or theoretical results, and Negele (1983) was forthright about the interpretation of the so-called experimental MFP values. Among his concluding remarks, he stated that:

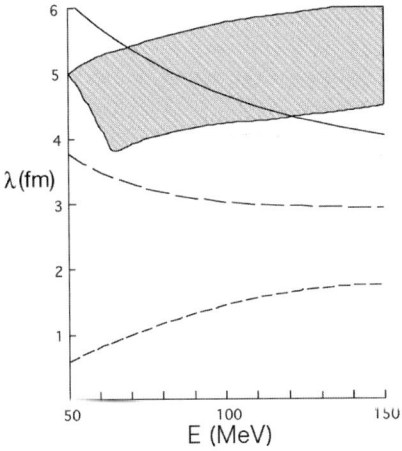

Fig. 5.4. Comparison of theoretical and experimental values (*shaded band*). The *short-dashed line* denotes λ calculated without a Pauli correction. The *long-dashed line* includes the Pauli effect; and the *solid line* denotes what Negele and Yazaki (1981) believe to be the correct result based on a non-local optical potential. The experimental band can be lowered by a factor of two by choosing alternative parameters (after Fig. 5.1 in Negele and Yazaki, 1981)

"It is tempting to conclude this Comment with the assertion that not only do we now understand how to calculate the mean free path microscopically, but also that the theoretical result is unambiguously confirmed by experiment. I believe the theoretical arguments summarized here stand on their own merits, but I have growing doubts that the present experimental situation is quite as unambiguous as suggested..."

"One hint of possible trouble is seen in the recent demonstration... that when sufficient flexibility is allowed in the parametrization, [experimental results, such as those presented by Bohr and Mottelson] can be fit by drastically different optical potentials, with interior values of [the imaginary part of the potential] which differ by a factor of 2 and correspond to $\lambda = 2.2$ or 5 fm respectively."

"Once one allows the possibility of significantly different surface shapes of the optical potential, determination of λ by total or reaction cross sections is also rendered ambiguous... Whereas the value $\lambda = 5.1$ fm is extracted from cross sections ... using $R = 1.37\, A^{1/3}$, essentially the same data ... using $R = 1.28\, A^{1/3}$ yield $\lambda = 3.2$ fm."

For these comments alone, Negele deserves a prize.

Inexplicably, Negele subsequently reverted to favorable citation of the Bohr and Mottelson MFP discussion that he had previously criticized – now rounding the Bohr and Mottelson value of 5.2 fm up to 6 fm (Negele and Orland, 1988, p. 259)!

Equally bizarre is the fact that Negele's earlier work that was skeptical about the lengthy MFP argument has been cited by others as an example of the *resolution* of the problem. Feshbach (1992), for example, reproduced the same figure (Fig. 5.4) from Negele and Yazaki (1981) and commented that it is

> "a result of importance for the understanding of the foundation of the mean field (shell model, optical model, etc.) approximation in nuclei" (Feshbach 1992, p. 355)

But is an MFP of 5 fm, not to mention an MFP of 3 fm, a solution to the problem?

Yuan et al. (1989) discussed the same topic, and showed how a specific set of parameters for the nuclear force (specifically, one of the Skyrme potentials) can produce a rather large (4.0 ∼ 7.3 fm) MFP at proton energies of 50 ∼ 170 MeV. They concluded that they had

> "demonstrated the usefulness of the extended Skyrme interaction in calculations of the mean free path [and had] resolved much of the discrepancy between the simple formula [for the MFP] and experimental data" (p. 1455).

Their theoretical results for Pb^{208} (Fig. 5.5) appear at first glance to coincide nicely with the experimental data, but two points should be noted. The "experimental" values were obtained by assuming a value for $r_0 = 1.35$ (Fig. 5.5; open circles). Equally valid "experimental" values obtained with $r_0 = 1.07$ tell a somewhat different story about the agreement between experiment and theory (Fig. 5.5; filled circles). The MFP with $r_0 = 1.07$ falls at 3–6 fm and leaves the MFP issue as the paradox it has always been.

Far more disastrous for the conclusions of Yuan et al., however, was the fact that the set of Skyrme parameters that produced the MFP curve shown in Fig. 5.5 was only one of 12 sets of parameters that have been reported by various groups and that were examined by Yuan et al. for their implications regarding the proton MFP. They were honest enough to note that:

> "The mean free paths obtained with the other parameter sets are too short (∼3 fm)" (p. 1455).

In other words, either Yuan et al. (1989) have hit upon the one and only correct set of parameters for the Skyrme potential or their results are not generally valid. Whichever the case may be, the discrepancy between theory and experiment is resolved at an MFP = 4 ∼ 7 fm (or at 3 fm using other Skyrme potentials). Such MFP values might be large enough to maintain the concept of "nucleon orbiting" inside of the alpha particle, but certainly not inside of a heavy nucleus.

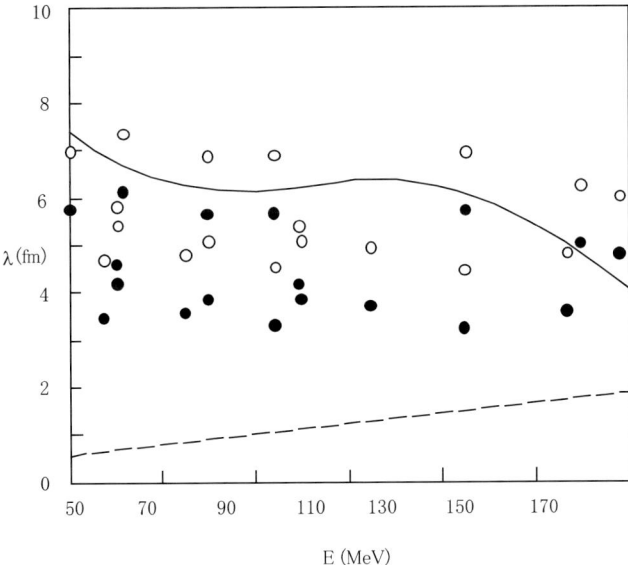

Fig. 5.5. The proton MFP in Pb^{208} (after Fig. 5.8 in Yuan et al., 1989). The *solid* line denotes the MFP obtained with one set of Skyrme interaction parameters. The *dashed* line shows the MFP calculated on the basis of the nuclear density. *Open circles* are experimental data using $r_0 = 1.35$ (as reported by Yuan et al.); *closed circles* are the same data using $r_0 = 1.07$

Finally, a word should be said about the optical model – which is the principal theoretical tool used for interpreting low-energy nucleon-nucleus scattering experiments. Often "experimental" MFP values are obtained by first obtaining a best fit between experimental data and optical model parameters, and then deducing the MFP from those parameters. That is entirely acceptable practice in nuclear theory, but it is important not to pretend that model parameters are like the empirical values you get from weighing a brick! Regardless of which set of parameters gives a "best fit," other fits are often quite good and a dramatically different set of parameters (maybe not a "best fit," but a "good fit") might be justified on theoretical grounds. Particularly for calculating the MFP using the optical model, the relative strengths of the adjustable real and imaginary potentials play a central role (e.g., Enge, 1966, pp. 409–415) and there is simply no unique set of parameters that can be unequivocally accepted as correct. It is therefore dangerous to treat values deduced from a rather malleable set of parameters as unambiguous facts, and quite incorrect to refer to such values as "experimental results."

Typical MFP values deduced from the parameter settings of the modern optical model are shown for representative nuclei in Fig. 5.6. In the 40–200 MeV range, MFP values are typically 3–6 fm, but it is at the lower end $E < 50$ MeV where MFP values relevant to bound nucleons are to be found.

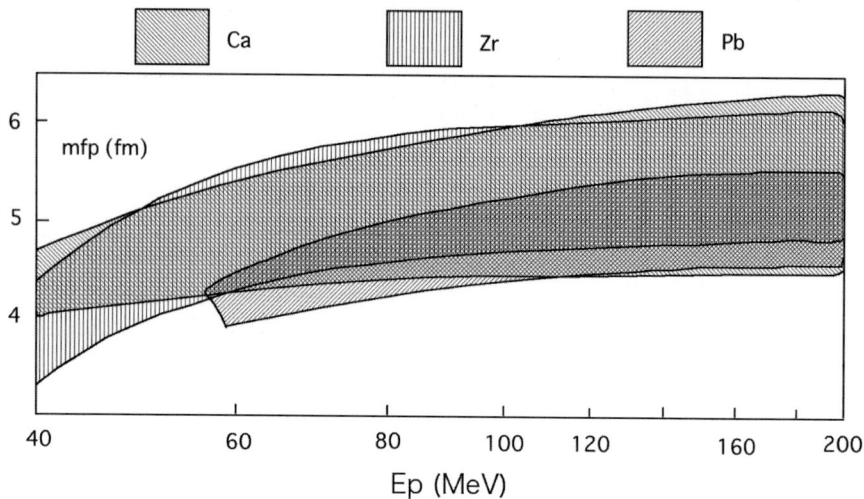

Fig. 5.6. The mean free path for calcium, zirconium and lead (after Fig. 13 of Nadasen et al., 1981). The *shaded bands* are the range of values calculated using optical model parameters

5.3 The Weisskopf Solution

As indicated by virtually all published calculations on the MFP, the question of the free-orbiting of nucleons remains today very much a problem for nuclear theory. In what sense, we must ask, is it an issue that can be totally overlooked in the textbooks or simply dismissed out of hand? In fact, most attempts at estimating the MFP point directly to the extremely high density of nucleons in nuclei and therefore to an extremely small MFP – features that are characteristic of solids and liquids, rather than gases.

Historically, what has happened to discussions of the MFP is that Weisskopf's early suggestion of a solution based on the exclusion principle was seized upon and accepted as a necessary and *sufficient* explanation by nearly an entire generation of nuclear theorists. Believing that theoretical justification of the shell model had to be found, a qualitative argument that suggested the possibility of nucleon orbiting has been uncritically accepted and the underlying paradox largely dismissed as "resolved." As a consequence, the picture in the back of the theorist's mind is one where nucleons with large hard-core dimensions magically slip and slide past their close neighbors without any interaction at all. Faced with an apparent contradiction between the simultaneously high- and low-density requirements of the nuclear models, this picture suggests that the nucleus is at the same time both a dense liquid and a diffuse gas. Unfortunately, quantitative work on the MFP has shown that the problem remains.

5.3 The Weisskopf Solution

Let us therefore go back to the origins of the Weisskopf solution to see how the current understanding of the MFP has evolved. In 1951, Weisskopf argued that, although the nucleus is a very dense substance with relatively large nucleons separated by relatively small distances (as described in the liquid-drop model), nucleons do not collide and interact because of the effects of the Pauli exclusion principle.

Old questions about the meaning of the exclusion principle itself will be raised below, but here it is worth recalling what this universally-accepted principle describes. Originally, Pauli proposed that the electron build-up of the periodic chart could be understood if it were assumed that only one (fermion) particle can enter a given quantum state at a time. If a particle is in a particular energetic state, with characteristic quantum numbers defining that energy state, then all other similar fermions are "excluded" from that state: they must therefore exist at a different energy level.

Weisskopf applied this idea to the nucleus, arguing that if each nucleon excluded other nucleons from its own energy state, it could by the very act of "exclusion" prevent collisions and therefore suppress energy exchanges among nucleons that are nonetheless tightly packed within the nuclear volume. It might then be the case that, even in the dense interior of a nucleus, nucleons can orbit several times before colliding and sharing their energy with other nucleons.

Weisskopf's own description of the dilemma and its resolution in 1951 went as follows:

> "It must be emphasized that this [shell model] picture is based upon a far-reaching assumption: The nucleons must be able to perform several revolutions on their orbits before they are disturbed and scattered by the interaction with neighbors. This condition is necessary for the existence of a well-defined energy and angular momentum in each separate orbit. The 'mean free path' within nuclear matter must be of the order of several nuclear radii in order to justify the existence of separately quantized independent states for each particle.
>
> "The strong interaction [liquid-drop] models are based upon the opposite assumption. They are all derived from the concept of the compound nucleus. Bohr has pointed out that, in most nuclear reactions, the incident particle, after entering the target nucleus, shares its energy quickly with all other constituents. This picture presupposes a mean free path of a nucleon that is much shorter than the nuclear radius. Nevertheless the compound nucleus picture is very successful in accounting for the most important features of nuclear reactions...
> "The two viewpoints seem to be totally contradictory."

Having set the scene, he suggested a solution:

> "It is very probable that the Pauli principle prevents the strong interaction from exhibiting the expected effects. The interaction cannot

produce the expected scattering within the nucleus, because all quantum states into which the nucleons could be scattered are occupied...."

"It may be useful to discuss in this connection an analogous situation that one finds in the theory of the electron motion in solids. The electronic properties of metals and insulators can be described very successfully by assuming that the electrons move in a common potential field... The interaction between the electrons is completely neglected...."

"The success of this description is perhaps also surprising in view of the fact that the interaction between electrons is by no means small... In spite of this fact, the mean free path of the metallic electrons is very much greater than the interatomic distances... The reason is again found in the Pauli principle, which does not admit any scattering of electrons by electrons, because all states into which the scattering process may lead are occupied."

"[In conclusion], the influence of the Pauli principle upon the mean free path of the electrons may serve as a useful analogy to understand the possibility of an independent-particle picture in the presence of strong interactions between nucleons."

The analogy here is certainly of interest, but it is relevant to note the obvious differences between the two realms. Electrons are always treated as *waves* or as *point*-particles, because they have no detectable hard-core radius, whereas nucleons have measurable sizes as space-occupying objects. Even if electrons have a hard-core structure as large as nucleons, the relative volume of space within the atom that the electrons (as particles) occupy is negligible in comparison with the relative volume that nucleons occupy within the nucleus. Whatever factors may influence the interactions of electrons in a metal solid, there is clearly much less electron substance within the atomic volume than there is nucleon substance within the nucleus. The analogy between atomic and nuclear structure may perhaps be useful, but if the analogy breaks down it is likely to fall apart on this point of the relative size of the particles involved.

The simplicity of the idea that nucleons do *not* interact because they obey a principle, together with the attractiveness of Weisskopf's analogy with solid state physics proved quite seductive for most nuclear physicists in the 1950s. Today, if the MFP problem is mentioned in textbooks on nuclear structure, it is invariably followed by Weisskopf's suggestion – usually presented as an established fact and with the relevant calculations rarely mentioned. Before final conclusions are drawn, however, it is worth pointing out that this "solution" was only a qualitative argument and, as seen above and discussed further below, numerical work in subsequent decades has simply not supported it. Depending upon the various assumptions that go into the calculations, an MFP of 3–5 fm for bound nucleons can be produced with the help of the exclusion principle, but even 5 fm accounts for much less than one revolution in a nucleon orbit within a large nucleus. It is therefore doubtful that the Weisskopf solution solves anything at all.

In other words, even if there are grounds for believing that the property of fermions that the exclusion principle summarizes helps them to move past one another without interacting, it remains the case that – the near-universal enthusiasm for this suggestion notwithstanding – no one has been able to demonstrate quantitatively that this is so. The idea that the exclusion principle automatically allows for *several* revolutions of bound nucleons has simply not been vindicated. As discussed above, DeVries and DiGiacomo (1981) found that relatively lengthy MFPs may exist in the nuclear skin region, but not in the nuclear interior where most of the nucleons are located most of the time.

Weisskopf himself was honest enough to write:

> "It remains to be proved whether this effect is sufficient to establish independent orbits in low-lying states of nuclei in spite of the existence of strong interactions" (Blatt and Weisskopf, 1953, p. 778).

But supporters of the shell model have not bothered to glance back.

That the MFP problem could not be solved by one wave of the Pauli wand was evident in terms of early nucleon-nucleus experimental results and on the basis of every serious attempt at theoretical study. But the possibility that Weisskopf's analogy could avoid a head-on collision between the theoretical basis of the liquid-drop model and that of the shell model has proven extremely attractive for most theorists. Since the contradiction between the experimentally-based conclusions about the short MFP, on the one hand, and the entire theoretical framework of the independent-particle model, on the other, is clearly disastrous for the model of nuclear structure, even a qualitative argument that has not found quantitative support has been widely embraced. As a consequence, despite internal contradictions, contrary empirical findings and embarrassing theoretical calculations, the exclusion principle "solution" has been quite uncritically accepted as an essential part of an understanding of the nucleus.

Although the problem can be traced back to Weisskopf's influential work, its exacerbation is the result of many hands. Most discussions of nuclear structure treat the Weisskopf argument as a known fact. For example, a summary of nuclear theory by Elliott and Lane in 1957 began with what has subsequently become a familiar refrain:

> "The central physical feature of a shell-model of a closed system of particles is that the motions of the various particles are largely uncorrelated, so that each particle moves essentially undisturbed in its own closed orbit. Another popular way of expressing the same thing is to say that the mean free path of a particle against collision with other particles is long compared to the linear dimensions of the system...."

"There are two main reasons why the nuclear shell-model has taken so long to become mature and respectable. The first is that the model has no apparent theoretical basis. In fact, until recently, arguments why the model should not be valid were more numerous and convincing than those in favor of the model. These former arguments almost invariably commence with the observations that the nucleus, in contrast to the atom, has no overall central potential and that, furthermore, the short range of nuclear forces means that one cannot use a smooth average potential to represent the actual potential felt by a nucleon, which will have strong local fluctuations. These facts clearly throw strong doubt on the validity of the nuclear shell-model. In the absence of any further feature, they would surely lead to its rejection. However such a feature does exist in the form of the Pauli Principle." (pp. 241–242)

Similarly, in a textbook published in 1967, Meyerhof stated that:

"the Pauli exclusion principle, which forbids two nucleons of the same kind, e.g., two protons, to occupy states with identical quantum numbers, produces effects which keep nucleons apart from each other." (p. 35)

And later, Eder (1968) went through the same exercise in discussing the theoretical basis for the shell model:

"The independent-particle model is based on the assumption that we can single out any one nucleon from the other A-1 nucleons and assume that the effect of the A-1 nucleons on the nucleon so chosen is described by a uniform single-particle potential".

"The usefulness of such a model in the atomic physics of the electron shells is understandable because the electrons move in a common force field, namely the Coulomb field of the nucleus... The situation is quite different in the nucleus because the nucleons themselves generate the force field in which they arose. Since nuclear forces are among the strong interactions, it would seem reasonable to assume, to begin with, that a nucleon has only a short mean free path in the nucleus; we would assume that nucleon-nucleon collisions occur at very small spatial separations, and that in these collisions energy is gained or lost, with the result that it would be meaningless to speak of a particular stationary state. In fact, however, the Pauli principle strongly restricts the number of possible transitions..." (p. 63).

Burcham (1973) also asserted:

"Historically, the development of the shell model in its modern form was retarded because it seemed difficult to envisage any nuclear structure of strongly interacting particles which could provide a strong central potential of the sort known in the atom. The feature of strong interaction was fundamental to the successful liquid droplet model and this model seemed to exclude the possibility of the long MFP for a nucleon in nuclear matter required by the shell model. Recently it has been realized that the Pauli principle operates to lengthen mean free paths because collisions in nuclear matter cannot take place if they lead to states of motion which are already occupied by other nucleons." (p. 2)

Later, discussing the short mean free path when nuclei are bombarded with nucleons, he restated the Pauli magic:

"The long mean free path [of an incident particle] is at variance with the strong interaction theory of Bohr, but it may be understood because the Pauli principle will inhibit collisions which lead to occupied momentum states." (pp. 406–407)

In Segre's text, *Nuclei and Particles* (1965), the issue of the MFP that troubled so many nuclear theorists in earlier years was dealt with as follows:

"Even without going into the complicated subject [of the shape of the potential well in the shell model], we must point out a serious difficulty in the shell model. How can a nucleon move in an orbit in nuclear matter?"

"Using free particle cross sections, we would expect a nucleon mean free path in nuclear matter to be short compared with the distance required before one can speak of an orbit. Pauli's principle gives a partial answer to this difficulty by inhibiting collisions within a nucleus when the final states that the colliding nucleons should reach are already occupied." (p. 281)

Williams expressed the same doubts in his text, *Nuclear and Particle Physics* (1991), and offered the same resolution:

"A question comes up at once: how can it be that a nucleon can occupy an orbit for a time sufficient to allow that concept to make sense when the nucleon is in an environment crowded with nucleons? Normally the mean free path of an energetic ($>10\,\mathrm{MeV}$) nucleon moving in nuclear matter during the course of a reaction is about 2 fm. If this applies to bound nucleons, then even a once-around-the-nucleus orbit without a collision is almost [!] impossible. It is the Pauli exclusion principle which rescues us from this difficulty...." (p. 134).

Antonov, Hodgson and Petrov (1988) made similar assertions:

"Due to the action of the Pauli principle the nucleus is not a dense system of nucleons and the strong nucleon-nucleon forces are reduced by the fact that the nucleons are quite far apart. The interactions due to the singular force at small nucleon-nucleon distances are infrequent and this makes the idea of the independent motion of the nucleons acceptable as a first approximation to the many-body problem" (p. 6).

And, again, Hodgson, Gadioli and Gadioli Erba (1997) frankly raised the issue of the apparent contradiction between gas and liquid models in their text, *Introductory Nuclear Physics*, but then side-stepped the inevitable crisis by simply asserting that an answer is to be found in the exclusion principle:

> "The answer to this problem is provided by the Pauli exclusion principle, which forbids any interaction that puts particles into states that are already occupied. This prevents many interactions that would have taken place in free space, and thus ensures that the mean free path of nucleons inside the nucleus is long enough for the shell model to be valid." (p. 315)

Unfortunately, whatever qualitative attractions that the "invocation of a principle" may offer, numerical estimates based on the effects of the exclusion principle also need to be examined. So, in spite of the fact that many authors since the early 1950s have concluded that the MFP problem is neatly resolved by the exclusion principle, the results published in the nuclear physics literature should also be examined.

5.4 Exclusion Principle "Correlations"

The idea that the exclusion principle keeps particles apart has been formalized – within the context of both atomic and nuclear physics – and generally expressed as the "correlations" between particles. Indeed, it is frequently argued that atomic and nuclear matter does not collapse under the force of the mutual particle attraction due principally to correlations among particles as a direct consequence of the exclusion principle. A more common-sense view that the hard-core of the nuclear force or the electrostatic repulsion between charge-bearing particles might play a role in keeping them apart is explicitly rejected. Indeed, some authors are adamant that the exclusion principle itself is responsible and that it is fundamentally unrelated to the physical forces acting between particles. For example, de-Shalit and Talmi (1963) argued as follows:

> "The 'Pauli repulsion' has nothing to do with the forces between the nucleons and does not result from any dynamical consideration. It is merely a manifestation of the Pauli principle, and exists in any system of fermions irrespective of whether the real forces between them are attractive or repulsive." (p. 138)

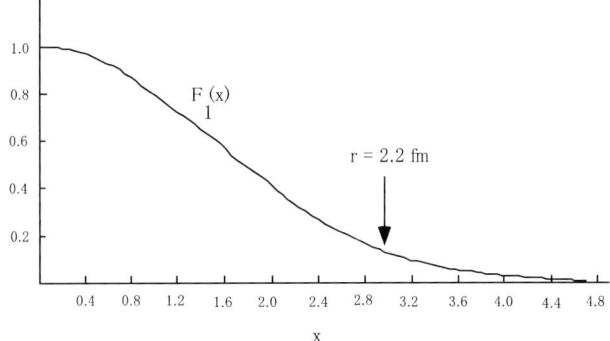

Fig. 5.7. The Pauli correlation function F_1 for an infinite medium (after Fig. 5.1 of Preston and Bhaduri, 1975, p. 249). The *horizontal axis* in the original figure indicates distance in units of "x"

In effect, the Pauli correlations are said to prevent particle interactions and thus to have effects that are as real as those exerted by the known forces of nature. This mysterious force that isn't a force ("Pauli repulsion... is a manifestation of the Pauli principle") begs for further exploration, but within the present context it is certainly relevant to inquire about the dimensions over which correlation effects are thought to act inside of a nucleus. If we have a nucleon at some location, and the exclusion principle suppresses the probability of finding another nucleon at a nearby location, then it is of interest to know how much suppression over how many fermi.

A numerical discussion of such correlations in the nucleus can be found in *Structure of the Nucleus* by Preston and Bhaduri (1975, pp. 248–252). Their line of reasoning concerning particle correlations goes as follows. Two nucleons classically have a reasonably high probability of being found in a given volume of space within the nucleus, but "quantum statistics [i.e., Pauli effects] alters this situation drastically." That is, the probability of finding two identical nucleons, 1 and 2, P_{Fermion} (1,2) in a volume is:

$$P_{\text{Fermion}}(1,2) = \rho^2[1 - F_1(k_F\rho)] \tag{5.3}$$

where ρ is the nuclear density and k_F is the particle momentum. Since ρ (0.17 nucleons/fm^3) and k_F (1.36 fm^{-1}) are constants determined by the density of nuclear matter, the probability of finding the two nucleons together, $P_{\text{Fermion}}(1,2)$, is a function of the distance separating the two particles.

This is illustrated in Fig. 5.7, reproduced from Preston and Bhaduri (1975). They note that $F_1(x)$ has the value of unity at $r = 0$, in other words, a probability of zero of finding the two particles at the same point in space. Moreover, $F_1(x)$

"remains large up to $x \sim 2$, and is small (less than 0.1) for $x > 3$. With $k_F = 1.36\,\text{fm}^{-1}$ corresponding to nuclear-matter density, these two values of x correspond to $r = 1.5$ and 2.2 fm. The short range part of the two-nucleon force between identical nucleons will therefore be suppressed because P_Fermion (1,2) is small where the force is large. It is apparent that the exclusion principle alone accounts for the possibility of using an effective force in nuclear matter, much weaker than the free nucleon-nucleon force, for collisions between pairs of identical particles." (pp. 249–250)

The calculation of the correlation function is straight-forward and Preston and Bhaduri's explanation is characteristically clear, but we must still be cautious about any preconceived conclusions. What precisely has the exclusion principle been invoked to accomplish? It has reduced the probability of like-particle interactions to 50% at an internucleon distance of 1.3 fm in infinite nuclear matter, but the reduction is only 16% at a distance of 2.2 fm. So, as real as the Pauli correlation effect may be, it adds only a small effect to that which the dynamic effects among nucleons are known to produce anyway.

In what sense, therefore can it be concluded that "the exclusion principle alone accounts for the possibility of using an effective force in nucleon matter, much weaker than the free nucleon-nucleon force" (p. 250)? Has the exclusion principle prevented the strong interaction between neighboring nucleons? Yes, but only to a small degree. Like most nuclear theorists, however, Preston and Bhaduri would like to reach the conclusion that nucleons do *not* interact strongly with each other and that the gaseous model is well justified, but they are too careful simply to declare an "infinite" MFP. Indeed, having examined the case of the MFP in nuclear matter, they then proceed to examine the suppression of nucleon interactions in a small nucleus. Although their preconceived notion of significant Pauli effects remains unchanged, their own calculations show that such effects are reduced even further in finite nuclei.

In infinite nuclear matter, the correlation effects were relatively strong at less than 1.6 fm, but a comparison of the correlation effects found in O^{16} and infinite nuclear matter shows diminished Pauli blocking (reproduced here as Fig. 5.8). They concluded that:

"It is seen that, at least as far as estimates of the Pauli correlation effects are concerned, calculations with an infinite medium would give a very good approximation to the finite case." (p. 252)

In fact, their own figure shows virtually no Pauli effect when the two particles are "each a distance $r/2$ from the center of the system" – with $r = 2.0$ fm. Let there be no confusion here! The value of r is the center-to-center distance between two nucleons, and *not* the distance over which the Pauli exclusion principle operates "outwards" from the center of one nucleon. If the latter were the case, then the exclusion effect would operate over a distance up to $2r$ (~ 4.0 fm) – effectively keeping nucleons far enough away from each other that the strongly attractive part of the nuclear force would be suppressed by 40% between neighboring nucleons. But, in fact, such calculations show that

5.4 Exclusion Principle "Correlations"

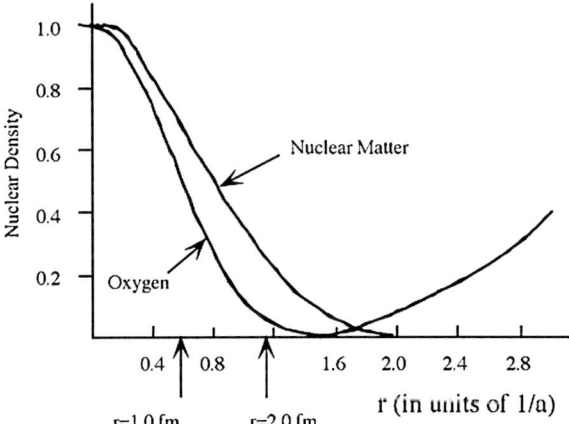

Fig. 5.8. The Pauli correlations in nuclear matter and in the oxygen nucleus. When two nucleons are separated by 2.0 fm, the Pauli effect is small. $a = 0.5769$ (after Fig. 5.3 of Preston and Bhaduri, 1975, p. 252)

the exclusion principle correlations act with decreasing strength only over a distance of $r/2$ or 0.75–1.1 fm from the center of any given nucleon. Exclusion principle effects are thus similar to the effects of dynamical considerations and tend to prevent nucleons from approaching closer than (center-to-center) about 1.6 fm.

It is also worth noting that if the exclusion principle kept nucleons 3.0–4.4 fm apart from one another, then the theoretical nuclear core density could not possibly be as dense as the known value of 0.17 nucleons/fm^3, but would be reduced to 0.01–0.05 nucleons/fm^3 depending on assumptions about the packing of nucleons in the nuclear interior. If that were the case, the 3~10-fold decrease in nuclear density alone would require rewriting all of nuclear theory.

As seen above, the quantitative effect of the exclusion principle is real, but quite modest. Unfortunately, while the numerical effect is small, the rhetorical reverberations of this idea have been tremendous. Without examination of the available theoretical calculations, the exclusion principle argument has allowed theorists to declare that, although nuclei are extremely dense, most of the expected physical implications of that high density have been "suppressed." This is often stated paradoxically in terms of the interactions of orbiting nucleons that, in the end, don't interact. In other words, the collisions that a short MFP implies do not lead to observable effects. Nadasen et al. (1981) put it this way:

> "it should be kept in mind that, in the simple transmission picture of a nucleon traversing nuclear matter, a nucleon-nucleon collision does not necessarily lead to a nuclear reaction" (p. 1040).

Such assertions are motivated by the desire to leave the independent-particle model in place as the central paradigm in nuclear theory, but they inadvertently push theoretical nuclear physics in the direction of counting angels on pinheads: "Sure they're there! It's just that there aren't any observable implications of their celestial presence!"

5.5 Further Doubts

Doubts about the validity of the solution to the weak-interaction/strong-interaction paradox on the basis of the exclusion principle also arise from evidence supporting the cluster model perspective of nuclear structure. MacGregor (1976) has noted that:

> "in order to account for single-particle orbitals in dense nuclear matter, the Pauli exclusion principle must be invoked, and this precludes the existence of any significant nucleon clustering in the center of the nucleus."

According to the shell-model view of nuclear matter, nearest-neighbor interactions of nucleons in the nuclear interior do *not* occur because of Pauli smoothing of the nuclear potential – thereby increasing the MFP and *preventing* local nucleon-nucleon bonding. The inevitable consequence of this suppression of nucleon interactions would be the suppression of the clustering of nucleons into alpha particles. But quite in contradiction to such a theoretical argument, clustering is known to occur. Clustering is a major, well-studied feature of the smallest nuclei (^{12}C, ^{16}O, ^{20}Ne, ^{24}Mg, etc.), and MacGregor cites more than 20 experimental studies (circa 1970) that indicate the presence of alpha clusters *deep in the interior of medium and large nuclei*. Moreover, the release of alpha particles from the large radioactive nuclei is generally thought to be indication that alpha clusters also have a significant presence in the nuclear skin region (where the MFP is relatively long and nucleon clustering might otherwise be expected to be weak).

Notable among such experimental findings are the quasi-fission and knockout experiments in which the nuclear fragments subsequent to high-energy collisions are examined. For MacGregor and the authors of those studies, the results indicate the presence of clusters and therefore show that the cluster model is justifiable, even for the medium-sized and heavy nuclei. But, if nucleon clustering is real, then we must go back and ask what has happened to the effects of the exclusion principle. Can they be invoked at will, and just discarded when they are not theoretically convenient? Clearly not. Either the effects of the exclusion principle are strong (and sufficient to allow nucleons to orbit inside the nucleus and not form clusters) or the effects are modest (and strong interactions occur among nucleons and consequently nucleon clustering is real). Quantum physics or not, we cannot draw opposite conclusions when discussing one and the same phenomenon!

5.5 Further Doubts 119

The implications are obvious: if clustering occurs, then the nuclear force must be operating locally between nucleons in the nuclear interior, as well as on the nuclear surface. But, if the interactions of nucleons are *not* suppressed, then the exclusion principle cannot be simultaneously acting to give nucleons long MFPs.

In summary, the experimental data indicating nucleon clustering in the nuclear interior must also be considered as an obstacle to the non-interacting, orbiting nucleon viewpoint and, therefore, as indication that Pauli blocking – at least on the scale imagined by Weisskopf – does *not* occur. If the concept of the independent-particle orbiting of nucleons is to be sustained, it appears that the Pauli principle is not going to be the mechanism.

Could invocation of the exclusion principle to rescue the independent-particle model be simply mistaken? Can the results of a large number of nucleon-nucleus scattering studies and clustering studies be taken at face value, as indicating an extremely short MFP of low energy nucleons in nuclei? Does the high-density, strong-interaction perspective on the nucleus suffice to explain nuclear phenomena? Can the realistic nuclear potential derived from nuclear scattering experiments explain nuclear structure – leaving only one further chore, that is, the reinterpretation of the successes of the shell model on the basis of a realistically-dense (realistically-short MFP, realistically short-range, strong nuclear force) nuclear model? Could it be possible that a genuinely unified model will emerge not from a manifestly-fictitious low-density, gaseous nuclear theory, but from a realistic high-density, strong-interaction view?

An answer to these questions will be attempted in later chapters, but let us ask one more question about the long MFP in relation to the nuclear force. The most common way of stating the contradiction between the strong-interaction perspective implied by the liquid-drop model, on the one hand, and the orbiting-nucleon perspective of the shell model, on the other hand, has been in terms of the diametrically opposed views on the effective range of the nuclear force. Turner (1977) acknowledged this frankly, but then took a quick exit through the Pauli back door:

"The single most important question that has plagued nuclear physicists for the past two decades is the following: can the detailed properties of nuclei be calculated accurately using interactions deduced from nucleon-nucleus scattering data? The answer, unfortunately, is inconclusive."

"One difficulty is that the nature of the realistic nucleon-nucleus interactions, exhibiting as they do strong short-range repulsive components, leads to poor convergence..."

"This observation pin-points the central dilemma: two-body scattering data exhibit a behavior characteristic of strong short-range components in the nucleon-nucleus interaction, whereas nuclear structure studies [i.e., the theoretical constraints of the independent-particle model] indicate the effective interaction in nuclei to be weak, long-range, and attractive. This dilemma was resolved when the significant role played by the Pauli principle in smoothing out short-range fluctuations was appreciated." (pp. v–vi)

But even when we are "rescued" by the "action" of the Pauli principle, even when the exclusion principle is enthusiastically "understood," "realized" and "appreciated," the contradiction does not disappear if numerical work on exclusion principle correlations does not effectively keep nucleons apart from one another. *Experimental* nuclear reaction physics is left with a nuclear force that is understood as short-range and strong, while *theoretical* nuclear structure physics is left with the shell model's paradoxical need to postulate the existence of a weak and long-range "effective" nuclear force, while short-range effects are "suppressed."

Whether phrased as "due to the exclusion principle," as "a consequence of Pauli blocking," or as "a result of the inability of more than one fermion to enter a given quantum state," the solution is essentially based on Weisskopf's analogy with solid state physics – an interesting theoretical suggestion, perhaps, but one that, quite simply, has not stood up to analysis by dozens of theorists over the course of 50 years.

Particularly in the light of calculations of the MFP in nuclei done in recent years, Weisskopf's suggestion simply cannot be considered a true solution to the paradox of nucleons orbiting in the dense nuclear interior. At best, the exclusion principle allows for a modest increase in the MFP that, nonetheless, remains smaller than the nuclear radius. In the long run, the Weisskopf hypothesis may become a useful stimulus to theorists because it effectively raises the important issue of just what the exclusion principle itself means, but clearly it has not resolved the issue of the nuclear texture. As is true for any valid descriptive principle, the exclusion principle does indeed describe important characteristics of fermion-particle interactions. Nevertheless, as valid as it is as a *description* of certain particle effects, the principle itself requires a dynamical explanation. This is in fact a long-standing, unresolved problem that is almost as old as quantum mechanics itself (see the next section). It must be stated, however, that, for the purposes of the present discussion of nuclear structure, the fact remains that, even when numerical calculations take the exclusion principle into consideration, a large (>20 fm) MFP is *not* produced. Never! No one has presented a numerical argument which indicates an MFP in the nuclear core that would allow for even *one* full intranuclear orbit, much less several orbits or the free-orbiting of nucleons. That one historical fact suffices to show that an unresolved problem remains in nuclear structure theory concerning the nuclear texture.

5.6 What is the Pauli Exclusion Principle?

For pointing out the contradiction of postulating a long MFP in a substance as dense as the nucleus, discussion of the meaning of the exclusion principle is unnecessary: all empirical indications and all theoretical estimates (including those based on the Pauli principle) point to the fact that the MFP of nucleons in nuclei is rather *short* and this fact argues strongly against the independent-particle model's orbiting nucleon conception of the nucleus. Nevertheless, the exclusion principle has been consistently invoked within the context of nuclear structure physics in an attempt to resolve the apparent contradiction between high- and low-density nuclear models. In so far as there is simply no doubt that the exclusion principle – as a summary of particle dynamics – is correct, how are we to understand the underlying mechanism of its action?

There is unfortunately no easy answer to questions about the meaning of the exclusion principle and there are surprisingly few queries about its meaning in the modern literature on quantum physics. It is now generally considered to be so fundamental that, like multiplication in the teaching of mathematics, little work in nuclear, atomic or molecular physics can proceed without it. It is learned at an early stage and invoked when needed. But this unquestioning attitude was not always the case and there was a time when theorists worried about the dynamical basis for "exclusion."

In his classic textbook entitled *Atomic Spectra and Atomic Structure* (1944), Herzberg noted that:

> "The Pauli principle does not result from the fundamentals of quantum mechanics, but is an assumption which, although it fits very well into quantum mechanics, cannot for the time being be theoretically justified" (p. 123).

In an equally-respected volume by Condon and Shortley, *The Theory of Atomic Spectra* (1935), the authors stated that:

> "It is to be observed that although we have found a natural place for the Pauli principle in the theory we have not a theoretical reason for the particular choice of the antisymmetric system. This is one of the unsolved problems of quantum mechanics. Presumably a more fundamental theory of the interaction of two equivalent particles will provide a better understanding of the matter, but such a theory has not been given as yet" (p. 167).

More recently, Yang and Hamilton (1996) have also noted the unresolved issue of an explanation of the exclusion principle:

> "What is the full physical basis of the rigorous repulsion expressed by the Pauli exclusion principle? While the Schrödinger equation can describe this mathematically, its physical basis remains an open puzzle." (p. 193)

In so far as the exclusion principle is an essential part of the description of the relationships among fermion particles in the build-up of atoms or nuclei, it is a useful short-hand, but caution must be used when invoking it as a causal mechanism. In a nuclear context, the most common confusion has been as follows. We assume that all of the low-energy quantum states are occupied in a ground-state quantal system, such as a nucleus. For description of what energy states are possible and their occupancy, the exclusion principle plays an essential role. Now we want to account for the nature of the interaction of two nucleons within the nucleus. If the nucleons collide and exchange energy, then we must expect that they will move from one energy state to another – states that, in agreement with quantum mechanics, can be described by a unique set of quantum numbers (n, l, j, m, i, s) The exclusion principle, however, is here invoked to maintain that most nucleons in the nuclear core have no available, unoccupied low-energy quantum states into which they can move because the states are already filled. If a nucleon moves to a state above a certain energy level, it would be ejected from the nucleus, so the high-energy states are clearly not available for low-energy interactions. The fact that at low excitation energies the nucleon is *not* ejected and the fact that all of the low-level energy states are already filled is then taken to mean (in line with an interpretation of the exclusion principle as a dynamic force) that the energy change must not have happened at all! The tortured logic of this type of application of the exclusion principle is therefore that, since the interacting nucleons have no place to go, then the preceding collision between them must not have occurred. The effect is "not allowed" by a principle, therefore the cause must not have happened. Clearly, this is a topic in need of further study.

5.7 Summary

The mean free path of nucleons in stable nuclei is of interest because it is a central concept in nuclear theory and yet, simultaneously, questions about the MFP are not easily answered either experimentally or theoretically. On the one hand, the independent-particle model of the nucleus *demands* a lengthy mean free path, in order that the starting assumption of the collisionless orbiting of nucleons in the nuclear interior can be justified. Unfortunately, all experimental measures and most theoretical estimates indicate an MFP of about 3 fm for most nucleons in most nuclei most of the time. Does this indicate that the independent-particle model is simply wrong? Clearly not, in so far as the model explains a great variety of experimental data regarding nuclear spins, parities and other nuclear properties. But the independent orbiting of nucleons within the dense nuclear interior is a fiction. It is this paradox of the independent-particle model that still requires resolution.

6
The Nuclear Size and Shape

In Chap. 5 the mean-free-path over which nucleons within bound, stable nuclei can move freely before coming into contact with other nucleons was examined *without* consideration of the size of the nucleons themselves. In the present chapter, the texture of nuclei will be considered – and here the relative size of nucleons and nuclei becomes an important factor. Again, our main concern is the historical record concerning how problematical issues related to the nuclear texture have been dealt with. Specifically, the topics of the nuclear density, the nuclear skin region and the nuclear radius will be discussed.

6.1 The Nuclear Density

The size of the nucleus and the density of nucleons inside are fundamental aspects of nuclear structure. From the density value, the average distance between nearest neighbors can be calculated, and therefore the range over which the nuclear force has its effects within the nucleus can be determined and, importantly, related to the dimensions of the nuclear force obtained from nucleon-nucleon collisions. Although each of the nuclear models has a realm of particular applicability, all hypotheses concerning nuclear structure and the nuclear force must be consistent with the basic facts about nuclear size. So, let us review what is well-known and well-established – and then see how those facts are interpreted.

Experimental work on the size of nuclei began with Rutherford's studies of the scattering of alpha particles from nuclei. He established the concept of the nuclear atom, and demonstrated the surprising fact that nuclei have diameters of less than 10^{-13} meters. The modern era of such work on the size and shape of the nucleus dates back to the late 1940s and early 50s, when the first accelerators were built. Experimental and theoretical techniques were developed for bombarding nuclei with electron beams and analyzing the scattering patterns of the electrons. Although many refinements have been made since then, the basic techniques and first-order conclusions remain valid.

Indeed, Hofstadter won a Nobel Prize in 1963 for his studies during the 1950s on electron scattering from various nuclei. His principal conclusion was that the experimental data for nuclei across the periodic chart are best explained by assuming a nuclear size constant, r_0, that is approximately 1.12 fm. Using that value to calculate the average nuclear density, ρ, a value of 0.17 nucleons/fm^3 was obtained.

Both values are, in fact, uncertain to a level of 5–10%, but such numbers are the uncompromising starting point for discussions of the nuclear texture. The density value leads directly and inextricably in the direction of liquid-drop-like conclusions concerning the nuclear interior. Of course, the liquid-drop model was the central concept in nuclear structure theory in the 1930s and 1940s, but, during the 1950s when Hofstadter was experimentally demonstrating the liquid-drop texture of the nucleus, theorists were engaged in the development of a gaseous theory of the nucleus in which nucleons were assumed to be free to travel independently of one another – and where nuclei were considered to be quite different from a dense liquid.

The contradiction inherent to these two directions was fully realized at the time and the paradoxical validity of both positions was widely discussed. Despite the fact that the basic conclusions drawn by Hofstadter and colleagues have never been challenged, ultimately, the independent-particle (shell) model won the hearts of most theorists. In the early 1950s, the new idea in town was the shell model and, in order to make it work, it was essential to assume that, regardless of the density value the experimentalists provided, the nuclear interior was a diffuse gas. Logically, the liquid-gas paradox raised a very real problem regarding the correct view of the nature of local nucleon interactions in the nuclear interior, but there seemed no way to resolve the problem other than to state the idea that both views were somehow valid. As a consequence, most textbooks acknowledge the high-density and the low-density views in separate chapters without attempting a reconciliation.

When reconciliation is attempted, inevitably problems arise. It is therefore rather unfair to single out those discussions of nuclear size where attempts have been made to deal directly with the problem of high- and low-density hypotheses, because *all* nuclear textbooks without exception embody this contradiction, even if never acknowledged. The problem does not therefore lie with individual authors, but is inherent to the study of nuclear structure physics. Some authors prefer not to address problems head-on when they simply cannot be resolved, while other authors prefer a direct approach and offer possible compromise positions.

The implicit or explicit contradiction about the nuclear texture that all discussions of nuclear structure encounter is not an easy issue, but let us approach the topic in the way that a curious undergraduate might take. Imagine a student motivated to understand what kind of substance the nucleus is. Having taken some courses in higher mathematics, introductory quantum mechanics and the philosophy of modern science, he might wonder what specifically nuclear physics is all about. The human genome, brain research and subatomic physics – those are the modern challenges! So the student might wander into the university library, and find several rows of textbooks on various aspects of nuclear physics. Most of those texts will be surprisingly old – dating from the 50s and 60s when the topic of nuclear physics was still hot, with relatively few additions from the 1990s and onwards. Invariably, the texts will be thick with details on experimental techniques and, in contrast to textbooks on atomic structure, rarely include a diagram depicting what a nucleus might look like. A sensible student might conclude that these dusty tomes are indeed the classics in nuclear physics and deserve serious study, but that jumping in at the deep end is not the best introduction to a new field of study. Beginning with a specialist text that was written even before one's parents were born might strike our curious undergraduate as being equivalent to beginning with Newton's *Principia*. So, while leaving open the possibility of returning to these library shelves later, a more likely first stop will be the local bookstore. University bookshops will inevitably carry a few books on quantum theory, and sometimes carry an introductory book, published in 1993 by Dover and entitled *Lectures on Nuclear Theory*. Priced at $5.95 and containing only 108 pages, this might be a good place to start.

Reading the back cover will reveal that the authors, Landau and Smorodinsky, were renowned Russian physicists who (despite the comments that follow) made important contributions to nuclear physics. Their book includes a modest dose of mathematics, is clearly *not* a simple-minded popularization, and is lauded on the back-cover as a "real jewel of an elementary introduction to the main concepts of nuclear theory". Having made the purchase and put this jewel in hand, the student is likely to discover only later that *Lectures on Nuclear Theory* were delivered as lectures in 1954, first published in 1959, and here reprinted in 1993 in unaltered form.

Despite the publisher's blurb that, "Throughout, the emphasis is on clarity of physical ideas," our undergraduate is in for a rude surprise. Although "clarity of physical ideas" is routine in nuclear experimentation, it is virtually unheard of in nuclear theory. Lectures One, Two and Three are concerned with the nuclear force – with surprisingly little about the physical objects known as nuclei, but, doggedly in pursuit of the hidden jewel, the student will be glad to find that Lecture Four turns to the problem of nuclear structure. In Lecture Three, the authors had noted that "the proton radius is found to be 0.45 fm" (p. 28) – and in Lecture Four they began their explanation of the independent-particle model. Unfortunately, the discussion of nuclear structure begins with a disclaimer concerning why a self-consistent understanding of the nucleus has not been achieved:

> "It is clear that no simple model can represent all the properties of an extremely complicated quantum system such as the nucleus. Any model, of necessity, must have limited application. We should not be surprised if different effects require different models for their

description; sometimes these models may even have mutually exclusive properties...." (p. 33)

Indeed, this has come to be a common view concerning nuclear modeling, and must be considered as orthodoxy insofar as various models are actually in use in nuclear theory. That type of metaphysical argument will not, however, circumvent basic numerical problems within any given model and should not deter us from squarely facing internal contradictions. The authors proceed by suggesting a rough estimate of the nuclear density, as follows:

"We have already directed attention to the fact that the nucleon dimensions are of the order of 0.45 fm. The distance between nucleons in a nucleus is approximately 1.8 fm. Thus, roughly speaking, the nucleons occupy 1/50 of the volume of the nucleus." (p. 33)

Our student may find the change in terminology from "proton radius" to "nucleon dimensions" a bit curious, but the basic point seems clear enough: Nuclei are apparently rather insubstantial entities – only 2% filled with nucleons – the rest being empty space through which the nucleons orbit.

But where did these numbers come from and is 1/50th the correct conclusion to draw? In 1954, when these Lectures were actually delivered as lectures, it was already known that a radial value of 0.45 fm is rather small as an estimate of nucleon size. Hofstadter's early electron scattering work (Hofstadter, Fechter and McIntyre, 1953) on nucleon and nuclear radii had been published, and by 1959, when these lectures were published as a book, he and others had summarized this work several times (Yennie, Ravenhall and Wilson, 1954; Hahn, Ravenhall and Hofstadter, 1956; Hofstadter, 1956), wherein both the electrostatic and magnetic RMS radii of the proton was reported to be 0.88 fm. By using a radial value that is about 1/2 the known value, Landau and Smorodinsky had effectively chosen a nucleon *volume* that is only 1/8th of the more realistic volume based on the known proton radius. Today the best estimates of the charge radius of the proton are 0.866 fm (Borkowski et al., 1975) and 0.8521 fm (Angeli, 1998). A radial value of 0.45 fm is a rather poor start at determining the correct value for the nuclear density, but that small value for the nucleon radius alone does not explain the astounding conclusion that nucleons occupy only 1/50th of the nuclear volume.

The other numerical value used for calculation is the average distance between nucleons, 1.8 fm. The authors did not state how the 1.8 value was arrived at, but it is likely to have been obtained from the widely accepted value of the nuclear density published by Hofstadter and colleagues in which a rough estimate of the average internucleon distance can be obtained by assuming that nucleons are in a simple cubic lattice. That is, if nuclei are thought of as the cubic packing of nucleons, a cubic edge length of 1.8 fm gives a nuclear density of about 0.17 nucleons/fm^3 – which is the experimentally determined density value obtained from the same electron scattering studies of Hofstadter mentioned above. As illustrated in Fig. 6.1a, if this cubic packing

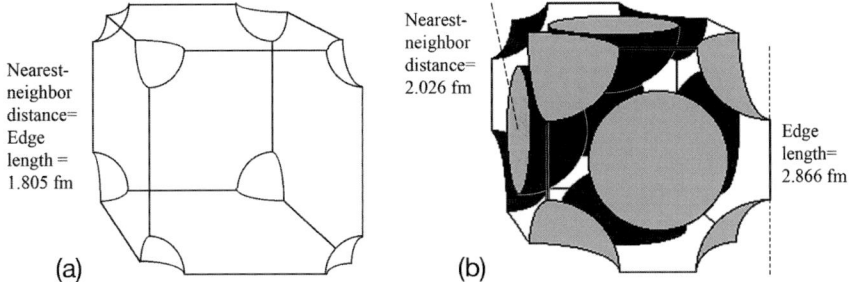

Fig. 6.1. Ballpark calculations of nuclear density. (**a**) Landau and Smorodinsky's calculation of the number of nucleons per unit of nuclear volume assuming simple cubic packing of nucleons. The edges of the cube are 1.8 fm and each corner contains 1/8th of a nucleon, giving the known nuclear density. (**b**) Calculation of the nuclear density assuming the close-packing of nucleons inside of nuclei. The figure is drawn to scale in which each of the nucleons has a radius of 0.862 fm and there is a core density of 0.17 nucleons/fm³

scheme for nucleons is assumed, then the nuclear density is easily calculated: one octant of a nucleon at each of the eight corners of the cube gives one complete nucleon per unit cube. If the cube length is 1.805 fm, then we get a nuclear density of one nucleon per 5.882 fm³, or 0.170 nucleons/fm³.

Of course, neither Hofstadter nor Landau and Smorodinsky argued that nuclei are in reality simple cubic arrays of nucleons; cubic packing was assumed solely to obtain some ballpark figures on nuclear size. The simplest assumption is simple cubic packing, allowing for an elementary and unambiguous point about the nuclear texture.

But let us see where precisely the 1/50th result came from. This, after all, was the main thrust of Landau and Smorodinsky's introductory section on the independent-particle model: they maintained that a very simple geometrical argument implies that nuclei have a rather low density. If they have established that fact, then there are many implications concerning nuclear structure and the nuclear force. So, using a "nucleon dimension" of 0.45 fm, they proceeded to calculate the approximate volume that each nucleon occupies inside its average cubic volume within the nucleus:

$$(0.45\,\text{fm})^3/(1.805\,\text{fm})^3 = 0.016 \text{ or } \sim 2\% \quad (6.1)$$

which, as they say, means that only about 1/50th of the nuclear volume is filled with nucleonic material. And such a figure seems to justify the starting assumptions of the independent-particle model...

But, what about the change in terminology from nucleon "radius" to "dimension"? They had stated just five pages earlier that the proton *radius* is 0.45 fm, so what is the significance of doing calculations on the basis of the nucleon "dimension"? It seems that the meaning of "dimension" here is actually "diameter", so they have effectively reduced the nucleon radius from 0.45

to 0.225 fm for their density calculation. In other words, by placing a cubic nucleon with a 0.45 fm edge length at each corner of the cube, the nucleonic cube has one octant of its volume inside each of the eight corners. So, we see that the volume of the octant, calculated as $(0.225)^3$, is the key to obtaining the 1/50 value. A better calculation using 0.45 as the nucleon radius would therefore be the same as (6.1), but multiplied by 8 to include the volume of each of the nucleon octants at the eight corners:

$$8 \cdot (0.45 \,\text{fm})^3/(1.805 \,\text{fm})^3 = 0.124 \quad \text{or} \quad \sim 12\% \qquad (6.2)$$

This is not quantum physics, but grammar school geometry and much too simple to make a mistake if we do not have preconceived notions about what the independent-particle model demands as the correct answer.

And what if we use the *experimental* nucleon RMS radius, known since the early 1950s, and assume a spherical nucleon volume?

$$4\pi(0.862 \,\text{fm})^3/3/(1.805 \,\text{fm})^3 = 0.456 \quad \text{or} \quad 46\% \qquad (6.3)$$

The correct answer to such a ball-park calculation is that nuclei are about half filled with nucleon matter – in first approximation, a dense liquid or a solid, rather than a diffuse gas! Suddenly we realize how difficult the rest of Landau and Smorodinsky's Fourth Lecture on the independent-particle model would have become. If the nucleus is one half filled with nucleon matter, it is obvious that justification of the free-orbiting of nucleons will not be an easy task.

The other ballpark calculation that should be checked before concluding that the nucleus is a diffuse gas is a similar estimate in which the *close-packing* of nucleons is assumed. Simple cubic packing is a rather loosely packed lattice configuration, so it is useful to do a similar calculation on a more-densely packed configuration.

Unlike hexagonal close-packing (for which an identical packing fraction of 0.74 is obtained, but which is geometrically more complex), face-centered-cubic close-packing provides a cubic unit (Fig. 6.1b), from which density calculations can be easily made. Portions of 14 different nucleons can be seen in the unit cube: six hemispheres on each of the six faces of the cube and eight octants at each of the corners. In order to obtain the known nuclear density of 0.17 nucleons/fm^3, it is necessary for the cube to have an edge length that is considerably greater than that assumed with simple cubic packing. Specifically, an edge length of 2.866 fm gives the correct density value, and implies a nearest neighbor distance of 2.026 fm. Clearly, the percentage that such a cube is filled with nucleonic matter (a cube containing four nucleons [6 hemispheres and 8 octants]) will be the same as that obtained in (6.3):

$$4 \text{ nucleons} \cdot 4\pi(0.862)^3/3 \,\text{fm}^3/2.866 \,\text{fm}^3 = 0.456, \quad \text{or} \quad 46\% \qquad (6.4)$$

but what is of interest is that the distance between neighboring nucleons is significantly larger than that in a simple cubic lattice. So, while there are

6.1 The Nuclear Density

important differences concerning the inter-nucleon distance between simple cubic packing and close-packing, the percentage of the nucleus filled with nucleon matter (46%) is the same in both cases. In either case, this density figure – based on the *known* nucleon radius – tells a very different story about the nuclear interior than the 2% figure presented by Landau and Smorodinsky.

Unfortunately, when a scientist with the authority and reputation of Landau makes a hasty mistake, it is likely to echo through the halls of academia. Two decades later in 1975, Sitenko and Tartakovskii published a monograph, *Lectures on the Theory of the Nucleus*, in which the same erroneous values for the nucleon radius (later referred to as a nucleon "dimension") and the nuclear volume were used to come to the same conclusion that nuclei are approximately 1/50th filled with nucleon matter. Their text is similar to that of Landau and Smorodinsky:

> "Knowing the density of the nuclear substance, it is not difficult to determine the mean distance between the nucleons in the nucleus: this amounts to about 1.8×10^{-13} cm. But the spatial dimensions of the nucleons are equal to 0.45×10^{-13} cm. Consequently, the nucleons occupy only about one-fiftieth of the whole volume of the nucleus. Therefore, despite the saturation of the nuclear interaction, nucleons in a nucleus retain their individual properties." (p. 75)

Our unfortunate undergraduate who wanted only a rough idea of the nuclear texture is clearly off to a disastrous start. But are the calculations by Landau and Smorodinsky and Sitenko and Tartakovskii unrepresentative of modern estimates of nuclear density? Unfortunately not. When the topic of the nuclear density is discussed, various ploys are routinely used to suggest that nuclei are somehow *not* extremely dense entities, since the answer demanded by the independent-particle model is apparently the ultimate goal of all such calculations, regardless of the experimental facts. Some textbooks attempt a numerical estimate, but most authors simply state Hofstadter's figure for the density of nucleons at the nuclear core of heavy nuclei – 0.17 nucleons/fm^3 and leave it at that. Just a fraction of a nucleon per cubic fermi seems so small! If there remains some doubt about the interpretation of such numbers by naïve and impressionable students, blunt assertions might be employed:

> "the nucleus is not a dense system of nucleons and ... this makes the idea of the independent motion of the nucleons acceptable" (Antonov, Hodgson and Petrov, 1988, p. 6).

> "The average distance between the nucleons is much larger than the radius of the nucleon hard core" (Povh et al., 1995, p. 223)

Such statements notwithstanding, if the 0.17 nucleons/fm^3 density figure and 0.86 fm radial value are correct, a little bit of solid geometry leads to the conclusion that nuclei are rather filled with their nucleon constituents.

In 1991, Williams argued that nucleon matter takes up less than one percent of the nuclear volume. Instead of the direct attack on the density issue used by Landau and Smorodinsky (1959), Williams reviewed the properties of the liquid-drop, cluster and shell models before addressing the possibility that problems might remain. As is always the case when nuclear models are compared, some contradictions had emerged in the course of the chapter, so just before the closing section, he included a section entitled "Reconciliation." Here we find that:

> "The shell model and the liquid drop model are so unlike that it is astonishing that they are models of the same system." (p. 155)

Quite so, but that is apparently not cause for worry, because the shell model is the only real player in town, and the liquid-drop characteristics are a bit misleading since

> "the fraction of the nuclear volume that is hard core [nucleon matter] is... 0.009" (p. 156).

A similar calculation has been suggested by Jones (1986):

> "...from the range of the repulsive force as required from nucleon-nucleon scattering, it is possible to attach an effective size to each nucleon ($r_{\text{eff}} \sim 1/2\,x$ range, since two nucleons cannot approach closer than this range). The effective volume of the nucleons in a heavy nucleus is then found to be $\sim 1\%$ of the nuclear volume."

Just one percent filled with nucleon matter is even more convincingly gaseous than the estimate provided by Landau and Smorodinsky, but again we must ask where such figures come from. Williams calculated the nuclear density using a standard formula:

$$\rho = (4\pi r_0^3/3)^{-1} \tag{6.5}$$

for which an r_0 value of 1.2 fm was used. Values of r_0 ranging from 1.07 to 1.4 are sometimes used in nuclear theory, but this 30% leeway must be handled with care or else numerical results will be equally approximate. A value of 1.2 fm for r_0 produces a nuclear density of 0.138 nucleons/fm^3, whereas the usual value for the nuclear core (0.17) where most nucleons are located is almost 25% larger. In so far as we are interested in an approximate figure for the nuclear density, it is worth noting that a value for r_0 of 1.07 gives the best fit to nuclear radii across the periodic chart, whereas a slightly larger value of 1.12 gives the core nuclear density of 0.17 nucleons/fm^3. By choosing a still larger value for r_0, Williams calculated the nuclear density by reducing the amount of nucleon material by 23%.

Clearly, the section of Williams' text entitled, "Reconciliation," was designed to resolve the apparent problem of the gas and liquid models implying very different nuclear textures, but the contradiction between liquid and gaseous models was resolved by introducing yet another model – this time,

6.1 The Nuclear Density

a model of the nuclear force. He introduced the concept of the nucleon size by noting that, according to one model of the nuclear force, the impenetrable "hardcore" interaction between two nucleons occurs at $C = 0.5$ fm – which implies a hardcore per nucleon of only 0.25 fm. On the basis of that model of the nuclear force, it was then an easy matter to show that there isn't much nucleon substance in the nuclear volume. This reduced nucleon radius was then used, together with the slightly inflated r_0 value, to calculate the percentage that nuclei are filled with nucleon matter:

$$(C/2)^3/(r_0)^3 = 0.009, \quad \text{or} \quad \text{about} \quad 1\% \tag{6.6}$$

The reason for such a remarkably small value is that the nucleon hard core was not taken to be the electromagnetic RMS radius (~ 0.8 fm) and not taken to be the hard core of the nuclear force in the well-known Paris or Bonn potentials ($0.4 \sim 0.5$ fm), but rather half of that size. Since the calculation is of a volume, the effect is to reduce the nucleon volume by 1/8th. Together with the 11% increase in r_0, the numerical result leads to a familiar, but dubious conclusion:

> "Thus hard core collisions occur very infrequently and the motion of an individual nucleon can be described by assuming it moves independently in a smoothly varying potential well provided by the remaining nucleons." (p. 156)

A similar estimate was given by Bohr and Mottelson (1969, p. 255), but clearly all such conclusions are only as valid as the assumptions that go into the calculations. The essential question is whether this specific value for the hardcore of the nuclear force is an appropriate estimate of the size of the nucleon. Specifically, is the *infinitely* impenetrable radius of the nucleon really the relevant number to use for estimating the volume of nucleon substance?

Figure 6.2 (left) shows the experimental radial charge distributions of the proton and neutron. Here we see that there is considerable charge present at a radial value of 1.0 fm in both types of nucleon. It may not indicate an "infinitely impenetrable core" at 1.0 fm, but it certainly suggests the presence of a significant amount of nucleonic matter extending from the center of one nucleon until it reaches the midway point between neighboring nucleons (separated center-to-center by ~ 2.0 fm). Similarly, most nuclear potentials (Fig. 6.2, right) become *infinitely* repulsive at ~ 0.5 fm, but there is already strong repulsion at $0.7 \sim 0.8$ fm. The difficult problem of the nuclear force cannot be neatly resolved here in a discussion of nuclear density, but it is important to realize that, by the same token, controversial interpretations of the nuclear force cannot be constructively utilized in resolving problems of the nuclear density. Using debatable numbers concerning nucleon size to justify arguments concerning the nuclear force (á la Landau and Smordinsky) is as mistaken as uncritically borrowing numbers from nuclear force models to justify conclusions about the nuclear density (á la Williams).

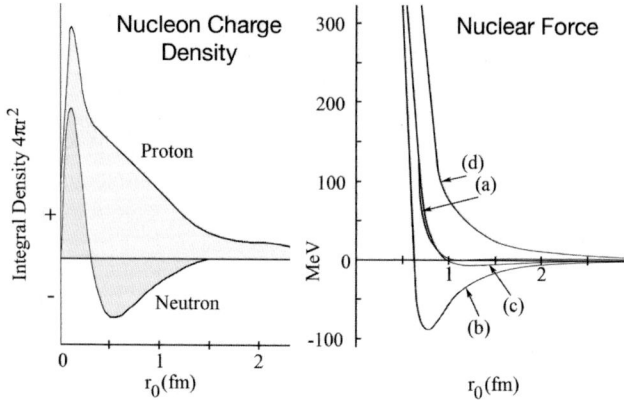

Fig. 6.2. On the *left* is shown the experimentally-determined charge distribution in the proton and neutron plotted against distance from the center of the nucleon (Littauer, et al., 1961). On the *right* are shown the dimensions of nuclear force effects (Bohr and Mottelson, 1969, p. 253). Curves (**a**), (**b**) and (**c**) represent various parametrizations of the nuclear force, while (**d**) shows the energy of a hard-sphere gas

Textbooks normally discuss the nuclear force and nucleon size in separate chapters, and rarely make comparisons of the distribution of charge in protons and neutrons (Fig. 6.2, left) and the known dimensions of the nuclear force (Fig. 6.2, right), but a direct comparison is instructive. Clearly, measures of both the nucleon and the nuclear force show consistent results indicating that the bulk of the substance of the nucleon lies within 1.0 of the nucleon center, and all significant interactions with other nucleons occur at less than 2.0 fm. Typically, the "infinite repulsion" of the nuclear core is shown at some radial value, such as $0.4 \sim 0.5$ fm in the Bonn or Paris potentials, but what is important is the fact that the attractive part of the nuclear force turns repulsive at ~ 1.0 fm. The repulsion is not strong, much less "infinite" at that point, but the slope is steep, indicating an ever-increasing repulsion as the distance between nucleons decreases below 1.0 fm (Fig. 6.2, right).

There is in fact no easy answer for our student trying to understand the first-order texture of the nucleus, because all of nuclear theory is built on numbers such as those discussed above. A nuclear core density of 0.17 nucleons per cubic fm is the most commonly quoted value, with references to the work of Hofstadter and colleagues, but values of 0.16 and occasionally 0.138 are also used. Clearly, the lower the density value, the easier it is to justify the independent-particle model approach to nuclear structure, but even a value of 0.138 (nuclei 37% filled with nucleon matter) is far from being unambiguous support of the concept of freely orbiting nucleons in the nuclear interior.

Let us return to the experimental data. The most widely cited findings on nuclear sizes in all of nuclear physics are the results of Hofstadter and reproduced in most nuclear textbooks (Fig. 6.3). The density of proton charge in the nuclear core was determined to be about $1.1 \sim 1.2 \times 10^{19}$ Curies/cm^3 – for most nuclei. Electron scattering off of nuclear protons does not reflect the distribution of neutrons in the nuclear interior, but subsequent work using uncharged pions indicates that the neutron distribution is fundamentally the same as the proton distribution. That assumption implies that the nuclear core contains 0.17 nucleons/fm^3.

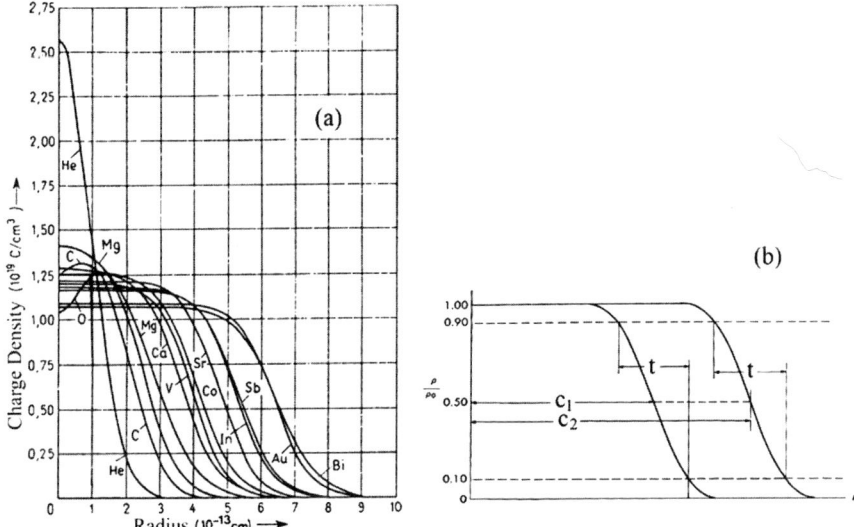

Fig. 6.3. (a) The charge density curves for various nuclei. Most nuclei have a constant density core at 1.1~1.2, but ^{4}He has a core density that is more than twice that of other nuclei (Hofstadter, 1956). Why do most textbooks show these data with the Helium curve absent? (b) The theoretical Fermi density curve used to approximate nuclear dimensions; c_1 and c_2 are the 50% density radii of two hypothetical nuclei; t denotes the skin thickness

A great deal of information is summarized in Fig. 6.3a, but three points are outstanding and have subsequently been confirmed many times: (i) the charge density of the nuclear core is roughly constant (with the exception of Helium); (ii) the radii of nuclei increase with the number of nucleons; and (iii) at the periphery, the density falls to zero over a roughly constant distance (2.4 fm) regardless of the size of the nucleus (therefore producing the many parallel diagonal lines in the figure). In other words, both the core density and the skin thickness are constant. Similar data are also presented in Table 6.1 for nuclei whose skin thickness has been studied. The accuracy is reported to be within 2% for the radii and 10% for the skin thickness. Clearly, there is a gradual increase in radius with the number of nucleons (with some anomalies

Table 6.1. Charge Distributions for Representative Spherical Nuclei (Überall, 1971)

Nuclide	Skin Thickness	RMS Radius (Experiment)*	RMS Radius (Theory)[+]	RMS Radius (Theory)[++]
$^{4}\text{He}_2$	–	1.67	2.15	1.88
$^{6}\text{Li}_3$	–	2.51	2.34	2.07
$^{12}\text{C}_6$	2.20	2.47	2.74	2.45
$^{14}\text{N}_7$	2.20	2.54	2.85	2.55
$^{16}\text{O}_8$	1.80	2.71	2.94	2.64
$^{24}\text{Mg}_{12}$	2.60	3.04	3.25	2.94
$^{28}\text{Si}_{14}$	2.80	3.13	3.38	3.06
$^{40}\text{Ca}_{20}$	2.51	3.49	3.70	3.37
$^{48}\text{Ti}_{22}$	2.49	3.60	3.80	3.55
$^{52}\text{Cr}_{24}$	2.33	3.65	3.89	3.63
$^{56}\text{Fe}_{26}$	2.50	3.74	3.97	3.71
$^{58}\text{Ni}_{28}$	2.46	3.78	4.05	3.74
$^{89}\text{Y}_{39}$	2.51	4.24	4.42	4.23
$^{93}\text{Nb}_{41}$	2.52	4.32	4.49	4.29
$^{116}\text{Sn}_{50}$	2.37	4.63	4.74	4.57
$^{139}\text{La}_{57}$	2.35	4.86	4.92	4.82
$^{142}\text{Nd}_{60}$	1.79	4.90	4.99	4.85
$^{197}\text{Au}_{79}$	2.32	5.44	5.39	5.34
$^{208}\text{Pb}_{82}$	2.33	5.51	5.45	5.43
Mean ±SD	2.36 ± 0.25			

* The RMS charge radii are the currently accepted best estimates (Firestone, 1996).
[+] These theoretical values are obtained using $R = r_0 \, Z^{1/3} + 0.8$, where $r_0 = 1.07$ fm.
[++] These theoretical values are obtained using $R = r_0 \, A^{1/3} + 0.58$, where $r_0 = 0.82$ fm (Hofstadter, 1967).

among the smallest nuclei), whereas the nuclear skin thickness remains quite constant.

The RMS radial data find a ready explanation in the liquid-drop model (see Sect. 6.3). If nucleons are space-occupying particles each with a constant volume within the nucleus, then a steady increase in nuclear volume with A would be predicted, and is experimentally found. In contrast, if the nucleons within stable nuclei were diffuse clouds in analogy with electron clouds, then their spatial overlap – similar to the overlap of electron clouds – would not lead to the linear dependency of the mean radius on the number of nucleons (Figs. 2.25 and 2.26). Other factors (the effects of the exclusion principle, the hard core of the nuclear force, etc.) might be brought into the shell model picture, but clearly a gaseous phase model must be developed in ways that push it in the direction of more closely resembling the liquid-drop model in order to explain these size effects.

The results on nuclear sizes generally support the liquid-drop model, but two notable anomalies remain to be explained. First of all, the extremely high density of Helium is a factor of two greater than the value for nuclear matter (at the nuclear core). In the liquid-drop conception, nucleons move at random within a volume defined by the number of particles involved. To begin with, there are neither spin nor isospin effects, and independent-particle model shells are not an intrinsic part of the liquid-drop analogy. Although an explanation in terms of the magic number 2 in the shell model may come to mind, that logic is not consistent with the complete absence of magic shells elsewhere in the data on nuclear sizes (Figs. 2.25 and 2.26). The extreme compactness of the 4-nucleon system is clearly unusual and suggests the reality of nuclear substructure beyond the general features of "nuclear matter" that are the usual focus of discussion. Despite the fact that several of the extremely light nuclei have relatively large radii (e.g., ^{2}H [2.1 fm] and ^{6}Li [2.51 fm]), ^{4}He is extremely small (1.67 fm) for the amount of charge it contains. This fact is well-known, clearly anomalous, and yet rarely commented on. Indeed, the Hofstadter data shown in Fig. 6.3 are frequently reproduced in the textbooks with the troublesome Helium curve removed.

The second anomaly within the liquid-drop model account of nuclear sizes is the nuclear skin region.

6.2 The Nuclear Skin

As evident from the above, most findings on the size of nuclei (excepting Helium) provide general support for a liquid-drop view of the nucleus. As outlined in Chap. 2, nuclear radii show no signs of shell structure, but instead a gradual increase in size with the number of nucleons. Other gross properties of nuclei – notably, the over-all trend in binding energy – are rather well explained in terms of the liquid-drop model. The model starts with a plausible assumption that nucleons interact only with their nearest-neighbors, with each nucleon occupying a characteristic volume. This implies that the binding energy will be proportional to the number of particles present, as is empirically known. It also gives each nucleus a constant density core, as is empirically known, and a nuclear radius that is dependent on the number of protons and neutrons present, as is empirically known. The fit between theory and experiment is not in fact perfect, but it is good indication that the liquid-drop model is not too far off in its first-order description of nuclear size and shape.

It is worth noting, however, that the liquid-drop model has certain unavoidable implications regarding the nuclear surface. Assuming that nucleons interact predominantly with nearest neighbors, any nucleon in the nuclear interior will have a maximum of 12 neighbors. Regardless of the packing scheme in the nuclear interior, the nucleons on the surface of a nucleus will have proportionately fewer nearest neighbors. This means that the contribution of individual nucleons to the total binding of a nucleus will differ somewhat depending on their location in the nuclear interior or on the nuclear surface. In a close-packed interior, each nucleon will have about 12 neighbors to which it is bound, and on the nuclear surface, each nucleon will have 6 ~ 9 nearest neighbors.

So, regardless of the exact number of nearest-neighbor bonds that an individual nucleon has, all nucleons in the liquid-drop are a part of a constant density "nuclear droplet". Any nucleon "sticking out" from the surface will be pulled toward the nuclear center until a state of equilibrium – a more-or-less spherical liquid-drop – is established. And therein lies the nuclear skin problem of the liquid-drop model. It is experimentally known that the density of nucleons is roughly constant in the nuclear center, but decreases rather *gradually* from a maximal value of the nuclear core to zero density at the surface. In fact, the information obtained from electron scattering experiments is most accurate for the nuclear skin region and it has been shown that the gradual decrease in nuclear density occurs over a distance of approximately 2.3 ~ 2.4 fm, *regardless of the size of the nucleus*. For both medium-size and large nuclei, the shape of the surface region remains much the same (see Table 6.1 and Fig. 6.3).

Unfortunately, a naïve liquid-drop model implies a sharp fall in nuclear density at the surface. Starting with a typical spherical nucleus in the liquid-drop conception (Fig. 6.4a), the dependence of the density on the radius can be shown as in Fig. 6.4b. That is, the density is constant from the center to the surface and then drops immediately to zero – as would be the case for a spherical billiard ball with constant density. Since most nuclei are thought to be spherical or only modestly deformed, the density plot shown in Fig. 6.4b should be the usual case.

Prolate or oblate deformation of a nucleus (Fig. 6.5a) results in only a small change in the density plot. Again, there will be a constant density spherical core, but the average density will gradually drop near the surface as we travel from the core toward the equatorial regions of the ellipsoid shape (Fig. 6.5b). The fall in nuclear density will be less precipitous than that for a perfect

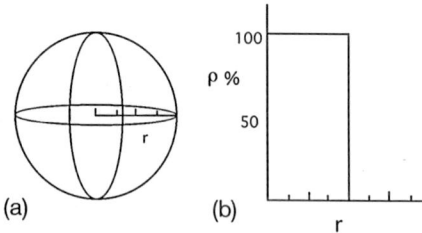

Fig. 6.4. A perfectly spherical liquid-drop (**a**) would show a rapid fall in nuclear density (**b**) with essentially no skin region

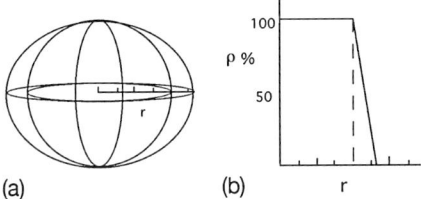

Fig. 6.5. A prolate or oblate deformation of a spherical liquid-drop with a typical deformation of 0.3 (**a**). A static deformation of this kind shows a narrow nuclear skin region of roughly 1 fm (**b**), whereas a skin of 2.3 ∼ 2.4 fm thickness is experimentally known

sphere, but the fall is still sharp for a nucleus with either an oblate or a prolate deformation of a magnitude typical of that known for the non-spherical nuclei (deformation parameter, $b = 0.3$). In other words, the majority of the non-spherical nuclei are deformed to such a small extent that there will be a nuclear skin region whose density falls from 100% to 0% over a distance of little more than 1.0 fm.

The majority of nuclei are approximately spherical or have very small oblate or prolate deformations ($b < 0.2$), so that a nuclear skin of >2 fm remains anomalous in a strictly liquid-drop formulation (Fig. 6.4). Even for the minority of nuclei that are thought to be deformed into ellipsoids – with the ratio of the lengths of the semi-major and semi-minor axes, b, being between 0.2 and 0.4 (a value of 0.3 is illustrated in Fig. 6.5), the liquid-drop model underestimates the experimental skin thicknesses. For the largest nuclei with ellipsoid deformations, most b-values are about 0.3 and the nuclear radius will be upwards of 6.0 fm for $A > 220$. This implies a nuclear skin thickness within the liquid-drop model for even the most-deformed large nuclei of only 1.8 fm.

As illustrated in Fig. 6.6, deducing the charge (mass) distribution of nuclei directly from experimental data is more difficult than an indirect method, by which theoretical form factors are compared against the data. The oscillatory patterns shown in Fig. 6.6b are typically found for all medium and large nuclei and the form factor that most nearly reproduces that pattern is a homogenous sphere with a diffuse surface. There is consequently little doubt that the so-called Fermi curve (illustrated in Fig. 6.3 and shown as the "sphere with a diffuse surface" in Fig. 6.6a) is a rather good approximation of nuclear density for all medium and large nuclei. The many strengths of the liquid-drop model notwithstanding, the thick nuclear skin implied by experimental findings argue against the idea of a "homogenous sphere".

6.3 The Nuclear Radius

Although Hofstadter's experimental work from the 1950s is the foundation of the current understanding of nuclear size, today there are experimental values

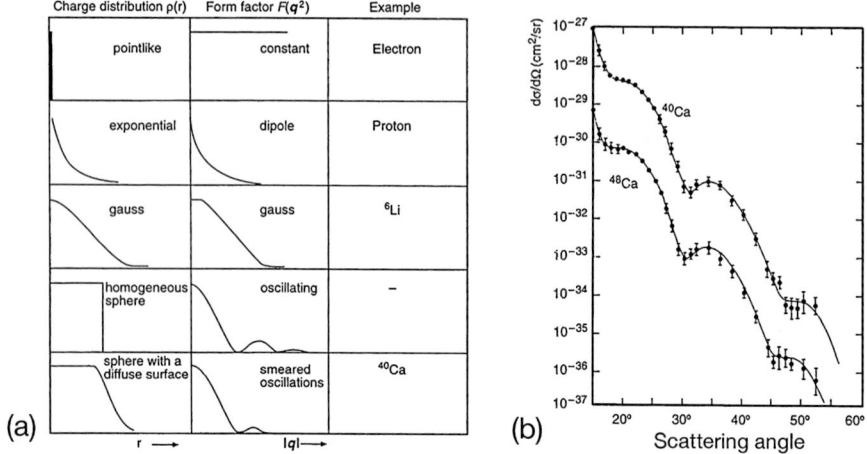

Fig. 6.6. A comparison of theoretical curves (a) and experimental data (b) suggest the reality of a diffuse nuclear skin region (from Povh et al., 1995, pp. 63–64)

for 623 isotopes ranging from $^1H_1$ to $^{243}Am_{95}$. These data are well explained using an equation that takes both the number of protons (Z) and the number of excess ($N > Z$) neutrons into consideration:

$$\text{RMS radius} = r_c(2Z)^{1/3} + r_{ne}(A - 2Z)^{1/3} \tag{6.7}$$

where r_c is the size of the $N = Z$ core nucleons, and r_{ne} is the size of the excess neutrons that lie external of the ($N = Z$) core (Fig. 6.7). The fit is excellent ($R^2 > 0.99$) and slightly better when the nucleons in the nuclear core ($N = Z$) and the neutron excess ($A - 2Z$) are considered separately, than when a regression is done using only the number of protons (Z) or only the number of nucleons (A) without distinguishing between the core and skin.

The linear dependence of the nuclear charge radius on the number of nucleons tells us that all core nucleons take up a constant volume, whereas the excess (electrostatically neutral) neutrons have a much smaller influence on the charge radius. When the analysis is done excluding the smallest nuclei ($Z < 12$), the coefficient for the core nucleons r_c in the multiple regression gives 0.98 fm and that for the neutron excess r_{ne} is 0.05 fm). Both of these values are what might be expected from the charge distributions of the proton and neutron (Fig. 6.3a).

It is seen that the only anomalies are among the small nuclei ($Z < 12$), most notably the deuteron and ^6Li, both of which being significantly larger than expected on the basis of the liquid-drop model.

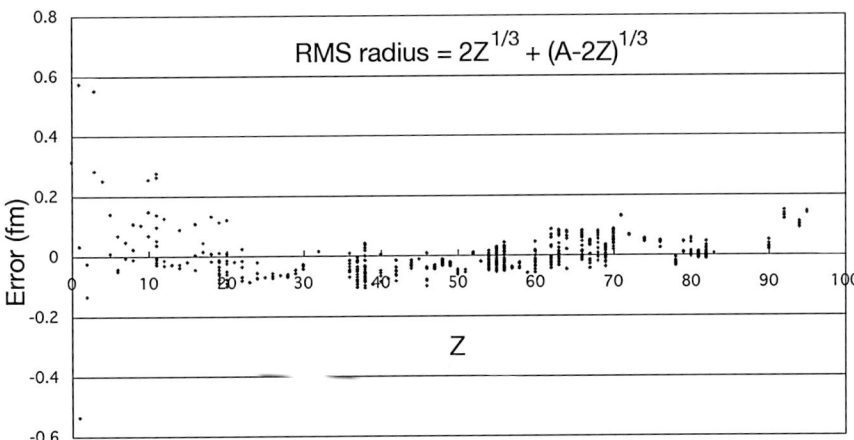

Fig. 6.7. The linear dependence of the nuclear radius on the number of nucleons is notable for all but the smallest nuclei ($Z < 12$)

6.4 Summary

The experimental values for the radii and density of nuclei are rather well accounted for in the liquid-drop model (Chap. 4) and are indication that the liquid-drop analogy is not a bad first approximation. Unfortunately, the extremely high density of Helium and the low density of Lithium are anomalous, and the thick nuclear skin for all nuclei is not consistent with the liquid-drop analogy.

7
The Nuclear Force and Super-Heavy Nuclei

The core dilemma of nuclear physics concerns the nuclear force. On the one hand, no topic during the 20th Century was the focus of as many PhD theses and no topic in all of natural science has launched as many upper-echelon careers in university life, publishing and the government bureaucracy as those beginning with graduate work on the nuclear force. Clearly, many bright people have spent many long hours studying the interactions among nucleons, but it is an open secret that the underlying force remains a puzzle. Progress has been made and insights have been obtained, but what is known about nucleon-nucleon interactions simply does not "plug into" any of the models of nuclear structure to give self-consistent quantitative answers concerning nuclear shape, size and stability.

Kirson (1997) has stated the modern dilemma, as follows:

> "The longstanding and well-documented failure of existing models of nucleon-nucleon interactions to reproduce the empirical binding energy and equilibrium density of infinite nuclear matter is a serious problem to those who would compute effective interactions for shell model calculations from realistic nuclear potentials. Existing many-body results on nuclear matter are a genuine triumph of fundamental nuclear physics, but the quantitative discrepancy alluded to is of great significance for computation of finite nuclei." (p. 290)

Similar sentiments were expressed by many of the principal players in nuclear structure theory at the time of the 40th anniversary of the shell model. J.P. Elliott commented that:

> "Much has been accomplished in forty years but there are still many unresolved problems and difficulties." (Elliott, 1990, p. 15)

and again:

> "I may be pessimistic, but I think it is unlikely that a sufficiently accurate effective interaction can be derived from the nucleon-nucleon force." (Elliott, 1989, p. 19)

7.1 The Nuclear Force

One of the first theoretical tasks in nuclear structure physics is to devise a model that explains how much energy is needed to hold the nucleons together in stable nuclei. In principle, this is not a difficult problem – and the successes of the liquid-drop formula show that an approximate solution can be obtained (Chap. 4). In detail, however, the quantitative description of the nuclear force that has emerged from nucleon-nucleon reaction studies is not compatible with what is known about nuclei: the short-range strong interaction that is known from experimental work simply cannot be used to deduce the properties of nuclei. This is arguably the central unanswered question in all of nuclear physics and the reason why experimentalists speak of a "realistic" nuclear force – known with great precision from nucleon-nucleon experiments – while theorists speak of an "effective" nuclear force – the sum of nuclear effects that emerges from complex many-body systems.

The modern understanding of the two-body nucleon-nucleon interaction dates back to the 1950s and can be summarized as in Fig. 7.1 (left). Here we see that the strength of the interaction between two nucleons depends principally on two factors: (i) the distance between the particles, and (ii) the relative angular momentum (spin) of the two particles. A third factor (isospin) plays a smaller role. Two more recent descriptions of the nuclear potential are referred to as the Bonn potential and the Paris potential (Fig. 7.1, right). In all important respects, they are identical – and highly similar to the Hamada potential from 1962.

Despite the complexity of the various components that contribute to the interaction of two nucleons, the force is believed to be fairly well characterized by the curves shown in Fig. 7.1. The obvious question is therefore: Why can this known force not be used to explain the binding of many-nucleon nuclei? Despite the universally-recognized validity of the above two-body potentials, they are not the basis for explaining nuclear structure: rather, an "effective" central potential-well is the starting point for quantitative work on the energy states of nuclei. There are in fact many variations on the effective force that are employed for nuclear calculations, but none resembles the potentials shown in Fig. 7.1. On the contrary, the three most popular *effective* potentials used in nuclear structure theory (Fig. 7.2) show little overlap with the *realistic* nuclear potential.

Noteworthy is the spatial extent of the effective force. For a small nucleus, the potential-well can be constructed to act over a range of 2–3 fm, but for a large nucleus, such as ^{208}Pb, the force has effects up to $7 \sim 8$ fm. In order to obtain agreement with experimental data, small deformations of the

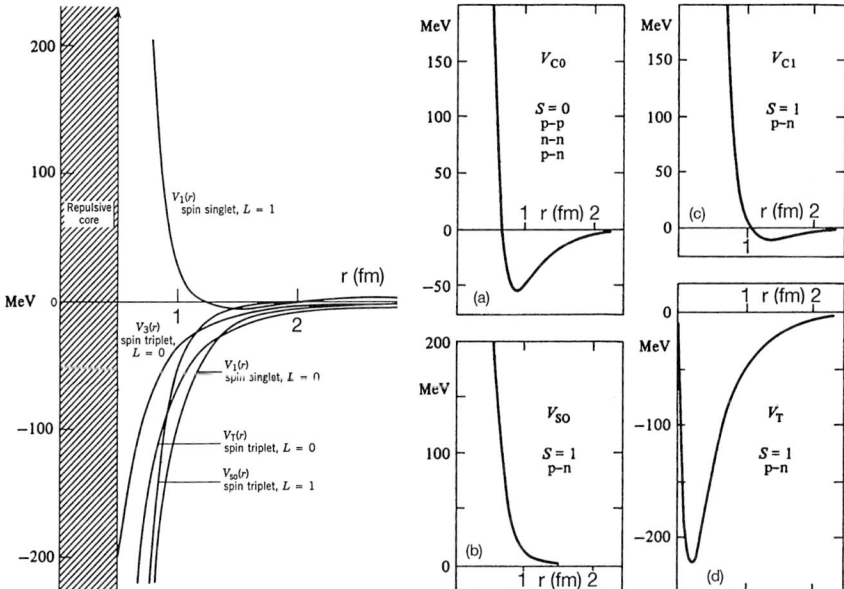

Fig. 7.1. Some of the most important nucleon-nucleon potentials. *Left*: The Hamada & Johnston (1962) potential. Note the repulsive hard-core at 0.5 fm and the disappearance of all effects at >2.0 fm. *Right*: Isolated components of the Paris potential (after Lacombe et al., 1980). (**a**) When spins sum to 0, the potential has a typical Lennard-Jones shape – repulsive at $r < 0.7$ fm, but attractive up to $r = 2.0$ fm. (**b**) The spin-orbit potential is repulsive between protons and neutrons with the same spin ($r < 1.0$). (**c**) The central potential between protons and neutrons with the same spin is only weakly attractive between 1.0 and 2.0 fm, but (**d**) the tensor potential has an angular dependence of two magnetic dipoles and is strongly attractive

potential-well can be later introduced by temperature changes, non-spherical shapes, etc., but the most remarkable aspect of the effective force is the *lack* of agreement with the actual nuclear force. As indicated by the arrows in Fig. 7.2, there is a small region where the realistic nuclear force of the Paris potential and the effective force of the shell model overlap, but it is hard to believe that both of these potentials are describing the same physical force.

The use of an effective force for all theoretical work on real nuclei ($A > 4$) can be understood as a necessary approximation to the complex effects of multiple, time-averaged, two-body nuclear force effects and it bears emphasizing that a great many nuclear properties can indeed be accurately described in this way. What cannot be claimed, however, is that nuclear properties have been deduced from the known two-body nuclear force. On the contrary, the known features of the strong, local nuclear force have been explicitly set aside when effective forces are in use.

144   7 The Nuclear Force and Super-Heavy Nuclei

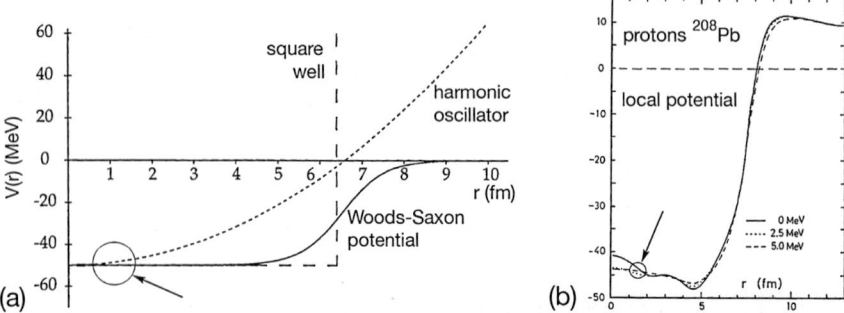

Fig. 7.2. Effective nuclear forces. (**a**) The three most common textbook potential-wells are the square well, the harmonic oscillator and the Woods-Saxon potential (from Greiner & Maruhn, 1996, p. 238). (**b**) The highly-popular Hartree-Fock potential used for the ^{208}Pb nucleus shown at three different temperatures (after Brack & Quentin, 1974). The arrows indicate the only regions where the *presumed* effective force overlaps with the *known* nuclear force

7.2 Super-Heavy Nuclei?

The modern version of the binding energy formula is remarkable for its numerical capabilities based upon a relatively small number of parameters (Möller et al., 1992) (see Chap. 4). Since the late 1960s, however, calculations based on this description of nuclear binding has consistently predicted "islands of nuclear stability" and the existence of so-called super-heavy elements at values of Z and N well beyond the known range of nuclei in the periodic table (Fig. 7.3). Experimental searches for the possible existence of super-heavy elements have been made for more than three decades and, in recent years, experimental techniques have been devised for the possible creation of super-heavy nuclei. A great many interesting phenomena and short-lived nuclei have been discovered, but the results regarding long-lived super-heavies have been disappointing.

It is relevant to point out that more than twenty transuranic elements ($92 < Z < 114$) have been synthesized and a variety of extremely short-lived isotopes have been found. Much knowledge has been obtained concerning these very large nuclei and, for this reason alone, the investment in super-heavy research has been worthwhile. Whatever future research may yet reveal, however, it is already clear that (i) super-heavy nuclei do not exist in any detectable abundance in nature, and (ii) stable or quasi-stable super-heavy nuclei cannot be constructed simply by colliding one large nucleus into another and letting the compound settle into a stable system. For anyone interested in the fundamentals of nuclear theory and for an understanding of why the chemist's Periodic Table is finite, those are important results. Nevertheless, crucial questions in nuclear theory remain to be answered: If the current understanding of the nuclear force is more-or-less correct in predicting the

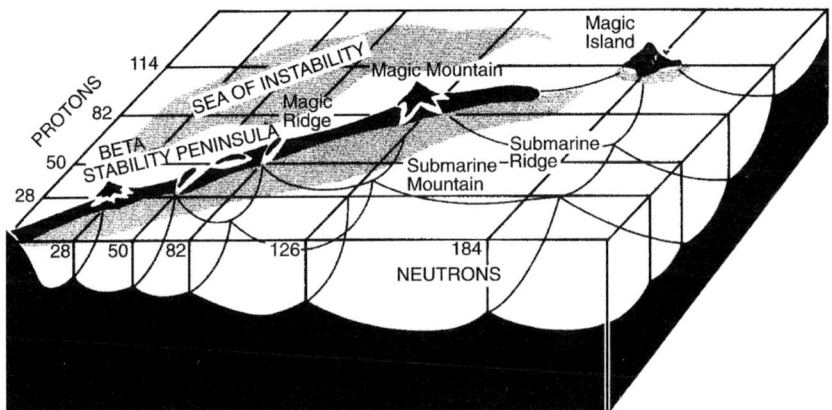

Fig. 7.3. A magic island of stability is predicted beyond the region of known nuclei (after Seaborg & Bloom, 1969)

possibility of super-heavy nuclei, why cannot the rather modest extrapolation beyond the known heavy nuclei produce unambiguous theoretical results concerning the existence or non-existence of the super-heavy nuclei?

Thus far, after a considerable research effort in terms of dollars and man-years, the search for the predicted stable super-heavy elements has been unsuccessful: *all empirical findings indicate that there is attenuation of the periodic table* (i.e., decreasing isotopic half-lives) extending only as far as $Z \sim 114$. Precisely where the periodic chart may come to a complete dead-end depends on how one defines attenuation. Improved experimental techniques allow for the detection of nuclei with extremely short half-lives, but every empirical indication points to the attenuation of the periodic chart somewhere in the vicinity of element 114 without the appearance of a magic island of stability (Fig. 7.4).

In contrast to the bleak picture painted by the experimentalists, theorists have remained convinced that super-heavies do exist and/or can be constructed. Clearly, something is amiss. Either better experimental techniques are needed to produce the elusive super-heavies or extrapolation to the binding energies of extremely large nuclei on the basis of conventional theory is not correct. It is important to point out that the prolonged search for super-heavies has been justified primarily by the *unanimity* of theoretical models in predicting the existence of super-heavies with $Z = 110, 112, 114, 118$ or 126 and $N = 164, 178$ or 184 (and indeed the predictions, although not the confidence of theorists, have changed somewhat over the decades).

Seaborg and Loveland (1987) argued strongly that:

> "One fact should be emphasized from the outset: while the various theoretical predictions about the superheavy nuclei differ as to the expected half-lives and regions of stability, all theoretical predictions

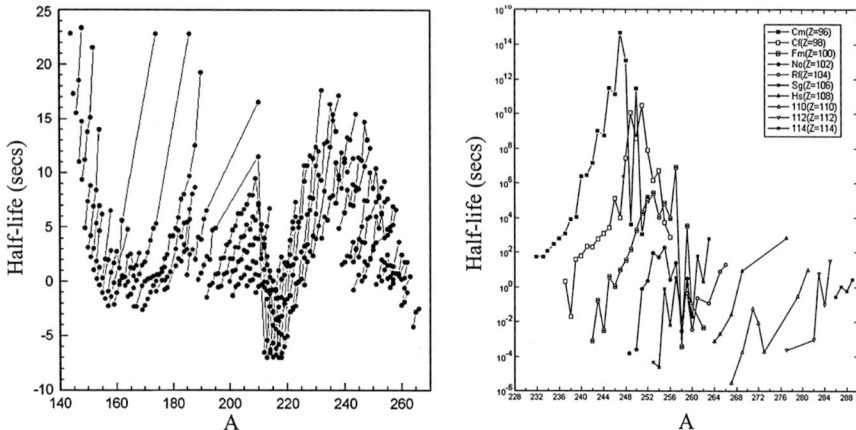

Fig. 7.4. Half-lives of the heavy nuclei. (**a**) shows the half-lives of all unstable nuclei with $A > 140$. A trough of very short half-lives near $A = 220$ is followed by the relative stability of Uranium and other actinides, but thereafter a gradual decrease. In (**b**) the half-lives for all known even-Z nuclei (>94) are shown. Another upturn in half-lives at $A > 260$ indicative of an island of stability has not been found

are in agreement: superheavy nuclei can exist. Thus, the search for superheavy nuclei remains as a unique, rigorous test of the predictive power of modern theories of the structure of nuclei."

More recently, Greiner (1995) has noted that:

"The simple extrapolation of the shell model shows the existence of new magic numbers 114, 124 (a subshell), and 164 for protons and 184, 196, 236 (a subshell), 272 (a subshell) and 318 for neutrons. Various parameter sets yield different subshell structures, but the principle shell structure remains." (p. 3)

To the unbiased observer, the flexibility of the theory is amazing. Most even numbers greater than 100 have at one time or another been proposed as "magic" or "semimagic" and it is hard to imagine any experimental indication of relative stability that could not, with *post hoc* clarity, be advertised as demonstration of the correctness of the theoretical calculations. But, clearly, if the existence of super-heavy elements is in any sense an empirical test of conventional nuclear structure theory, then there must be at least the *possibility* of a failure to confirm the theory. Since the null case, the non-existence of the super-heavies, can never be logically proven, this "unique, rigorous test" must be carried out on two fronts: (i) the creation of experimental conditions that might reveal the existence of super-heavy elements; and (ii) the examination of theoretical assumptions that lead to predictions about the attenuation or non-attenuation of the periodic table.

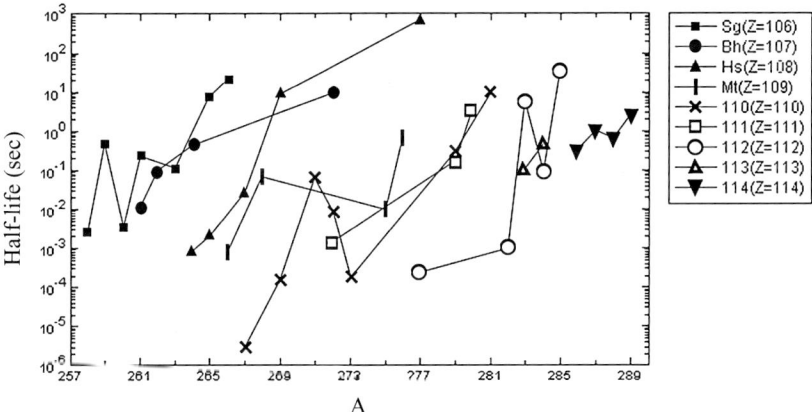

Fig. 7.5. The experimental half-lives for all isotopes $Z > 105$ (from the Nuclear Reactions Video database, 2005, www.nrv.jinr.ru/nrv)

The discrepancy between experiment and theory prompted Kumar (1989) to introduce his book, *Superheavy Elements*, devoted to the problem of the super-heavy nuclei, as follows:

> "Nuclear scientists face a major crisis at present. According to the currently predominant paradigm, there should be a 'major island' of stability surrounding the doubly magic superheavy nucleus containing 114 protons and 184 neutrons." (p. 1).

But this is the same "major crisis" that was around in the late 1960s, when a host of unstable transuranic nuclei were created, and is still with us in the 21st Century. For experimentalists and theorists alike, this is all very interesting work, but, as far as the existence of stable or semi-stable super-heavy nuclei are concerned, more than three decades of research have produced only negative results.

Nonetheless, in 1992 Möller and Nix revised their previous calculations – now predicting a super-heavy island around $^{282}110$ with fission half-lives on the order of a few milliseconds. That became the newly predicted "island of stability". Since the theoretical calculations were so inherently flexible with an abundance of adjustable parameters, Möller and Nix concluded their 1992 study on an upbeat note:

> "For odd systems specialization energies can lead to huge increases in spontaneous-fission half-lives, with up to 10 orders of magnitude possible" (p. 100)

With 10 orders of magnitude leeway, no one can be certain about the stability or instability of these isotopes, but two years later Möller and Nix (1994) maintained the same level of optimism with revised predictions:

> "We now have a much better appreciation of the experimental difficulties that have so far prevented us from reaching the spherical superheavy island around $^{294}110$ [a different island!]. However, we are confident that this superheavy island exists because the models that predict it have now been tested in several different ways and are sufficiently reliable for extrapolations to the spherical superheavy region." (p. 1744)

They in fact predicted a half-life of 10^{15} years for $^{272}\text{Mt}^{163}$ (element 109), in stark contrast to the experimentally-determined half-lives of most of the nuclei in this region (Fig. 7.5). Note that these are all artificially created elements and therefore testament to an unprecedented technological feat in creating and measuring new isotopes, but there is yet little indication of magic islands.

As the years pass without empirical verification of these theoretical certainties, the theorist's certitude has become mixed with signs of realism. The theoretical techniques behind the shell model predictions of super-heavies were developed by Nilsson and colleagues, but even they have become discouraged:

> "It seems that theoretical calculations can only indicate that the chances for success are so large that it is worth going on [searching for superheavies] but also that the uncertainties are large and we can neither expect nor exclude that any superheavy elements will be synthesized in the near future." (Nilsson and Ragnarsson, 1995, p. 175)

What is remarkably missing here is scientific skepticism: if there is only one unanimous theoretical conclusion, but no empirical verification even after several decades, theorists still find no grounds for re-examination of the starting assumptions of the theory.

The essential problem lies in the strange alliance between the shell and liquid-drop models – as embodied in the semi-empirical mass formula. On the one hand, the liquid-drop analogy provides a good, if rough, description of nuclear binding, but, for precise predictions of the characteristics of specific nuclei beyond the range for which the mass formula was developed, the empirical input concerning the balance of protons and neutrons (the symmetry term) and the special effects of certain numbers (the shell corrections) completely dominate the formula. The symmetry effect is expected to change gradually with gradual increases in nuclear charge, but the shell corrections change drastically with changes in the potential well and the closure of shells and/or subshells at various numbers. The conceptual contradiction between the strong, local nuclear force and the weak, long-distance potential-well is clear enough, but the confidence of the theorists remains unwavering because gaps in the energy levels of both protons and neutrons can be produced at various numbers by appropriate parameter selection.

7.3 Summary

The puzzle of the nuclear force is genuinely difficult and there is yet no solution to the problem of how the complex, but well-characterized force known from 2-body nucleon reactions can be reconciled with the effective force used in the independent-particle model of nuclear structure. Is the atomic nucleus dominated by a force that has significant effects up to about 2 fm and no effects at greater distances, or is it a force that has effects up to $7 \sim 8$ fm? No clear answer has been forthcoming. Meanwhile, efforts at extrapolating from the semi-empirical mass formula in light of shell effects in order to predict the stability/instability of super-heavy nuclei have led to the *incorrect* inference that super-heavy nuclei can exist. Experimental work has demonstrated that short-lived isotopes containing up to 289 nucleons can be constructed and detected, but the "island of stability" unanimously predicted by conventional ideas about the nuclear force appears to have been a mirage. If this abject failure is not a crisis in nuclear theory, what is? What level of discrepancy between theory and experiment is required to indicate that a problem exists – that the "effective" nuclear force is a red-herring leading away from, not toward, physical insight?

8

Nuclear Fission

The defining physical phenomenon of the Nuclear Age, the single most important fact that has transformed the nature of military combat, and therefore all major political discussions, is nuclear fission. Although the specifically *nuclear* aspects of the modern world impinge somewhat less directly on our daily lives than many other modern technologies, the importance of nuclear phenomena – both as an actual source of energy and as a potential source of mass destruction – is nonetheless pre-eminent and lingers in the background of every major economic and political decision.

Because of the over-riding importance of the technological applications of nuclear fission, the empirical facts and the history of its discovery and exploitation are well-established and, as historical facts, not controversial. The story of nuclear fission has been the focus of many books and the essential developments can be outlined in a few pages. What is not so easily summarized is the physical mechanism underlying the break-up of large nuclei. Although nuclear fission has been, for both military and economic reasons, a central concern to the physics community for more than 60 years, unexplained problems remain. Already, nearly a half century ago, Halpern (1959) was frank in stating this fact:

> "It is in some ways awesome to compare the tremendous development of the applications of nuclear fission in this [20 year] period with the slow growth of our understanding of its fundamental features." (p. 245)

Seven years later, Fraser and Milton (1966) were equally straight-forward:

> "The phenomenon of nuclear fission was first identified by Hahn and Strassmann in 1938. Today, a quarter of a century later, we know a great deal about fission but understand rather little." (p. 379)

More recently, Moreau and Heyde (1991) have stated that:

"The theoretical description of the fission process ... is one of the oldest problems in nuclear physics. Much work has been done to understand this process and many aspects have been clarified, but it appears that a consistent description of fission is still very far away." (p. 228)

Why do specialists in this subfield within nuclear physics insist on saying that there is a lack of understanding? If the technology needed to produce nuclear power is well-understood, what are the theoretical issues that remain troublesome? Let us review the basic facts of fission, its history, and its treatment in modern physics texts to see where the problem lies.

8.1 Basic Facts of Fission

Many artificial and a smaller number of natural isotopes are known to be unstable and decay into other kinds of nuclei. The instability is a direct consequence of an imbalance in the numbers of protons and neutrons, and can most often be remedied by the transformation of one or several protons into neutrons, or vice versa. Because of the large amount of positive charge contained in the small volume of heavy nuclei, any nucleus with more than 83 protons spontaneously finds ways of reducing its total charge. One way is to release a charge in the form of a positron ($\beta+$ decay). This transformation leaves the total number of nucleons the same, but, with one fewer proton and one more neutron, so that a small increase in nuclear stability is gained. For some of the largest unstable nuclei containing excessive positive charge, however, internal adjustment of the balance of protons and neutrons cannot give stability unless some number of protons is first expelled from the nucleus. One such mechanism for proton expulsion is the release of an alpha-particle (α decay), thereby reducing the nuclear charge by two units. Another mode of decay for nuclei with excessive positive charge is fission, in which the parent nucleus breaks into two relatively large pieces, together with the release of several neutrons. This is known as binary fission (whereas the much rarer breakdown into three pieces is known as ternary fission).

The cause of such nuclear breakdown is the large amount of charge. For the issues of nuclear technology and the harnessing of nuclear power, the energy calculations based on the liquid-drop model are necessary and sufficient. What has remained unexplained is the fact that nearly all of the so-called fissionable nuclei – that is, nuclei that spontaneously or with the addition of only a small amount of energy – break into two *unequal* fragments. This is called asymmetric fission. One of the two fission fragments contains about 140 nucleons, whereas the remaining nucleons (minus a few neutrons released from both the large and small fragments) are contained in the small fragment. For example, in the fission of Uranium-235 induced by a low-energy "thermal" neutron, the average size of the large fragment is 139, whereas the average

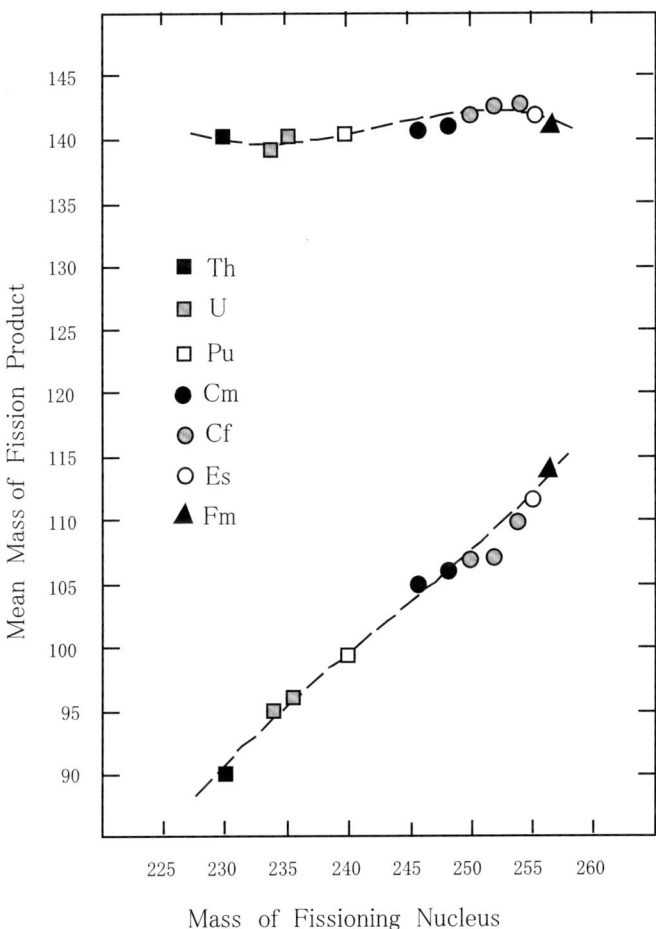

Fig. 8.1. A summary of light and heavy fragment masses (prior to the release of several neutrons) (after Krane, 1988)

size of the small fragment is 97 (prior to the release of neutrons). It is of interest that the liquid-drop model, which suffices to account for the energy released in fission, predicts that the dominant mode of fission should be symmetrical, i.e., 118 nucleons in each fragment, with half of the protons in each fragment. That is, the model that explains the basic energy transformations in fission predicts that the droplet would split into two equivalent halves. This prediction is incorrect. For the most important fissioning nuclei, asymmetrical fission, not symmetrical fission, is the dominant mode of nuclear break-up.

How significant is the *asymmetry* of the mass fragments in low-energy fission? The data for the entire set of low-energy fissionable nuclei are summarized in Fig. 8.1 and Table 8.1. What is apparent here is that *asymmetric fission* is the overwhelming mode of low-energy fission. Only when the excitation

Table 8.1. Average fragment masses for low-energy fission (after Hoffman and Hoffman, 1974, pp. 158–159)

Fissioning Nucleus (a)	Light Fragment (b)	Heavy Fragment (b)	Fissioning Nucleus (a)	Light Fragment (b)	Heavy Fragment (b)
^{228}Th*	(89)	(138.5)	^{250}Cf	108.0	141.9
^{230}Th*	89.6	140.4		107.5	142.5
	(87.6)	(139.9)		107.7	142.3
			^{250}Cf*	108.2	141.8
^{234}U*	95.0	139.0		(105.8)	(139.8)
	(93.3)	(138.2)		(106)	(139.5)
^{236}U*	96.6	139.4	^{252}Cf	108.5	143.5
	96.5	139.5		108.4	143.6
	(94.9)	(138.2)		108.5	143.5
^{238}U	(95.5)	(140.5)		(106.1)	(142.1)
				(107)	(142)
^{240}Pu	101.6	138.4	^{252}Cf*	(107.3)	(140.7)
	(100)	(137)	^{254}Cf	110.9	143.0
^{240}Pu*	100.3	139.7		110.6	143.4
	100.8	139.2		110	144
	100.4	139.6			
	(98.9)	(138.1)	^{253}Es	111.3	141.7
^{242}Pu*	102.6	139.4		(105.9)	(142.4)
	(100.4)	(138.8)	^{255}Es*	112.7	142.3
				(110.6)	(140.2)
^{243}Am*	(100.9)	(139.1)			
^{242}Cm	(102.5)	(137.5)	^{254}Fm	111.5	142.5
^{244}Cm	104.5	139.5		(108.8)	(141.5)
	104.6	139.0	^{256}Fm	113.9	142.1
	(103.1)	(138.7)		(111.8)	(141.0)
^{246}Cm	106.0	140.0	^{256}Fm*	114.9	142
	105.9	140.1		(113.2)	(138.8)
^{246}Cm*	105.3	140.7	^{257}Fm*	128.5	128.5
	(102.8)	(139.2)			
^{248}Cm	107.3	140.7			
	107.0	141.0	mean	104.4	140.5
^{250}Cm	107.5	140.5	std. dev.	±6.4	±1.7

(a) The asterisks denote thermal neutron-induced fission, whereas those without an asterisk are spontaneous fission.

(b) Numbers in parentheses indicate measurements made after the release of neutrons, and are consequently 2–4 nucleons less than the measurements of fragment size prior to neutron release.

energy is raised (making possible a variety of fission modes for these and many other nuclei) or when some of the extremely short-lived, artificial ($A > 256$) isotopes are considered is there any indication of symmetrical fission. It is important to emphasize that this phenomenon is *not* one of the minor forgotten recesses in the ancient history of nuclear theory: low-energy *fission* is one of a very small number of phenomena that makes nuclear physics a matter of more than academic interest; and *asymmetrical fission* is the dominant mode.

An explanation of the size of the fission fragments is therefore essential to an understanding of the phenomena of nuclear fission, but this old issue remains problematical. Fraser and Milton (1966) referred to the asymmetry of fission fragments as "the perennial puzzle" of nuclear physics. Hoffman and Hoffman (1974) noted that:

> "The asymmetric distribution of mass has long been recognized as one of the most striking characteristics of all spontaneous and low-energy fission." (p. 154)

Bohr and Mottelson (1975) have also stated:

> "A striking feature of the fission process is the asymmetry in the masses of fragments observed for low-energy fission of nuclei in the region $90 \leq Z \leq 100$" (p. 369).

And Moreau and Heyde (1991) commented that:

> "The preference of most actinide nuclei to divide asymmetrically at low excitation energy has been one of the most intriguing puzzles ever since the discovery of fission." (p. 228)

First and foremost, a theory of fission must address the mechanisms that underlie these "striking" findings. While a complete theory should also explain the symmetrical fission of the short-lived (10^{-11} years) ^{258}Fm nucleus and the trend toward increasing symmetry of fragments with increasing energy of excitation, the first theoretical task is to explain why nearly all of the fissionable nuclei break into fragments with mass ratios of 3:2. It is a curious pattern. Although the number of protons in the mother nucleus ranges from 90 to 100 and the number of neutrons from 142 to 157, the number of protons in the large fragment is consistently found to be 52–56, while the average number of nucleons in the large fragment remains remarkably constant at ~140.

Partly because of the technological importance of fission and partly because of the unambiguous nature of the data themselves (integer numbers of protons and neutrons), there is no uncertainty about what has occurred in asymmetrical fission. The precision of the modern data is evident in Fig. 8.2. Here, the distribution of mass fragments in the thermal fission of ^{233}U, ^{235}U and ^{239}Pu is shown, together with the experimental uncertainty. The uncertainty is so small that it falls within the diameter of the points indicating the mean values. In other words, the substructure near the peaks of the mass

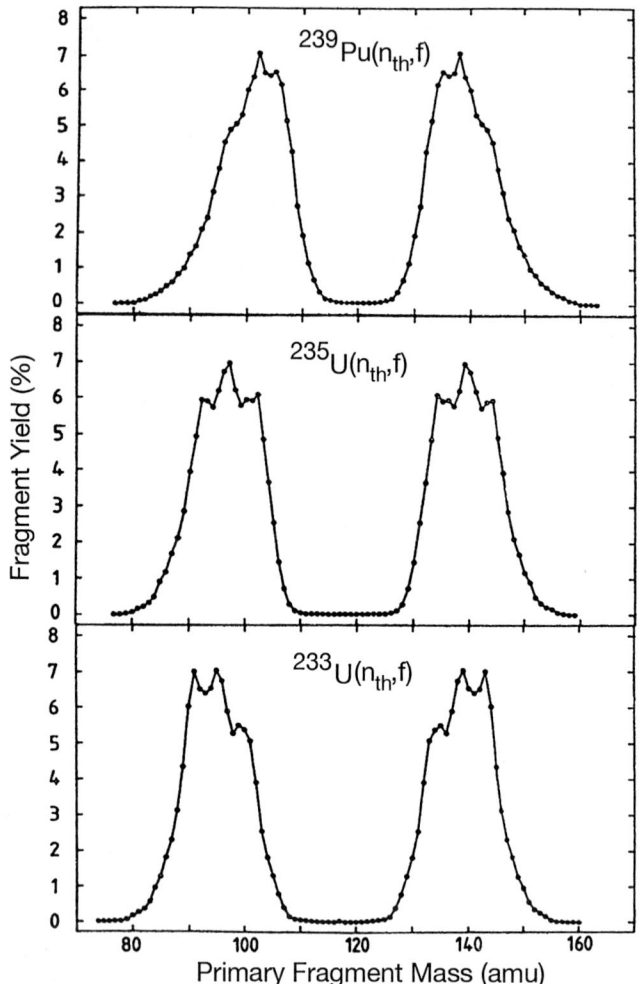

Fig. 8.2. Mass yields (prior to neutron release) vs. primary fragment mass for thermal neutron fission of ^{233}U, ^{235}U and ^{239}Pu (from Geltenbort et al., 1986). The two humps are mirror-images of one another until neutrons and occasionally other particles are emitted from the light and heavy fragments

distributions seen in Fig. 8.2 is real and not experimental "noise". The significance of the jaggedness of these curves lies in the fact that it is indication of nuclear substructure that survives the explosive break-up of the mother nucleus. In spite of the chaotic, stochastic events underlying fission, structure remains in the fragments.

The curves in Fig. 8.2 clearly indicate that the asymmetrical fission fragments – often summarized in terms of the roughly 3:2 ratio of nucleons in the heavy and light fragments – actually consist of three or more sharp, slightly

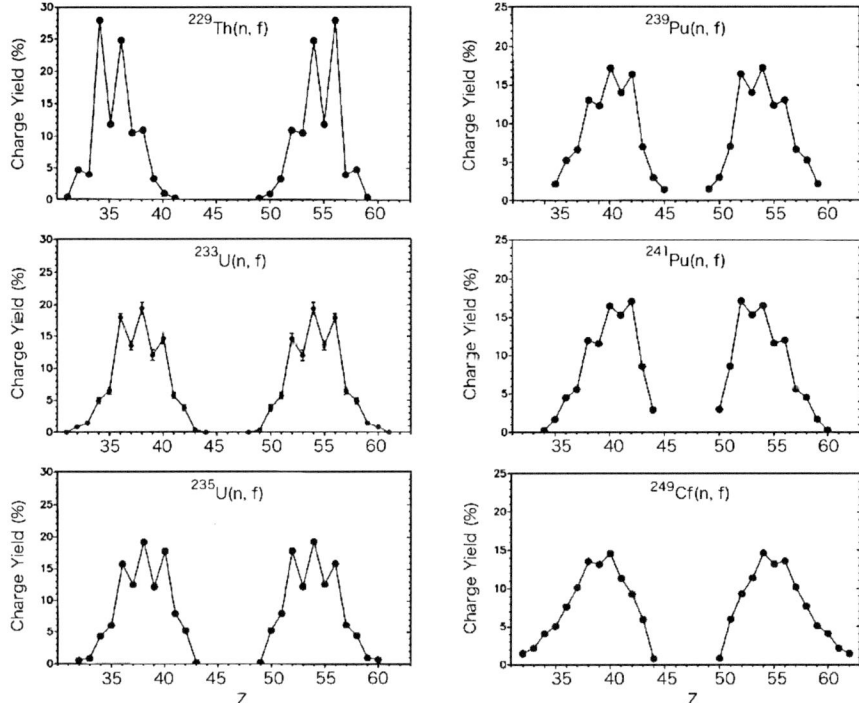

Fig. 8.3. Charge distributions in the light and heavy fragments for thermal neutron-induced fission reactions (after Fig. 92 from Gönnenwein, 1991). As was the case in Fig. 8.1, the light and heavy fragments are symmetrical until each fragment releases neutrons and undergoes internal transformations that bring it toward the optimal Z:N ratio

overlapping distributions within both pieces. The peaks for ^{233}U+n, ^{235}U+n and ^{239}Pu+n, respectively, lie at (^{234}U light: 91, 95, 99; heavy: 135, 139, 143), (^{236}U light: 92, 97, 102; heavy: 134, 139, 144) and (^{240}Pu light: 97, 102, 105; heavy: 135, 138, 143) nucleons. In fact, the number of protons within the fragments is also experimentally known (Fig. 8.3), providing exact constraints on what kinds of results a theoretical model must produce.

As is evident from Fig. 8.3, what is known about fission is known with great precision, but the experimental certainty has not led to theoretical clarity. On the contrary, the consistent failure of theory to explain the empirical facts about fission fragments has led to a curious neglect of fission – the hallmark of nuclear physics! – even in the textbooks. Before we address the curious mismatch between experimental precision and theoretical confusion, let us look briefly at the historical record.

8.2 The History of Nuclear Fission

The initial discovery of fission was made by Hahn and Strassmann in 1938. In experiments on uranium, they found that bombardment with neutrons resulted in the appearance of barium. Unlike uranium, barium is a medium-sized nucleus, suggesting that the large uranium nuclei were not merely transformed into other large nuclei, but were being broken into relatively small pieces.

Although this discovery was soon confirmed by Fermi and others, it was initially considered to be a curiosity without prospects of practical application. However, once Niels Bohr had noted that fission occurred only in one of the two naturally-occurring uranium isotopes and that ^{235}U was more likely to undergo fission when excited with low energy neutrons, the possibilities for practical uses became apparent and technological developments were rapid. In a matter of a few years, controlled chain reactions had been produced, and soon thereafter the first nuclear fission bombs were exploded.

Theoretical developments were also initially rapid. Bohr and Wheeler were able to explain the energy release in fission using the liquid-drop model and the compound nucleus reaction mechanism. They devised semi-empirical equations that could accurately predict the amount of energy release when large nuclei undergo fission. Subsequent developments of that theory remain central to the modern understanding of all fission phenomena:

> "The liquid-drop theory was ... systematically extended to non-spherical shapes to explain the process of fission in a, by now, classic paper by Bohr and Wheeler (1939). The ideas put forth in this paper still form a basis for the present theories of nuclear fission." (Nilsson and Ragnarsson, 1995, p. 29)

Initially, the fact that ^{235}U undergoes fission with only slight excitation, whereas other isotopes require significantly more excitation was considered a mystery. This was eventually understood as the result of two phenomena. Most importantly, low-energy (thermal) neutrons actually bind to ^{235}U, forming a compound nucleus. High-energy projectiles, in contrast, excite the target nucleus, but are themselves too energetic to remain bound within it. As a consequence, the lower energy projectiles can be more effective in producing fission than a somewhat higher-energy projectile. This understanding is based fundamentally on Bohr's compound nucleus model.

The second phenomenon that helped clarify the mechanism of fission concerns nuclear binding energies – and is based on the liquid-drop model. Because ^{235}U has an odd number of neutrons, the thermal neutron is strongly bound to the core nucleus, producing 6.4 MeV of energy. Most of that energy goes directly into the excitation of ^{236}U, rather than into externally-released radiation. Since the amount of energy required to induce fission – the so-called fission barrier – for ^{236}U is only 5.3 MeV, the probability of fission occurring is high when ^{235}U absorbs a neutron. In contrast, although other uranium isotopes can also undergo fission, the amount of energy internally released from the thermal neutron excitation of, for example, ^{238}U, is lower than its fission barrier (4.9 MeV neutron-binding versus a fission barrier of 5.5 MeV). As a consequence, ^{239}U is more likely to undergo other kinds of decay, rather than fission.

Similar calculations can be made for other nuclei and indicate which isotopes are likely to fission and how much energy will be released. For most practical applications of fission phenomena, the energy considerations are the most important – and in this realm the liquid-drop conception of the nucleus has proved invaluable.

8.3 Textbook Treatment of Asymmetric Fission

The world has been forever changed by the first fission explosion, so how do physics texts treat this revolution? Surprisingly, despite its practical importance, some nuclear textbooks do not even mention fission. McCarthy's *Introduction to Nuclear Theory* (1968) says not a word, nor does Heyde's *From Nucleons to the Atomic Nucleus: Perspectives in Nuclear Physics* (1998). Moreover, despite the fact that claims have been made that the fragment asymmetry in fission phenomena indicates the validity of the shell model (examined below), there is no mention of fission in Lawson's text, *Theory of the Nuclear Shell Structure* (1980). Of course, as every textbook author must know, many worthy topics require condensation or omission in books that summarize an entire field. If nuclear fission were not the defining phenomenon of the Nuclear Age, this omission could perhaps be understood as simply one of the many shortcuts in any textbook, but fission *is* the defining phenomenon of the Nuclear Age! Discussing nuclear physics without mentioning fission is like marriage without sex – technically possible, but missing something. And, as we will see below, nuclear theory without fission is a good example of academic prudery – a certain distaste for discussing matters that don't have easy answers within the accepted conventions.

Most textbooks do in fact discuss the energetic aspects of fission, but sidestep the issue of fragment asymmetry. In *Basic Ideas and Concepts in Nuclear Physics* (Heyde, 1994) and in *Particles and Nuclei* (Povh et al., 1995), the authors discussed a great many topics briefly, but not the asymmetry of fission fragments. In *Nuclear and Particle Physics* (1991), Williams discusses fission, but found the asymmetry of fission fragments quite unremarkable. The entire discussion is:

> "Another property of fission is that the production of equal or near equal mass fission products is unlikely and somewhat asymmetric fission is the usual outcome." (p. 78)

In a book with the same title, Burcham and Jobes (1994) condensed the discussion of asymmetric fission even further. After noting the reality of asymmetric fragments in thermal fission, the entire discussion of fragment masses was as follows:

> "The fragment distribution for fast fission is generally more symmetric than for thermal fission." (p. 211)

True enough, but the essential point that is not communicated in these textbooks is that the asymmetry of the fragments for both spontaneous and thermal fission is the *overwhelming* and *unexplained* essence of all the naturally fissioning nuclei.

In *Modern Atomic and Nuclear Physics*, Yang and Hamilton (1996) devoted 8 pages to the phenomenon of fission, including discussion of the basic energy predictions of the liquid-drop model, but did not mention the enigma of fragment asymmetry. And even in Mladjenovic's historical review in *The Defining Years in Nuclear Physics: 1932–1960s* (1998), the chapter on fission did not point out the unfinished business of explaining fragment asymmetry.

So, it is evident that some authors attempting to give an overview of 20th Century nuclear physics find nothing of interest in fission fragment asymmetry, but the specialist monographs devoted to fission have emphasized precisely this point over and over again. In a three volume work (*The Nuclear Properties of the Heavy Elements*, 1964), Hyde introduced the topic of fission, as follows:

> "Bohr and Wheeler developed a theory of the fission process in 1939 based on a conception of the nucleus as a liquid drop; Frenkel independently proposed a similar theory. Their application of this theory did not explain the most striking feature of fission, namely the asymmetry of the mass split, but it accounted satisfactorily for a number of features of the reaction.... No adequate theory of fission has ever been developed" (p. 7).

In a classic text on nuclear fission, Vandenbusch and Huizenga (1973) stated the situation as follows:

> "Asymmetric mass distributions have proved to be one of the most persistent puzzles in the fission process. Although many suggestions as to the origin of this effect have been offered, no theoretical model has been proposed which has been explored in a complete enough manner or has been sufficiently free of parameter fitting to be generally accepted." (p. 259)

And later:

> "The most significant failure of the [liquid-drop] theory is the failure to account for asymmetric mass distributions." (p. 273)

And again:

> "Although a large amount of experimental data on the mass distribution in fission for various nuclei under a variety of conditions has been

8.3 Textbook Treatment of Asymmetric Fission

available for a number of years, no suitable theory yet exists which explains all the observations." (p. 304)

In a textbook on nuclear structure, Preston and Bhaduri (1975) frankly admitted a theoretical failure:

"Up to the present time, there is no satisfactory *dynamical* theory of fission that describes the asymmetric mass split of the fissioning nucleus..." (p. 562)

And, in an equally solid text, Krane (1988) stated:

"Surprisingly, a convincing explanation for this [asymmetric] mass distribution, which is characteristic of low-energy fission processes, has not been found." (p. 484)

But most textbook authors have maintained that the asymmetry, although not fully understood, can be more-or-less explained on the basis of shell effects. The textbook argument is that nuclei with magic numbers of protons and/or neutrons (notably, 50 and 82) are so stable that, when nuclei break-up, they break into fragments, at least one of which has a magic number of nucleons. Already in 1955 with publication of *Elementary Theory of Nuclear Shell Structure*, Mayer and Jensen noted that:

"It has been repeatedly suggested that the asymmetry of fission of uranium and other nuclei might be accounted for by the discontinuities in the nuclear binding energy and stability limits. The great majority of fission events correspond to a division in which one of the fragments has not less than 82 neutrons, the other not less than 50..." (p. 37)

Bohr and Mottelson (1975) made similar comments:

"The liquid-drop model does not provide an explanation of this [asymmetric fission] phenomenon..., and it appears likely that the mass asymmetry must be attributed to shell-structure effects..." (p. 369)

And three decades after the emergence of the shell model, Nifenecker (1981) had similarly optimistic sentiments:

"While the liquid drop model is unable to explain these mass distributions, the latest remarks strongly suggest that shell effects are at work in determining the mass distributions" (p. 316).

Unfortunately, over the course of more than 50 years such suggestions have failed to become quantitative.

What is surprising is that the technical literature exhibits this same vague optimism. Strutinsky and colleagues labored explicitly on the problem of fragment asymmetries and published an authoritative review of their work (Brack et al., 1972). They presented extensive arguments on the theoretical manipulations of the nuclear potential well that are needed to produce asymmetrical fission, but did not deal with *any* specific cases of fission! Full of confidence, they nonetheless concluded their discussion as follows:

"It seems to us that these results may be considered as a basis for an explanation of the asymmetry in nuclear fission" (p. 380).

Despite the rather inconclusive nature of such conclusions, this theoretical paper is the definitive work to which reference is invariably made to indicate that the shell model explains the anomalies of asymmetric fission. Twenty years later, however, Moreau and Heyde (1991) let the cat out of the bag in a comprehensive 50th anniversary treatise on nuclear fission:

"Up to now, the microscopic and stochastic calculations have not been applied for predicting mass-yield curves. This is mainly so because of computational difficulties. Moreover, there are still major conceptual problems in these areas of nuclear research." (p. 231)

A liquid-drop, by its very nature, does not allow for nuclear substructure: it is an amorphous, structureless object and the random distribution of positive charges within the droplet inevitably leads to the prediction of *symmetrical* fission fragments. The shell model, on the other hand, implies distinct energy levels occupied by certain numbers of nucleons, and is therefore a reasonable place to turn in search of substructure in the nuclear interior that could lead to uneven fragmentation. So, the shell model theorist's optimism is perhaps understandable, but whatever theoretical model is employed, it is essential to make comparisons between theoretical predictions and experimental data – and to explain the role of model parameters in obtaining the results. The *possibility* that shell effects are important for fission is certainly worth examining (see below), but the conventional treatment of this problem is arguably one of the characteristic failures of the multi-model approach in nuclear theory: whenever one model fails (here, the otherwise successful liquid-drop model), the problem is relegated to a different model (the shell model).

The liquid-drop model is the main conceptual tool for understanding fission and is, according to all concerned, a quantitative success in explaining why some nuclei are stable, why some undergo fission, and how much energy is released. Many attempts have therefore been made since the early work of Bohr and Wheeler to tweak the model in such a way that the asymmetry of fission fragments could also be explained within the same formulation. Among these suggestions, some have argued that, as the liquid-drop becomes elongated under the influence of the Coulomb repulsion among protons, one lobe of the droplet might be larger than the other. Others have argued that, just before the two more-or-less equal lobes snap apart, the actual scission point in the "neck" connecting them might be closer to one lobe than the other. If all of the nucleons in the neck remain on one side or the other, then the mass asymmetry might be explained as due to the asymmetrical cutting of the neck.

Despite the effort spent on these theoretical manipulations, however, it is widely held that they have not been successful. If an adjustable parameter that explicitly produces the asymmetry is *not* introduced to produce asymmetrical

8.3 Textbook Treatment of Asymmetric Fission

structure in the liquid-drop or asymmetrical cutting of the neck, the liquid-drop analogy inevitably predicts that the nuclear droplet should split into two equivalent fragments. As a consequence, despite the fact that, within the framework of conventional nuclear theory, the liquid-drop model is the logical theoretical choice for explaining collective phenomena such as fission, when the model fails, the collective model theorists admit that this is one of the phenomena that one of the *other* models can undoubtedly handle. As Bohr and Mottelson (1975, p. 369) say, the mass asymmetry "must" be attributed to shell-structure effects. There is no other choice!

But is it the case that, the liquid-drop model having failed, the shell model is successful? This is what the textbooks maintain. For example, in *Neutrons, Nuclei and Matter*, Byrne (1994) asserted that:

> "The most surprising feature of the fission process with thermal neutrons is the marked tendency of the fissioning nucleus to disintegrate into two unequal fragments...
> "The explanation for this peculiar state of affairs did not come to light until about 25 years after the discovery of nuclear fission." (pp. 256–257)

And in *Atomic Nuclei and Their Particles*, Burge (1988) also suggested the solution lies with the shell model:

> "The mass distribution of the uranium fission fragments... revealed that the fission was predominantly asymmetric, a typical example being

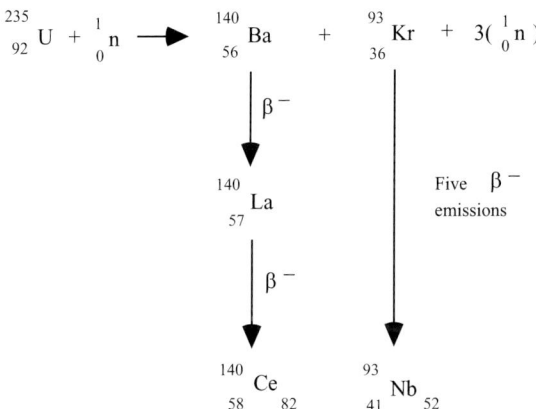

> The fact that ^{140}Ce has 82 neutrons and ^{93}Nb has 52 neutrons is significant since nuclei with Z or N values near certain 'magic' numbers are known to be particularly stable" (p. 109).

Indeed, one selected outcome of this reaction does produce the magic number 82, but the fission event itself produces $^{140}Ba_{56}$, with 84 neutrons and $^{93}Kr_{36}$ with 57 neutrons. Subsequent to the spontaneous readjustment of $Z{:}N$ ratios within the fragments over time, 82 neutrons remain in one stable nucleus, but the fact that 52 is "near" to a magic number does not have any real significance. In so far as the number 50 *is* "magic", it is magic relative to the "unmagicness" of 48 and 52, so the $^{93}Nb_{52}$ fission product has no power for explaining the role of the shell model in producing asymmetric fission. And even the magic number of neutrons in $^{140}Ce_{82}$ is not indication that a closed shell of neutrons was *causally* involved in the mass asymmetry. Certainly, among the many fission fragments there are magic nuclei, but can it be concluded that the shell model has explained anything about the mechanism underlying the asymmetric fission? Let us return to the experimental data to address this question.

8.4 The Empirical Data on Fission Fragments

The facts concerning the numbers of protons and neutrons in the large and small fragments following fission are best seen in the complete data set for any particular isotope, and not from selected examples (such as that above). First of all, it is important to note that, regardless of what kind of symmetrical or asymmetrical break-up occurs, a heavy nucleus such as Uranium begins with a large excess of neutrons over protons (3:2). As a consequence, nearly all fission fragments have neutron excesses that are huge for middle-sized nuclei and are therefore unstable to β^--decay. It is therefore not a surprise that, as the fission fragments themselves undergo internal transformation toward greater stability, some of the fission fragments will end up containing magic numbers of protons or neutrons. In other words, if fission occurs such that a neutron-rich fragment has a few more neutrons than a magic number or a few less protons than a magic number, then β^--decay will inevitably proceed in the direction of those magic numbers. The eventual appearance and stability of those magic nuclei are a simple consequence of (i) the size of the initial fission fragments, (ii) the neutron excess in both fragments, and (iii) the continuing process of β^--decay until stability is reached. The frequency with which certain numbers of protons and neutrons are then observed certainly indicates their relative stability, but it does not explain the asymmetry of the initial fission fragments. It is therefore the sizes of the initial fission fragments that are the basic empirical data in need of examination.

The essential question does not concern precisely where β^--decay comes to a halt, but concerns why the initial fission event was asymmetric (90–100:130–140) in the first place. So, we must go back and ask: what is the abundance of Z and N values *at the time of fission*, not subsequent to β^--decay? We can be sure that the eventual transformation of highly unstable fission fragments into stable nuclei will include a slight excess of "magic" nuclei, but – prior to

8.4 The Empirical Data on Fission Fragments

that drift toward stability – what was the nature of the fission fragments at the moment of fission? Do we find the presence of many "pre-formed" magic nuclei that somehow influence the fission process itself?

To answer this question, the known abundances of the fragments at the time of fission can be examined. The proton data for ^{235}U+n (and other isotopes that undergo thermal fission) were shown in Fig. 8.3. Significant proton peaks are seen at 52, 54 and 56 in the heavy fragment (and 36, 38 and 40 in the light fragment), but not at 50 (or other magic numbers). The neutron data are plotted in Figs. 8.4 and 8.5 for the heavy and light fragments of ^{235}U+n. For neutrons, some heavy fragments (24%) have a magic, closed shell of 82 neutrons at the time of fission, but the remaining 76% of nuclei contain non-magic numbers of neutrons – notably, 83 (13%), 84 (14%), 85 (19%), or 86 (14%) neutrons. The light fragment shows no peaks at magic numbers of neutrons (Fig. 8.5).

Despite the fact that the electrostatic repulsion between protons is the ultimate cause of fission, the data show that, when the most important radioactive isotope ^{235}U undergoes fission (and likewise for all other asymmetrically fissioning nuclei), there are no indications that proton shells cause asymmetric fission. For neutrons, a shell of 82 neutrons is found in about 1/4th of the large fragments following the break-up of ^{235}U, but magic numbers play no role in the majority of cases. It is clear that asymmetric fission is *not* determined by the "break away" of nuclear fragments that contain magic numbers of protons and/or neutrons.

Note that the distribution of nucleons at the moment of fission consistently indicates nucleon numbers that are slightly above the magic numbers (apparent in Figs. 8.3, 8.4 and 8.5; proton distributions in the light fragments show peaks at numbers that are far from any magic shell). Is it possible therefore to fashion an argument along the lines of the special stability of shells of "magic"

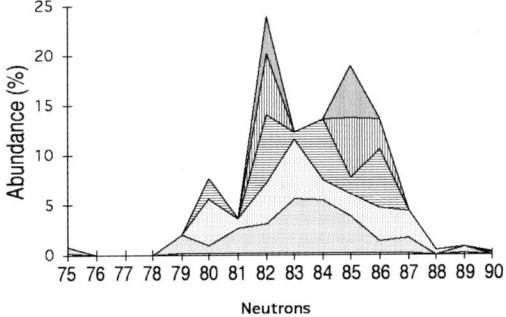

Fig. 8.4. The early abundance of heavy fragments prior to the decay process that leads to stable or semi-stable fragments. The various patterns in the figure represent different isotones

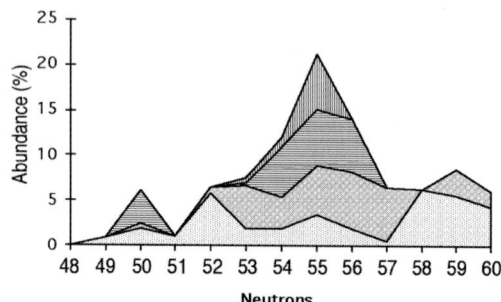

Fig. 8.5. The early light fragments show little sign of $N = 50$ magicness, but the abundance of $N = 55$ is quite strong. The various patterns in the figure represent different isotones

and "magic-plus-a-few nucleons"? This was the implication of the remark by Mayer and Jensen (quoted above), where they noted that fission fragments have nucleon numbers "not less than" the magic numbers. The logic of this argument is, however, not altogether clear. On the one hand, the concept of "magic" stability (for both nucleon and electron shells) is based on the idea that the particles outside of magic shells are loosely bound, while those just within a magic shell are, conversely, strongly bound. Indeed, the experimental data indicate the validity of this idea (Figs. 2.22 and 2.23 showing neutron and proton separation energies). If, therefore, magic stability were a factor contributing to asymmetric fission, the early fission fragments should have tightly bound structures with numbers of nucleons that are *less than or equal to* the magic numbers, while valence nucleons (the nucleons hanging on to magic cores) would be easily stripped away. This would imply a ratio of protons in the heavy and light fragments following fission of Uranium of 50:42 and that for neutrons of 82:62. To the contrary, the data show that when fission occurs in the general vicinity of 50 or 82 neutrons or in the general vicinity of 50 protons, there is an abundance of fragments with neutrons or protons *greater* than the magic numbers (95% of large fission fragments with proton numbers greater than 50 vs. 5% at or below the magic number; 93% of small fragments with neutron numbers greater than 50 vs. 7% at or below the magic number; and 65% of large fission fragments with neutron numbers greater than 82 vs. 35% at or below the magic number). In other words, the presence of extra nucleons beyond the various closed shells that are found in fission fragments is the exact *opposite* of what a naïve shell model would predict. However these numbers may be explained, the empirical data give rather weak support for the idea that tightly-bound closed shells at the magic numbers play a dominant role in fission.

The textbooks that gloss over the fission story with a sentence or two implicating shell effects are not giving an abbreviated form of a well-understood

8.4 The Empirical Data on Fission Fragments   167

phenomenon, they are avoiding an unresolved issue in nuclear theory. A diagram that is commonly reproduced in the textbooks (Fig. 8.6) has some superficial attraction, but is not in fact relevant to the question of why fission is asymmetric. For example, Krane (1988) has noted:

> "Fission is generally treated as a collective phenomenon according to the liquid-drop model, and the analogy with a charged drop of liquid is not only helpful analytically, but also provides a useful mental image of the process. It is therefore perhaps surprising to learn that shell effects play an important and in many cases a decisive role in determining the outcome of fission. As a clue to the importance of shell structure we consider in somewhat more detail the asymmetric mass distribution of the fragments..." (p. 493)

That comment is then followed by a discussion of a figure similar to Fig. 8.6.

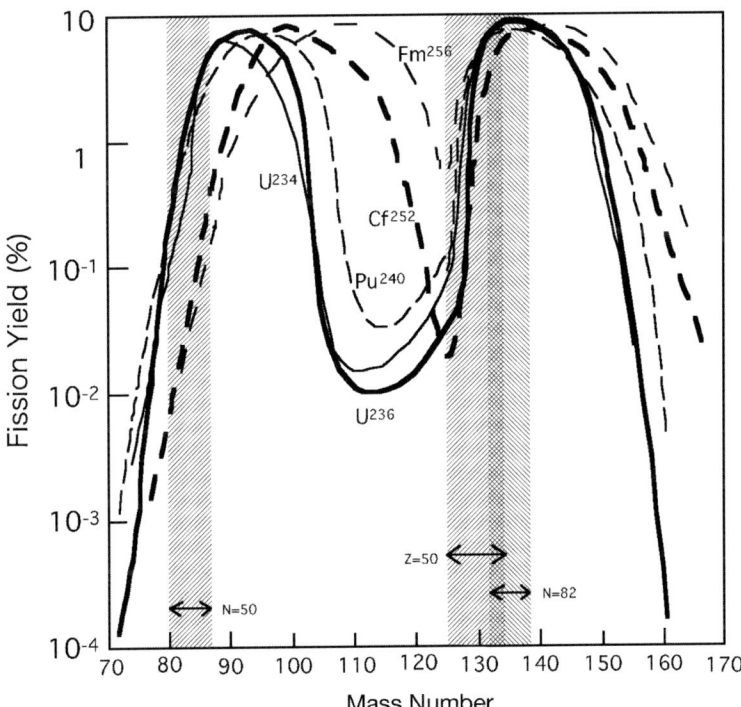

Fig. 8.6. Mass distribution of (post-neutron release) fission fragments showing the predominance of large fragments with $Z = 50$ and $N = 82$ (after Fig. 13.14 in Krane, 1988)

The suggested explanation is that, regardless of which of several parent nuclei we examine, the most common fragments have 50 or 82 protons or neutrons: "this unusual observed behavior lies with the shell model" (p. 494). There is no doubt that among the *final* heavy fragments, there is an abundance of nuclei with proton or neutron numbers *near to* certain magic numbers, but tin ($Z = 50$) is not a major fission product [<5% in ^{235}U+n fission, as compared to 19.5% for barium ($Z = 56$)] and the high frequency of fragments with $A = 132$ is *not* due to an overwhelming abundance of doubly magic $_{82}$Sn^{50} (trace amounts, as compared to 4.3% for $_{88}$Xe^{54}). In any case, the double magicness of $A = 132$ is illusory, because the peaks of the large fragment at the time of the fission event (prior to the release of slow neutrons) is at $A = 134$–144.

Why then is there such eagerness to attribute asymmetrical fission to the shell model when the magic numbers clearly do not play a major role in the initial break-up of the mother nucleus? The answer is the same as it was in the early 1950s. Regardless of the experimental data or the theoretical calculations, since the liquid-drop model inaccurately predicts the *symmetrical* break-up of the fissionable nuclei, the *only* theoretical alternative to the liquid-drop, i.e., the shell model, must be invoked. Examination of the fragment sizes inevitably shows, not an abundance, but a scarcity of traditional magic effects. Rather than admit an unresolved issue, the textbooks hint that a qualitative solution can be found in the shell model, whereas the technical literature suggests that the shell model itself must be revamped to produce different magic numbers (see below).

8.5 Adjusting the Nuclear Potential-Well to Produce Asymmetry

The textbooks are simply in error in arguing that fragment abundances show the importance of magic numbers, but in fact there are more sophisticated shell model arguments that do not rely on the invalid arguments outlined above. In a summary of extensive study of asymmetric fission, Brack et al. (1972) introduced the topic in a manner later taken up by the textbooks:

> "The asymmetry in the distribution of the fission fragments, i.e., the fact that the fissioning nucleus breaks into pieces of unequal size, has early in the history of fission been related to shell structure (e.g., Meitner, 1950, 1952). The lighter fragments are indeed close to a doubly magic nucleus with $N = 82, Z = 50$" (p. 377)

But, as discussed below, their own conclusions pointed in a somewhat different direction. Manipulations of the nuclear potential, corresponding to deformations of the fissioning mother nucleus, can produce energy gaps at various numbers of nucleons – depending explicitly on the extent of a deformation parameter – but, as they noted, "the shells responsible for [fission]

8.5 Adjusting the Nuclear Potential-Well to Produce Asymmetry

have hardly much to do with the magicity of spherical fragments" (Brack et al., 1972, p. 380). Although certain numbers corresponding to theoretically stable "shells" emerge after changes are wrought in the nuclear potential-well – adjusted explicitly to reproduce the experimental mass asymmetries, they are *not* the magic numbers of the shell model – despite their own comments implicating magic numbers 50 and 82!

As with all shell model contentions, modifications of the nuclear potential-well produce interesting results that are difficult to evaluate because of the extreme flexibility of the model, but the essence of the argument can be summarized as follows.

Strutinsky, Nilsson, and others in the 1960s showed that distortions of the nuclear shape imply alterations of the (otherwise spherical) nuclear potential-well and therefore changes in the spacing and sequence of independent-particle states. In large nuclei, a tendency for the nuclear liquid-drop to assume an elongated and, during fission, eventually a barbell-like configuration would mean that the shape of the nuclear potential-well is drastically altered from its spherical shape in most stable nuclei. As a consequence of that distortion, the energy gaps between independent-particle model states will theoretically *not* occur at the numbers found in the traditional shell model using a spherical potential-well (see Fig. 8.7).

Specifically, using an adjustable distortion factor, ε, a *completely new set of magic numbers* can be calculated. When ε is zero, we are given the traditional numbers of the shell model ($\ldots, 28, 50, 82, 126$), but, as distortion increases or decreases ($\varepsilon \neq 0$), the magic numbers change. When $\varepsilon = 0.8$, the new numbers that signify unusual stability are ($\ldots, 24, 36, 48, 60, 80, \ldots$) and these values can then be compared against the experimental data. Most of the new magic numbers calculated in this way (for both protons and neutrons) arise at intervals of 10 or 12 nucleons (beginning at various even numbers, depending upon the value of ε), so that *post hoc* predictions can be produced at almost any even number that is outstanding in the experimental data.

Finding energy gaps at various non-magic numbers, Brack et al. (1972) concluded that the mass asymmetry of fission fragments might be explained by this new set of numbers, but, again, the numbers are *not* the shell model magic numbers. In other words, by changing the nuclear potential-well, as expressed by ε, any number of new magic shell closures can be obtained – among which the experimental values can of course be found. So, in place of the puzzle of the experimental mass ratio found in asymmetric fission, we are given the new puzzle of a distortion factor, ε. The problem has been shifted from one place to another, but in what sense has progress been made?

Most technical discussions that raise the asymmetry problem ultimately suggest that the Strutinsky method for generating new energy gaps in the potential-well – and therefore new sets of magic numbers – is the correct solution, but Strutinsky himself has noted that the asymmetry of fission is related to the magic numbers of traditional nuclear structure theory *only* in the sense that they are both produced using an adjustable nuclear

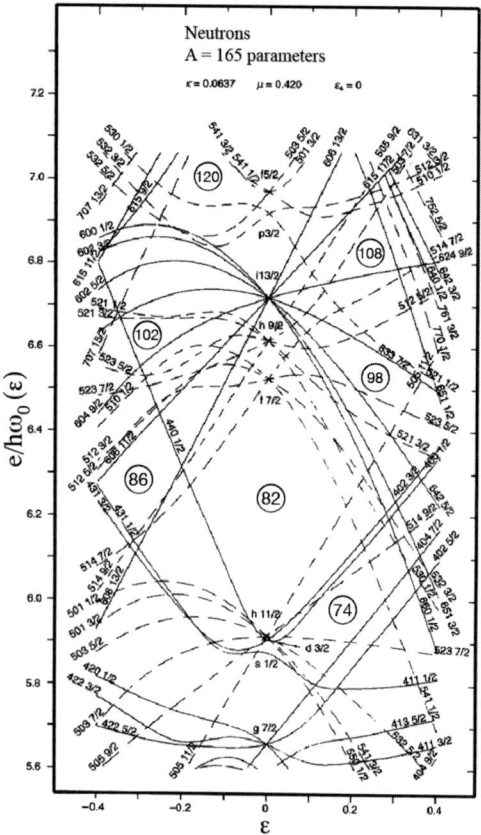

Fig. 8.7. A portion of the single particle levels of an anisotropic harmonic oscillator (Nilsson and Ragnarsson, 1995, p. 122). Either oblate or prolate distortion will produce energy gaps at various numbers that differ from the magic numbers of the shell model

potential-well with variables set to produce the experimentally-known energy gaps. The textbook arguments about traditional shell-model magic numbers are therefore, according to Strutinsky et al., fundamentally *incorrect*, and the band of high abundance around $A = 132$ (Fig. 8.6) has no significance with regard to a contribution by the magic numbers of 50 and 82.

Stated more positively, the Strutinsky argument has demonstrated that the nuclear potential-well technique that lies at the heart of the shell model can be manipulated into agreement with the experimental facts on nuclear fission. That is perhaps an important theoretical result for the shell model, and the work reported by Brack et al. (1972) is clear indication of the flexibility of the shell model approach. However, it cannot be said that manipulation of the shell model parameters to fit the experimental facts is equivalent to deducing the experimental facts from a theoretical model. Preston and Bhaduri's (1975) comment that inclusion of shell corrections leads "to a partial understanding

8.5 Adjusting the Nuclear Potential-Well to Produce Asymmetry   171

of mass asymmetry in low-energy fission" (p. 600) might be the appropriate level of skepticism for shell model explanations of asymmetric fission.

Whatever a "partial understanding" may be, almost two decades later, Brack and Bhaduri (1997) were more bullish on the successes of the shell model approach:

> "The mass asymmetry of the fission fragments, found experimentally for many actinide nuclei, had been a long-standing puzzle since it could not be explained by the liquid drop model.... It is only through the advent of Strutinsky's shell-correction method ... that a quantitative account could be given of the microscopic shell effects..." (p. 374).

But, even in a chapter devoted to such manipulations of the shell model potential in the Brack and Bhaduri (1998) textbook, the explanation of the asymmetry of fission fragments advocated by them does not include any concrete example! Having demonstrated that distortions of the nuclear potential-well produce energy gaps at a variety of numbers that depend on the size of the distortion, the authors declared this to be a "quantitative account". In fact, the actual achievement remains a modest "in principle" argument relying crucially on an adjustable distortion factor.

This rather inconclusive solution to the fragment asymmetry problem has found general acceptance simply because alternatives are lacking, and the promise of a shell model solution where the liquid-drop model has failed is alluring. Still more complex versions of the shell model approach have been developed (e.g., Möller et al., 2001), but always require parameters that directly reflect the "predicted" fragment asymmetry. In the Möller model, fission modes were examined over a 5-dimensional parameter space that included more than 2.6 million parameter permutations! The authors optimistically concluded that:

> "all of these observed fission phenomena can be understood in terms of nuclear potential-energy surfaces calculated with five appropriately chosen nuclear shape degrees of freedom" (p. 786)

But the trick is of course what is meant by "appropriately chosen". This is the same story (with some of the same authors) that lies behind the prediction of super-heavy nuclei. Manipulations of the same multidimensional nuclear potential-well is at the heart of both kinds of calculation, with the principal difference being that asymmetric fission is an established reality (to accommodate which, parameters must be manipulated), whereas the existence or non-existence of the super-heavy nuclei is uncertain (thus making the weighting of model parameters an open issue). Results would be convincing if there were independent criteria for parameter selection, or at least indication that the values chosen for the parameters that predict asymmetrical fission could then predict super-heavy phenomena, or vice versa.

Be that as it may, the computational resources, institutional support, manpower and indeed brain-power behind this kind of computational *tour de force* are impressive and should warm the hearts of Ptolemy and all shell-model-centric theorists of modern times. But reports that the shell model "explains" asymmetric fission – a view already touted in the early 1950s! – understandably leave theorists scratching their heads. On the one hand, most theorists would be happy to believe that the shell model has again (or is it finally?) saved the day, but, on the other hand, there is no realistic possibility of objectively evaluating a model with unverifiable adjustable parameters, individually set for each specific isotope. An unbiased summary of such work might conclude that the shell model is not inconsistent with asymmetrical fission, but prediction of same on the basis of first principles has not yet been achieved.

8.6 What Needs to be Explained?

The experimental data most in need of theoretical explanation were those shown in Figs. 8.2 and 8.3. A portion of those data has been redrawn in Fig. 8.8 to emphasize the discreteness of the proton constituents of the heavy fragments. There are in fact other, extremely precise experimental data that a comprehensive theory of fission should explain (the symmetrical or asymmetrical fission of many other fissionable nuclei [Table 8.1], the distribution of fragments in high-energy nuclear fission, and interesting findings on the numbers of neutrons released during the fission event), but the most precise data on the most important isotopes are shown in Fig. 8.8. Remarkable is the fact that, regardless of which mother nucleus is the source of the fragments, there emerges at the moment of fission (prior to the readjustment of Z:N ratios) three dominant fragments (among the large fragments) containing 52, 54 and 56 protons (often with small contributions at 50 and 58 – and variable numbers of neutrons). There is no comparable regularity regarding neutron constituents or the Z and N numbers of the light fragments.

Assuming that there is a small probability that any fragment will lose or gain one proton in the fission process, the jagged lines in Fig. 8.8 can be represented as the summation of several underlying Gaussian curves with peaks at 52, 54 and 56 protons in the heavy fission fragments. Depending on the initial charge of the mother nucleus, the number of protons in the light fission fragments will vary (32–48 protons), and the final proton/neutron ratios will depend on the various transformations of the neutron-rich fragments over time. Since, however, the repulsion among protons is the ultimate source of the nuclear instability of these nuclei, it is clearly the numbers 52, 54 and 56 that are the only meaningful "magic" numbers in the fission story. If those numbers can be explained, then the charges of the small fragments are necessarily understood, and the variable number of excess neutrons at each of these proton numbers might also find easy explanation.

8.6 What Needs to be Explained? 173

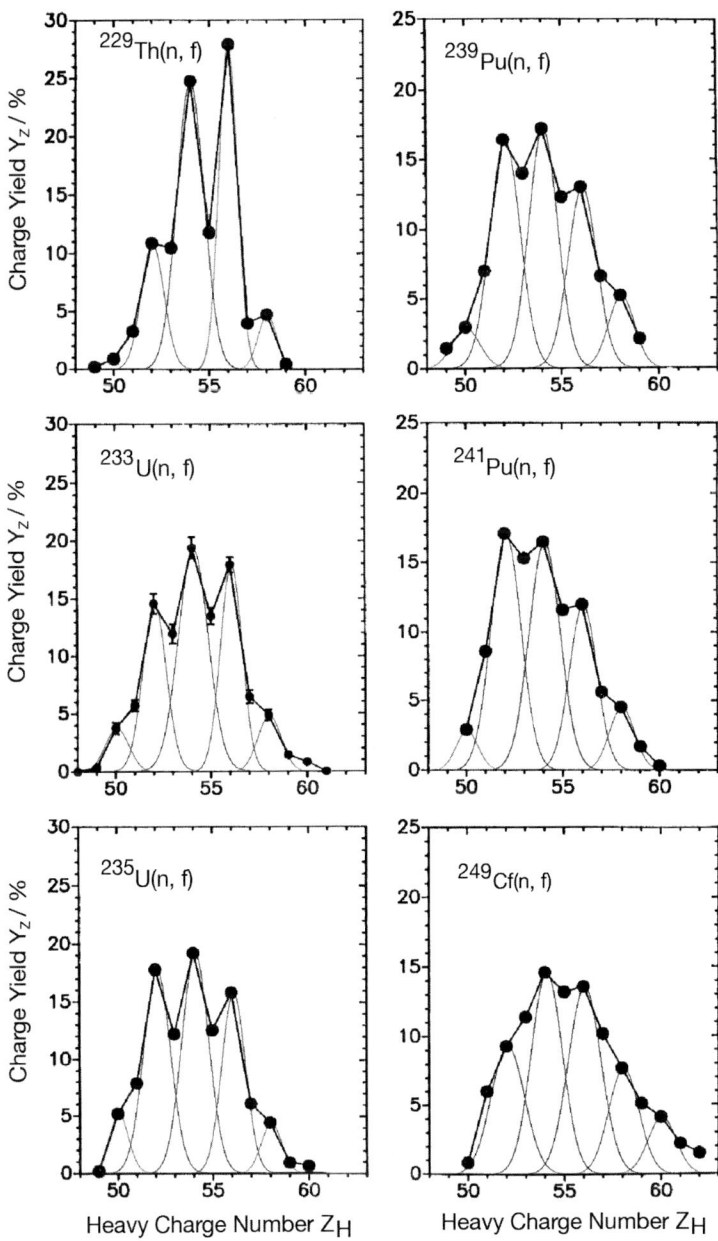

Fig. 8.8. The data on the most important fission fragments (as also shown in Fig. 8.3), with emphasis drawn to the 4 or 5 even-Z fragments that account for the charges of the heavy fragments. Why do large nuclei fission primarily into fragments containing 52, 54 and 56 (± 1) protons? (experimental data from Gönnenwein, 1991, p. 409)

8.7 Summary

Elaborations on the liquid-drop and shell models have not solved the problem of asymmetrical fission and the theoretical explanation of fission products remains incomplete. Because of the technological importance of fission, the precision of the available experimental data is excellent and provides theorists with the kinds of empirical constraints that are normally welcomed. Unfortunately, manipulations of the conventional models have not led to an understanding of why uranium and most other fissionable nuclei break into uneven pieces. Despite hopeful suggestions in the textbooks that the shell model can explain the asymmetry, most experts on fission acknowledge the continuing dilemma presented by the asymmetrical fragments and insist that theoretical work remains to be done.

If it is surprising that the mechanisms of a phenomenon as important as fission are still uncertain, it should be remembered that, as long as fundamental aspects of nuclear structure and particularly of the nuclear force remain unclear, inevitably certain details of nuclear dynamics will also remain ambiguous. Given that the empirical reality of fission and the technology used to control it are well-established, the danger of catastrophic nuclear accidents caused by this theoretical confusion is non-existent, but it is clear that scientific progress depends crucially on the identification and highlighting of those topics that current scientific theory cannot explain. In other words, the lack of understanding concerning asymmetric fission does not imply a crisis in applied physics, much less world affairs, but it is indication of where greater clarity is yet possible. Although the traditional magic numbers of the shell model do not, in fact, explain asymmetrical fission, the fragment asymmetry is strong indication of nuclear substructure playing a role in the fission process – substructure of a kind that apparently neither a nuclear gas nor a nuclear liquid can explain.

Part III

The Lattice Model

Chapters 1–4 outlined the basic issues in nuclear structure physics, and Chaps. 5–8 examined more closely four areas where significant theoretical problems have remained. Each of these problem areas has been addressed many times using one or the other of the nuclear models, but with only limited success. The short mean-free-path of intranuclear nucleons can be accounted for easily within a liquid-drop model of the nucleus, but if a liquid-drop is the approximate texture of the nucleus, then how can the experimental data indicating distinct shells and subshells of orbiting nucleons in the nucleus be explained? Nuclear total binding energies are also well accounted for by the liquid-drop model, but the experimental evidence for magic numbers indicates variations in nuclear stability that an amorphous liquid-drop cannot explain. The cluster models work quite well for explaining the structure and vibrational states of certain small nuclei, but how can a rigid cluster geometry be reconciled with the structureless texture implied by a liquid-drop or a Fermi gas? The radial values of nuclei are consistent with the liquid-drop model, but the low-density skin region appears to be more like a diffuse gas than a dense liquid. And, finally, the energy released in nuclear fission is accounted for by the liquid-drop model, but that model predicts symmetric fission. How can the mass asymmetry of fission fragments be explained within the context of the shell model without negating the energetic successes of the liquid-drop model? The reality of these conceptual difficulties has been emphasized in Part II without indication of how problems might be resolved. Here, in Part III, possible solutions and a unification of nuclear models is outlined within the framework of a lattice model.

The upshot of the two following chapters is that, despite the existence of the problems discussed in earlier chapters, nuclear structure theory is in fact not far from being a self-consistent, coherent body of thought. To achieve true coherency, however, the known problems must be squarely addressed and the relationship between individual nucleon states and nuclear states must be clarified. Among the known problems, the most important is the long-standing contradiction of maintaining the fiction of an "effective" nuclear force that is weak and long-range and, simultaneously, a "realistic" nuclear force that is strong and short-range. Nuclear physics cannot proceed toward unification until that paradox is resolved. Moreover, the "grand unification" of the forces of nature that is the goal of theoretical physics in general cannot be achieved if the nuclear component is incoherent.

In the next two chapters, the problems of nuclear theory will be addressed within the framework of a specific lattice in which the nuclear force is strong and short-range (essentially, the force known from two-body nuclear reactions and implied by the liquid-drop model), while the independent-particle properties of the shell model that have made it so important to nuclear theory are explained in terms of the geometric properties of the lattice. It is this combination of factors that will lead us toward a more coherent understanding of the nuclear system.

Several lattice models were briefly discussed in Chap. 4 to illustrate the merits of a lattice approach, in general, and to show the unique contribution of lattice simulations of high-energy heavy-ion multifragmentation data, in particular. Within that rather narrow field, lattices have been shown to be useful computational tools, but the lattice approach – more than the other nuclear models – has been advocated explicitly as a computational technique, rather than as a comprehensive explanation of nuclear structure. In the following two chapters, one variety of lattice model is developed as a unifying nuclear model. Given certain constraints concerning the internal structure and dynamics of the lattice, it can be shown that the lattice contains all of the principal strengths of the established nuclear structure models and, in addition, is capable of explaining various high-energy multifragmentation phenomena. In other words, I argue that the lattice might be the basis for a general model of the nucleus – not restricted to nuclei of a particular size, not restricted to nuclei near to or far from the magic numbers, and not restricted to phenomena of a certain energy range.

Before issues of unification are discussed, a less controversial claim about the lattice model will be addressed. That claim, in its simplest formulation, can be stated as follows: *The conventional application of quantum mechanics to the nuclear realm results in a pattern of shells, sub-shells and symmetries with specific particle-occupancies, and that entire pattern is reproduced in a particular lattice.* Based upon the isomorphism between quantum mechanics and the lattice, theoretical calculations made using the lattice will be discussed in Chap. 10 in relation to the experimental data. Prior to asking whether or not the lattice is a quantitative success as a nuclear model, however, the relationship between the lattice and quantum mechanics needs to be examined. This is the main topic of Chap. 9.

9

The Lattice Model: Theoretical Issues

9.1 The Independent-Particle Model Again

The independent-particle model has been first among equals in nuclear theory for more than 50 years because it is fundamentally quantum mechanical. As a realistic model of nuclear structure, it has strengths and weaknesses, but unlike the liquid-drop and cluster models, it is explicitly built from the Schrödinger wave equation and therefore has a theoretical "purity" that the other models do not have. As a consequence, the shell model does not rely solely on the fit between theory and experiment: insofar as quantum mechanics is the universally-acknowledged correct theory of reality at the atomic level, the shell model is theoretically sound from the ground up.

Let us return to the foundations of the shell model to examine what the potential-well of the nuclear force implies for nuclear structure. It is this concept that provides a quantum mechanical basis for the existence of nucleons with individual quantized properties and nuclei with distinct energy shells. Together with the assumption of spin-orbit coupling, many known properties of nuclei can then be reproduced within the framework of the theory.

The time-independent Schrödinger wave-equation for describing a particle can be written as:

$$-(h^2/2m)(d^2\Psi/dr^2) + V(r)\Psi(r) = E\Psi(r) \tag{9.1}$$

and expresses the idea that the wave, $\Psi(r)$, has an energy, E, which is a function of the particle's mass, m, acted on by a potential, $V(r)$. The simplest one-dimensional potential is the infinite well, such that the particle either feels the potential, $r = 0$, and oscillates back and forth over some range, or it does not feel the potential, $r = \infty$, and is unaffected. If the particle feels the potential, then there is a series of discrete oscillatory states, the energy of which can be defined as:

$$E_n = (h^2/2m)n^2 \tag{9.2}$$

178   9 The Lattice Model: Theoretical Issues

The particle described by the Schrödinger equation is generally thought of as existing as a standing wave, and its energy states are indexed with the quantum number, n, that can be any non-negative integer.

The wave-equation can be stated in various forms for ease of computation. Assuming the particle is confined to a 3D box, (9.1) can be rewritten to define the particle's position in Cartesian coordinates:

$$-(h^2/2m)(d^2\Psi/dx^2 + d^2\Psi/dy^2 + d^2\Psi/dz^2) = E_n\Psi(x,y,z) \qquad (9.3)$$

When using more realistic potentials that change gradually with distance, the mathematics of the wave-equation becomes more complex, but the basic idea is straight-forward: there are certain oscillatory states (essentially, integral numbers of sine waves) that are stable and all in-between, non-integral waves cancel out. The solutions that satisfy this equation in the one-dimensional case have a quantum number, n, that specifies the number of sine-waves. Using the idea of a harmonic oscillator, the energy of each state can then be defined as:

$$E_n = h\omega_0(n + 1/2) \qquad n = 0, 1, 2, 3, \ldots \qquad (9.4)$$

The solutions of the three-dimensional harmonic oscillator can be described in terms of three numbers, n_x, n_y and n_z. The total energy is then 3-fold the one-dimensional case:

$$E_N = h\omega_0(n_x + n_y + n_z + 3/2) = h\omega_0(N + 3/2) \qquad N = 0, 1, 2, 3, \ldots \qquad (9.5)$$

In other words, the energy states of the wave-equation in three dimensions depend only on the sum N, which is consequently called the principal quantum number. Different combinations of n_x, n_y and n_z that give the same total N-value denote spatially distinct "degenerate" states, with the same energy. By defining the allowed states as all odd-multiples of sine waves, a set of oscillatory states within the box are found to have occupancies related to the known energy levels of the nucleus. The first seven shells are listed in Table 9.1, together with the occupancies produced by the so-called degeneracy of spin and isospin.

In the framework of the shell model, energy states are generated using a harmonic oscillator potential (Fig. 9.1). The 3D geometry of the wave-equation becomes highly complex (analogous to the convoluted orbitals of electrons, Fig. 2.4), but the multiplicity and occupancy of nucleon states are the same as those shown in Table 9.1. Several general points about the energy levels in Table 9.1 and Fig. 9.1 are worth mentioning. As the nucleon moves to higher shells, its energy level increases in discrete jumps (3/2, 5/2, 7/2, ...). At the same time, the number of nucleons that can reside at each level increases due to the fact that the number of permutations of the coordinate indices (or n and l) increases. Prior to consideration of the degeneracies brought by spin and isospin, there is only one state possible at the lowest level, but 3, 6, 10, 15, 21, etc. states for higher levels. In other words, gradually more and more

9.1 The Independent-Particle Model Again

Table 9.1. Energy levels of a particle confined in a 3D box

Energy $(E/\hbar\omega_0)$	N-shell $(n_x+n_y+n_z-3)/2$	Wave-functions (n_x, n_y, n_z) *	No. of Distinct Wave-functions	Spin Degeneracy	Isospin Degeneracy	Total Occupancy
15/2	6	(13 1 1)(1 13 1)(1 1 13) (11 3 1)(11 1 3)(3 11 1) (1 11 3)(3 1 11)(1 3 11) (951)(915)(591) (195)(519)(159) (933)(393)(339) (771)(717)(177) (753)(735)(357) (537)(573)(375) (555)	28	56	112	336
13/2	5	(11 1 1)(1 11 1)(1 1 11) (931)(913)(391) (193)(319)(139) (751)(715)(571) (175)(517)(157) (733)(373)(337) (553)(535)(355)	21	42	84	224
11/2	4	(911)(191)(119) (731)(713)(371) (173)(137)(317) (551)(515)(155) (533)(353)(335)	15	30	60	140
9/2	3	(711)(171)(117) (531)(513)(351) (153)(315)(315) (333)	10	20	40	80
7/2	2	(511)(151)(115) (331)(313)(133)	6	12	24	40
5/2	1	(311)(131)(113)	3	6	12	16
3/2	0	(111)	4	1	2	4

* The entire set of wave-functions can also be defined using all positive integers, with the N-shells defined as $N = (n_x + n_y + n_z - 2)$, but using the odd-integers shows a more direct connection to the fcc lattice model (described below).

distinct states at the same energy level become possible as a consequence of the quantization and the increasing number of permutations that sum to N.

The changing occupancy of shells due to the integers, n_x, n_y and n_z, (or n and l) is the first topic noted in all discussions of the shell model, but there is further degeneracy caused by spin and isospin. For every distinct state, each nucleon can exist in either a spin-up or spin-down state *and* there are two types of nucleon – protons and neutrons (isospin-up or isospin-down) – that can co-exist in each state. Such degeneracy means that four times the number of states are allowed in each N-shell of the Schrödinger equation.

Although nearly all textbooks on nuclear physics reproduce a table similar to Table 9.1, they usually list only the spin, and not the isospin degeneracy.

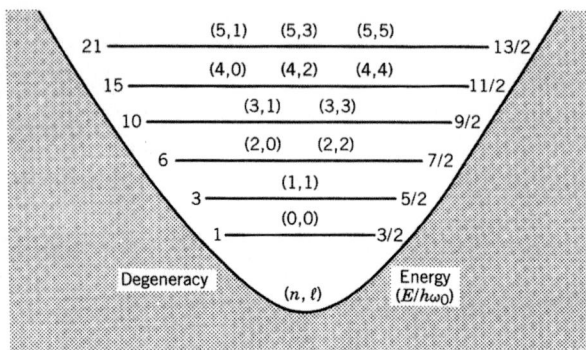

Fig. 9.1. The energy levels and degeneracy of the harmonic oscillator potential that is the basis of the shell model (Krane, 1988, p. 33). Note that the occupancy of the various energy levels is obtained through combinations of n and l, but the results are identical to those obtained with n_x, n_y and n_z using an infinite Cartesian well, as summarized in Table 9.1

Instead, the comment is made that a similar build-up procedure applies to both protons and neutrons. Since, however, we are interested in the issue of *total occupancy* of various energy levels, it is somewhat misleading to deal with isospin in a manner different from spin, as if protons and neutrons somehow occupied distinct realms within the nucleus. They do not, and they are both defined by the same Schrödinger equation.

The reason for the initial neglect of isospin degeneracy in textbooks is that the discussion of the particle in a box and the harmonic oscillator is usually a prelude to an explanation of shell closure and the idea of magic numbers for protons or neutrons. Empirically, the concept of magicness applies to either protons (Z) or neutrons (N), but *not* to total nucleons (A), so that the summation of protons and neutrons – even when doubly-magic nuclei are produced – is not of major theoretical significance. It should be noted, however, that as important as the empirical issue of proton or neutron magicness may be, the first order of business is to examine the quantum mechanical nature of nucleon states – all forms of which, including isospin, are relevant.

The patterns outlined in Table 9.1 and Fig. 9.1 mean that, given the wave-equation, a series of distinct energy levels and their occupancies can be deduced and compared with experimental data. As described briefly in Chap. 4, the independent-particle model requires one further assumption, spin-orbit coupling, in order to explain the empirical shell structure of the nucleus. That is, the total angular momentum of each nucleon is assumed to be the sum of the nucleon's intrinsic spin and its orbital angular momentum, $j = l + s$, that is experimentally observable. The known sequence of j-levels then matches well with the shell model sequence (Fig. 9.2).

While it remains to consider the spatial configuration of nucleons in nuclei, it can be unequivocally stated that the discrete j-levels are the foundation

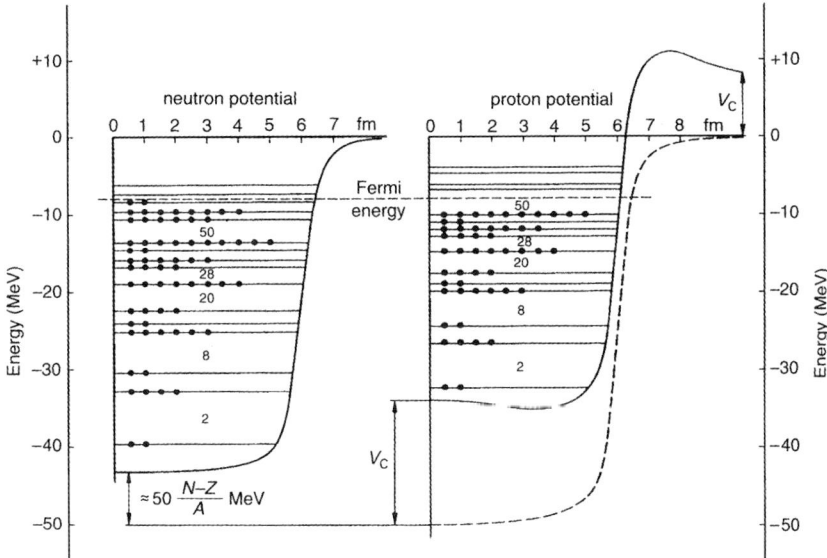

Fig. 9.2. The build-up sequence for protons and neutrons implied by the independent-particle model (Nilsson & Ragnarsson, 1995, p. 64)

from which the shells in the independent-particle model are explained. Each nucleon resides in an n-shell, with a fixed total angular momentum, j, that is the sum of its orbital (l) and intrinsic (s) angular momenta (spin-orbit coupling), and with an orientation, denoted by m. Together with its intrinsic spin quantum number, s, and its intrinsic nucleon character quantum number (isospin), i, each nucleon can be described as being in a unique state, specified by the five quantum numbers, $njms$ and i. Individual nucleons can exist in any state, where: $n = 0, 1, 2, \ldots$; $j = 1/2, 3/2, 5/2, \ldots$; $m = \pm 1/2, \pm 3/2, \pm 5/2, \ldots$; $s = \pm 1/2$; and $i = \pm 1/2$, with certain restrictions on the various permutations. In principle, any nucleus can therefore be described as the summation of the quantum values of its "independent" nucleons.

The extreme simplicity of the independent-particle model and the excellent agreement between theory and experiment concerning the nuclear build-up procedure are merits that theorists are understandably unwilling to abandon. Despite a variety of theoretical problems generated by the starting assumption of the central potential-well, this comprehensive scheme provides a quantum mechanical foundation for nuclear theory and leaves no doubt that nuclei are reasonably described as the summation of their independent-particle states.

9.2 Reproduction of the Independent-Particle Model in an fcc Lattice

The regularities of nuclear energy levels, as outlined in Table 9.1, are the foundations of modern nuclear theory. The question that arises is whether a self-consistent geometry of the nucleus emerges from this description of nucleon states. There certainly is a self-consistent description in an abstract 5-dimensional "quantal space" – as embodied in the Schrödinger equation and exploited in the independent-particle model – but what about the Cartesian geometry of the nucleus itself?*

The surprising answer is that the geometry of the n_x, n_y and n_z values in Table 9.1 produces a regular lattice that exhibits exactly the same energy levels and occupancies as the wave-equation (Cook and colleagues, 1976–1999; Dallacasa & Cook, 1987). Indeed, the wave-function subscripts in Table 9.1, n_x, n_y and n_z, are the absolute values of the integers corresponding to one octant of Cartesian space, but can be generalized to all nucleon coordinates, using permutations of positive and negative coordinates depending on the shell number. The geometry of this isomorphism between the symmetries of the Schrödinger equation and the symmetries of the lattice is described below.

It will be seen that the entire pattern of nucleon states in the independent-particle model, outlined above, is reproduced (without exceptions or special rules for light or heavy nuclei, deformed or undeformed nuclei, ground-state or excited-state nuclei) *in an antiferromagnetic face-centered-cubic (fcc) lattice with alternating isospin layers (the fcc model)*. The correspondence is precise and unambiguous, and is the basis for all model predictions. Until the isomorphism between known characteristics of nuclear "quantum space" and this particular lattice is understood, however, the theoretical possibility that a lattice might be the basis for a realistic model of the nucleus cannot be appreciated. Once that connection is clear, then many questions concerning model predictions and empirical data arise, and they will be discussed in Chap. 10. The theoretical arguments in the present chapter, however, should leave no doubt that there exists a lattice representation of the known quantum mechanical symmetries of the nucleus – a one-to-one mapping between abstract

---

*The 3D structure of the electron orbitals is well-known (e.g., Fig. 2.4), but already in the case of electrons (that occupy only 1/100,000th of the atomic volume) the shapes of the textbook orbitals are strongly distorted by the presence of other electrons. The best example of such local effects can be seen in the carbon atom, whose 4 valence electrons nearly always take on a tetrahedral geometry (107° bond angles), despite the fact that the uncorrelated electron orbitals would predict two strongly-overlapping spherical *s*-orbitals and two orthogonal *p*-orbitals forming a 90° bond angle. In other words, the spatial configuration of electron orbitals is distorted by local interactions. In the nuclear case, again the wave-equation predicts spatially overlapping nucleon orbitals, but the relatively strong local effects among nucleons will inevitably alter the geometrical configuration of nucleon states.

9.2 Reproduction of the Independent-Particle Model in an fcc Lattice

quantum mechanical symmetries and a 3D lattice that went unnoticed for the first half century of research in nuclear structure physics.*

Let us examine the geometry of the quantum numbers of the nucleons in a close-packed lattice, and then return to issues of nuclear structure.

The Principal Quantum Number, n

The principal quantum value, n, indicates the basic energy shell within which the nucleon is found. The n-value in the Schrödinger wave equation can in principle be any non-negative integer, $n = 0, 1, 2, 3, \ldots$, and the number of nucleons with each value (the total occupancy) can be defined as:

$$\text{occupation } (n) = 2(n+1)(n+2) . \tag{9.6}$$

The numbers of nucleons per n-shell in the wave-equation are therefore: 4, 12, 24, 40, 60, 84, 112, 168 and so on – implying a total occupancy at the closure of each shell of 4, 16, 40, 80, 140, 224, 336, and so on (see Table 9.1).

Given that the shell model implies a nucleon build-up procedure that produces these numbers (and finds abundant experimental support in the known properties of nuclei), the validity of (9.6) is well-established. Let us now look at how that nucleon build-up sequence might be represented in a lattice of nucleons. If the origin of the coordinate system is taken as the center of a tetrahedron of particles and all additions of nucleons to the central tetrahedron are in accordance with a face-centered-cubic close-packed lattice, then it is found that the closure of each consecutive, symmetrical ($x = y = z$) geometrical shell in the lattice corresponds precisely to the numbers of nucleons in the shells derived from the three-dimensional Schrödinger equation, as listed above.

---

*It is likely that the isomorphism between nucleon states and the fcc lattice has been discovered independently many times. The earliest "near discovery" was made by Eugene Wigner and published in *Physical Review* in 1937. There, he noted the patterns of quantum numbers in an early version of the independent-particle model in relation to the close-packed geometry of spheres in a 2D array. Unfortunately, Wigner never took the final step of stacking up the 2D layers – which would have given him the fcc geometry. Not until 1974 was the 3D argument published in the physics literature – in an obscure journal, *Atomkernenergie*. There, and in subsequent papers, Klaus Lezuo, a nuclear physicist in Mainz, Germany, argued that the close-packing of "Gaussian nucleon probability-spheres" – both face-centered-cubic and hexagonal-close-packed lattices – reproduces various nuclear features. At least three other independent discoveries of the fcc nuclear lattice have been reported. Peter S. Stevens of Harvard University circulated a preprint in 1972 – a copy of which reached me following distribution of a similar paper of my own on the "geometry of the nucleus" in 1971 (eventually published in 1976). And in the early 80s, Andrew Bobeszko, a Romanian physicist living in London, developed a similar model based on the fcc lattice. Those are the discoverers of the fcc nuclear symmetries that are known to me personally, so there are undoubtedly more.

A build-up procedure that places particles uniquely at the sites of an fcc lattice allows for a definition of the n-value of a particular nucleon, k, simply in terms of its x, y and z coordinate values:

$$n_{\text{nucleon}(k)} = (|x_{\text{nucleon}(k)}| + |y_{\text{nucleon}(k)}| + |z_{\text{nucleon}(k)}| - 3)/2 \tag{9.7}$$

assuming only that the x, y and z coordinates are odd-integers. These fcc shells are shown in Fig. 9.3.

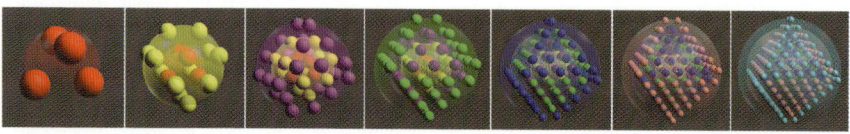

Fig. 9.3. The first seven $x = y = z$ symmetrical structures in the fcc model: ^{4}He, ^{16}O, ^{40}Ca, ^{80}Zr, ^{140}Yb, ^{224}Xx and ^{336}Xx. The smaller n-shells can be seen within each shell (although the overall sizes of the nuclei have been scaled to occupy approximately the same volume)

As shown in Fig. 9.3, these structures have octahedral ($x = y = z$) symmetry. The number of nucleons in the first three n-shells correspond to doubly-magic ($Z = N$) nuclei. For the larger structures, the closed n-shells do *not* correspond to magic numbers, but rather to the closed shells of the particle in a box (or harmonic oscillator). The empirical issue of deciding on what numbers are "magic" is rather complex (Chap. 2) and will be addressed again below, but, for now, the main point is that there is a correspondence between the symmetries of n that are inherent to the Schrödinger equation and this particular lattice structure. By referring to Table 9.1, it is seen that the correspondence for n-shells is exact for $n = 0 \sim 6$, and in fact continues indefinitely for any value of n.

What does the correspondence between particle build-up in (9.6) and (9.7) and the geometry of the fcc lattice mean? Clearly, if a lattice theory of nuclear structure makes any sense, the quantal properties deduced from the Schrödinger equation (and reproduced in the independent-particle model) should be reproducible in the lattice. The fcc model achieves this for the n-shells on a simple geometrical basis. Alone, such an identity may not be a sufficient grounds for elaborating an entire theory of nuclear structure, but an identity between n-shells and geometrical symmetries is certainly a *necessary* condition. Stated contrarily, all other lattices and alternative build-up procedures fail at this first step. Whether one starts with a simple-cubic, body-centered-cubic, diamond, hexagonal-close-packed or other build-up of polyhedrons (e.g., Wefelmeier, 1937; Anagnostatos, 1985), geometrical shell structures are obtained at various numbers, but never at those indicated by (9.6). The fcc lattice, as shown in Fig. 9.3, produces closed ($x = y = z$) shells uniquely at those numbers.

9.2 Reproduction of the Independent-Particle Model in an fcc Lattice

Below, we will see that the closure of *magic* shells deviates from the n-shells of the Schrödinger equation, principally because of: (i) the need for excess neutrons ($N > Z$) to hold the mutually repulsive protons in the nucleus for all nuclei with more than 20 protons, and (ii) the influence of protons on the neutron build-up sequence, and vice versa. In any case, the important point is that there is a correspondence between the symmetries implied by the n-subscript of the equation that lies at the heart of all quantum mechanics and the fcc lattice. Stated contrarily, if a lattice build-up produced uniquely the doubly-magic numbers $(4, 16, 40, 56, 100, \ldots)$, then it would *not* produce the symmetries of the Schrödinger equation. Clearly, the more fundamental reality is the wave-equation, while the issue of magic numbers is an empirical issue that must be dealt with separately.

The fcc coordinate system can be defined as the sites generated by three odd-integers whose product is a positive or negative number (depending on the shell number). The first n-shell contains four nucleons whose coordinates are ($\pm$) permutations of 1:

$$111, \quad -1-11, \quad 1-1-1, \quad -11-1$$

forming a tetrahedron at the origin of the coordinate system, as shown in Fig. 9.4a. The product of the coordinates of each of the lattice sites in the first shell is a positive number. If all combinations of ± 1 were allowed, the first shell of the lattice would contain 8 sites, and form a simple cubic packing (scp) lattice with twice the occupancy for each n-shell. (This turns out to be an important issue in defining the lattice-*gas* model, and is addressed in Sect. 9.4, below.)

The second shell ($n = 1$) consists of 12 lattice sites that are permutations of ± 1 and ± 3 (whose sum is five and whose product is a negative number).

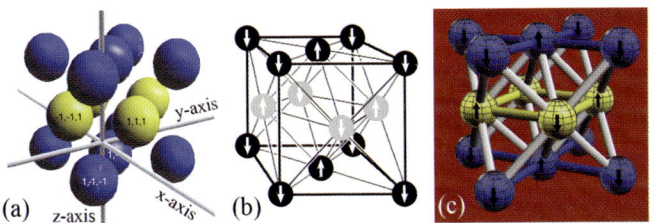

Fig. 9.4. On the *left* is shown the fcc lattice in relation to Cartesian coordinate axes (protons are yellow, neutrons are blue). Note that there is no nucleon at the origin of the coordinate system. Rather, straddling the origin is a tetrahedron of 4 nucleons (coordinate values shown). Although the unit cube of the fcc lattice is not symmetrical around the origin, this "misalignment" of the cube has no real significance. The cubic shape plays no role except as the conventional label for this particular close-packing scheme. In the *middle* diagram, lines are drawn between both first- and second-nearest neighbors, thereby emphasizing the unit cube. On the *right* is shown the same cube with only nearest-neighbor bonds drawn

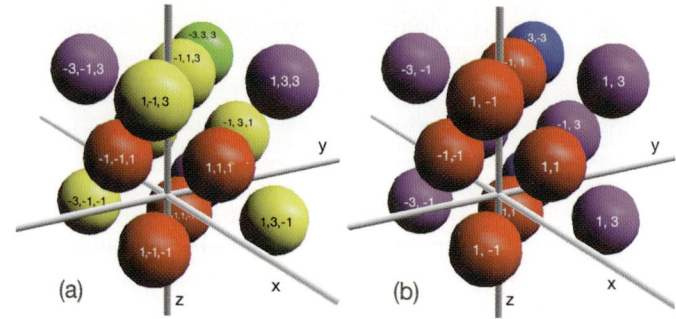

Fig. 9.5. (a) shows that each nucleon's n-value in the fcc lattice is a function (9.7) of its x-, y-, and z-coordinate values. (b) shows that the nucleon's j-value is a function of the x- and y-coordinate values only (9.9). The colors denote different n-shells (a) or j-shells (b)

Specifically, the next shell of fcc lattice coordinates are:

$$31-1, \quad 3-11, \quad -311, \quad -3-1-1, \quad 1-31, \quad -131, \quad 13-1,$$
$$-1-3-1, \quad -1-13, \quad 11-3, \quad 1-13, \quad -11-3$$

The third shell ($n = 2$) contains 24 sites that are permutations of ± 1, ± 3 and ± 5 (whose sum is seven and whose product is a positive number again), and so on. The coordinates for the 14 nucleons of the fcc unit cube can be seen in Fig. 9.5a.

The dependence of n on x, y and z means that the greater the distance of a nucleon from the nuclear center (Fig. 9.6), the greater the value of its principal quantum number, n.

As illustrated in Figs. 9.3 and 9.6, the n-shells are approximately octahedral in shape (more precisely, tetrahedrally-truncated tetrahedrons). Two

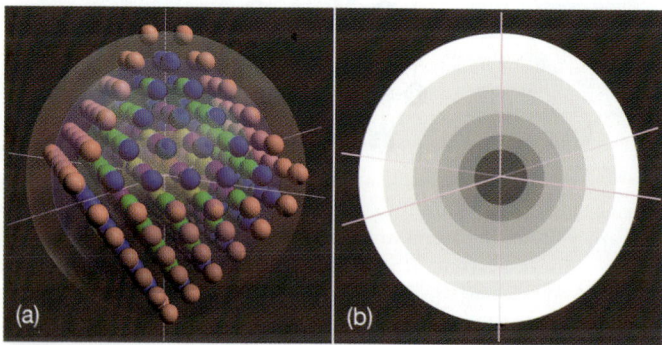

Fig. 9.6. The consecutive n-shells of the fcc lattice. The truncated tetrahedrons of nucleons in each shell are rotated 90 degrees around the vertical spin axis in each consecutive shell (a), giving an approximately spherical build-up of n-shells (b)

9.2 Reproduction of the Independent-Particle Model in an fcc Lattice

points should be emphasized about this symmetry: (i) all of the completed n-shells have $x = y = z$ symmetry (their geometry is identical viewed down the x-, y- and z-axes); and (ii) the occupation numbers are identical to those in the independent-particle model. The 3D symmetry of all of the closed n-shells is particularly hard to appreciate in 2D drawings, but is evident in 3D structures and in the NVS graphical software on the CD.

In other words, the "shells" of the independent-particle model as defined by the n-values of nucleons in the Schrödinger equation are literally geometrical shells in Cartesian space (Fig. 9.6). The entire sequence of closed n-shells of the lattice is not identical to the "magic" shells of the shell model for exactly the same reason that the harmonic oscillator produces the quantum mechanical degeneracy shown in Table 9.1 and Fig. 9.1, and not the magic numbers. Since the magic numbers are, however, found in the *subshells* of the harmonic oscillator (and in *subshells* of the lattice, see below), there is a means for explaining the special stability of the magic nuclei in the lattice that is analogous to the shell model explanation. On the basis of first principles, neither the harmonic oscillator nor the lattice predicts the experimentally-known magic numbers, but both models are consistent with unusual stability at certain combinations of N and Z when other factors are brought into consideration. Specifically, both models predict a quantal "texture" in the nucleus, with regularities of nuclear structure arising at various shells and subshells (see Table 9.2).

A Comment on the Magic Numbers

If the build-up discussed above reproduced not only the first three magic numbers 2, 8, and 20, but also the next four, 28, 50, 82, 126, then the lattice model would undoubtedly have long ago become a central pillar of nuclear structure theory – because the stability at the magic numbers would then be a simple consequence of the greater bonding of closed geometrical structures. In fact, the correspondence between the fcc lattice and magic numbers continues only up through the third energy level, and afterward requires consideration of subshells. Although maximization of the binding energy in compact structures appears to be the principle underlying magicity in all shells, the need for a neutron excess ($N > Z$) for medium and large nuclei means that there are no stable $N = Z$ nuclei larger than ^{40}Ca. The unequal numbers of protons and neutrons for all larger nuclei therefore make the geometry of bonding maximization more complex.

Consider, however, what the correspondence between the lattice and the wave-equation represents. The geometrical shells of the lattice unambiguously reproduce the basic energy (n) shells that are a direct implication of the Schrödinger wave-equation. Stated conversely, quantum mechanics itself implies doubly-closed shells at 4, 16, 40, 80, 140, 224, and so on. The fact that experimental data do not support this prediction for the larger n-values is not seen as grounds for discarding quantum mechanics, but rather grounds for

Table 9.2. The shells and subshells (and their occupancies) in the independent-particle model and in the fcc model (Cook & Dallacasa, 1987)

	n-shells $n=(|x|+|y|+|z|-3)/2$						j-subshells $j=(|x|+|y|-1)/2$							m-subshells $m=|x|/2$							Total
	0	1	2	3	4	5	$\frac{1}{2}$	$\frac{3}{2}$	$\frac{5}{2}$	$\frac{7}{2}$	$\frac{9}{2}$	$\frac{11}{2}$	$\frac{13}{2}$	$\frac{1}{2}$	$\frac{3}{2}$	$\frac{5}{2}$	$\frac{7}{2}$	$\frac{9}{2}$	$\frac{11}{2}$	$\frac{13}{2}$	
	2						2							2							2*
								4						2	2						6+
		6					2							2							8*
									6					2	2	2					14+
								4						2	2						18
			12				2							2							20*
										8				2	2	2	2				28*
									6					2	2	2					34
								4						2	2						38
Number				20			2							2							40+
of											10			2	2	2	2	2			50*
protons										8				2	2	2	2				58
or									6					2	2	2					64
neutrons								4						2	2						68
					30		2							2							70+
												12		2	2	2	2	2	2		82*
											10			2	2	2	2	2			92
										8				2	2	2	2				100
									6					2	2	2					106
								4						2	2						110
						42	2							2							112
													14	2	2	2	2	2	2	2	126*
												12		2	2	2	2	2	2		138

Each integer indicates the occupancy of protons or neutrons in a specific shell or subshell. Total occupancy is twice that value. Asterisks denote the magic numbers. Plus-marks denote semi-magic numbers (see Chap. 2 for discussion of the empirical criteria for magicness).

searching for mechanisms that will modify the quantum mechanical predictions. The mechanism that was postulated by Mayer and Jensen was spin-orbit coupling – similar to what was known from electron physics. The coupling effect allowed the n-shells of the wave-equation to be broken into subshells, among which the experimentally-known magic numbers are found. Quantitative justification of the spin-orbit force in the shell model has remained controversial (e.g., Bertsch, 1972; Bertsch et al., 1980), but n-shell splitting is an essential part of the model if the magic numbers are to be explained. The lattice model shows the same build-up sequence and consequently the same consistency with quantum mechanics – while also requiring further arguments to explain why certain of the j-subshells show relatively strong magicity.

9.2 Reproduction of the Independent-Particle Model in an fcc Lattice

The Total Angular Momentum Quantum Number, j

The independent-particle model postulates the existence of a total angular momentum value, j, that is due to the summation of the nucleon's intrinsic angular momentum (spin), s, and its "orbital" angular momentum, l: $j = |l \pm s|$ – spin-orbit coupling. Unlike the theoretical values of total energy level (n) and theoretical orbital angular momentum (l), the j-value is an observable quantity, and has been experimentally measured for thousands of ground-state and excited-state isotopes. For this reason, an explanation of j-values is absolutely crucial for any nuclear model – and was in fact an important factor in the general acceptance of the shell model in the early 1950s. Since l is defined as: $l = 0, 1, 2, 3, \ldots, n$, and spin is a binary value: $s = \pm 1/2$, j can take any positive half-integer value less than or equal to: $n + 1/2$, with again restrictions on the number of nucleons with any value of j for each energy shell. Specifically,

$$\text{occupation } (j) = (n+1) * 4 \tag{9.8}$$

Theoretical justification of spin-orbit coupling has remained a contentious issue, but the fact that the observed pattern of j-values corresponds closely with the theoretical values has made the independent-particle model indispensable. It is therefore of interest that the lattice model reproduces the exact same pattern of j-values. Based upon the same geometry as described for nucleon n-values, nucleon j-values can be defined in terms of the distance of the nucleon from the nuclear spin-axis (Fig. 9.5b). That is,

$$j_{\text{nucleon(k)}} = (|x_{\text{nucleon(k)}}| + |y_{\text{nucleon(k)}}| - 1)/2 \tag{9.9}$$

where again x and y can take all odd-integer values. The angular momentum of each nucleon is therefore dependent upon the nucleon's distance from the (z) spin-axis of the nucleus as a whole, analogous to the definition of angular momentum in classical physics. The relationship between j-values and the fcc lattice is illustrated in Figs. 9.5b, 9.7 and 9.8.

In the lattice model, the geometrical definition of j means that there are concentric cylinders of increasing j-values (Fig. 9.7). Just as the n-shells of the fcc lattice are triaxially ($x = y = z$) symmetrical (not spherically symmetrical), the j-subshells are biaxially ($x = y$) not cylindrically symmetrical structures. (It is worth noting that the entire lattice geometry could be deformed into a spherical system where n-shells are spherical and j-subshells are cylindrical, and the lattice dimension, d, varies with position in the lattice. The symmetries of n, j, etc. would, however, remain unchanged.) The labeling of j-subshells in the fcc model for ^{140}Yt is illustrated in Fig. 9.8, where it is seen that the cylinders have increasingly thick cylinder "walls" such that there is some overlap between neighboring j-cylinders.

How is total angular momentum to be interpreted within the lattice? Clearly, the classical orbiting of particles cannot occur in a solid lattice (nor,

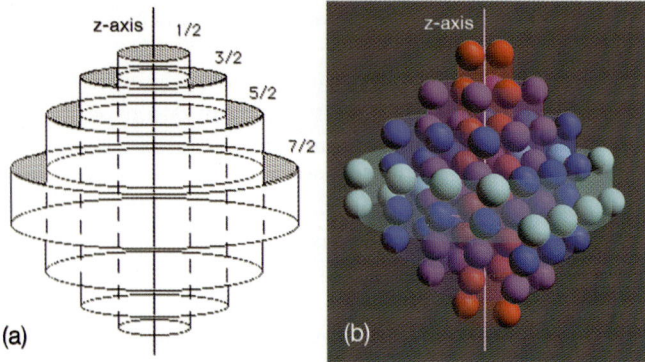

Fig. 9.7. The j-subshell cylinders in ^{80}Zr. Nucleons located closest to the vertical spin-axis of the nucleus as a whole have the lowest total angular momentum quantum values

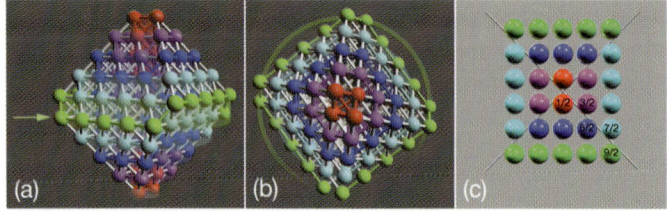

Fig. 9.8. The j-values in the fcc model for ^{140}Yt, as viewed from an angle that is orthogonal to the spin-axis (**a**) and down the spin-axis (**b**). It is seen that the j-value depends on the distance of the nucleon from the nuclear spin-axis. This is illustrated in (**c**) for the layer indicated by the arrow in (**a**). The 1/2 spin nucleons (red) are nearest to the spin-axis, followed by 3/2, 5/2, nucleons, and so on

as discussed in Chap. 5, in a nuclear liquid or gas!). Rather, the total angular momentum must be considered a property inherent to the wave-state of the individual nucleon – determined by the nucleon's position relative to the nuclear spin-axis, but not indicating intra-nuclear orbiting.

Any nucleon in the fcc lattice has a principal quantum number, n, that is a function of its distance from the nuclear center. Within any n-shell, there is a fixed number of nucleons for the given n-value, and that number is identical to the (proton + neutron) shell-filling implied by the Schrödinger wave-equation. By defining the total angular momentum quantum value, j, in terms of the nucleon's position in relation to the nuclear spin axis, every nucleon also has a fixed j-value in the lattice. For each n-shell, there is then a fixed number of nucleons with each allowed j-value. If j-values are assumed to be the basis for defining subshells, as in the shell model, then we find that both the sequence and the occupation numbers for every j-subshell deduced from the spin-orbit coupling model are reproduced exactly in the lattice (Table 9.2).

9.2 Reproduction of the Independent-Particle Model in an fcc Lattice

In other words, prior to discussion of the relative energy levels of the subshells, it can be said that the entire pattern of the splitting of n-shells into angular momentum subshells that was *the principal success of the shell model circa 1950* is reproduced in the lattice. The question of the precise energy differences among the subshells turns out to be problematical not only for the standard shell model, but also for the lattice model, principally because the energy levels are strongly influenced by the relative numbers of protons and neutrons in the given nucleus.

Despite such irregularities (known as configuration-mixing and intruder states), the (approximate) sequence and the exact occupation numbers of the angular momentum subshells are well-known experimentally. Figure 9.2 summarizes the approximate sequence of build-up for protons and neutrons, but particularly for closely-spaced subshells, the default sequence is frequently not followed. Both models predict n-shells and j-subshells, but neither model unambiguously predicts the empirical build-up sequence. The most conservative prediction of either model is simply that some sort of subshell "texture" should be evident in the build-up of nuclei, and that the texture should be sensitive to the relative numbers of protons and neutrons.

The Azimuthal Quantum Number, m

The so-called azimuthal quantum number, m, is defined in quantum mechanics as the projection of the orbital angular momentum along one axis, taken here as the x-axis (Fig. 9.9). In the lattice model, the height of the m-value cone is defined by the nucleon's lattice position along the x-axis.

$$|m| = |x|/2 \qquad (9.10)$$

The sign on m is determined by the nucleon spin (Fig. 9.10).

$$m = (|x|/2)(-1)^{(x-1)} \qquad (9.11)$$

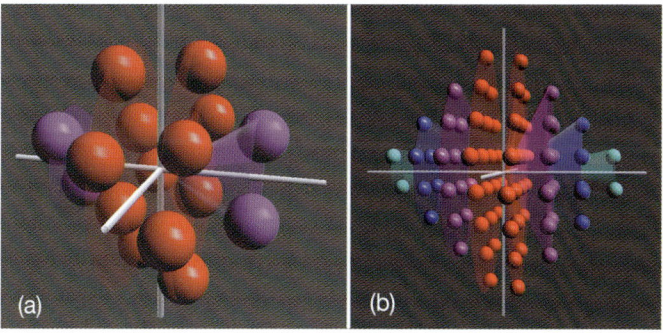

Fig. 9.9. The geometry of the azimuthal quantum number, m, for a small nucleus, ^{16}O, and a larger nucleus, ^{80}Zr

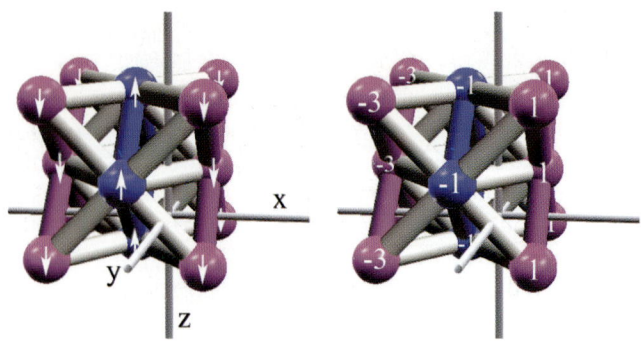

Fig. 9.10. The m-values in the lattice are a function of the direction of spin (*left*) and the absolute value of the x-coordinate (*right*)

The Spin Value, s

The intrinsic spin of nucleons in the lattice model is defined in terms of the orientation of the spin in relation to one axis of nuclear Cartesian space.

$$s = (-1)^{x-1}/2 \qquad (9.12)$$

As seen in Fig. 9.10, the spin-planes alternate between spin-up and spin-down along the x-axis, meaning that all nearest neighbors within each horizontal plane are of opposite orientation. This gives the lattice a spin-structure known as antiferromagnetic.

The Isospin Value, i

Finally, in order for the lattice model to reproduce the pattern of quantum numbers for both protons and neutrons, as shown in Table 9.1, the model requires protons and neutrons to be in alternating layers (Fig. 9.11). It is a remarkable fact that the antiferromagnetic fcc lattice *with alternating isospin layers* was also found by Canuto and Chitre (1974) to be the lowest energy solid-phase configuration for nuclear matter ($N = Z$), possibly present in neutron stars.

The isospin of nucleons is defined relative to the z-axis.

$$i = (-1)^{z-1}/2 \qquad (9.13)$$

Such layering is a means of separating protons from one another, but it also implies that, within the lattice, there are always precisely two lattice sites that have an identical set of n, j, m, and s quantum numbers – one in the northern hemisphere of the lattice and one in the south.

Due to the antiferromagnetic arrangement of nucleons in each layer and the alternating proton and neutron layers, the geometry of the intrinsic spin

9.2 Reproduction of the Independent-Particle Model in an fcc Lattice

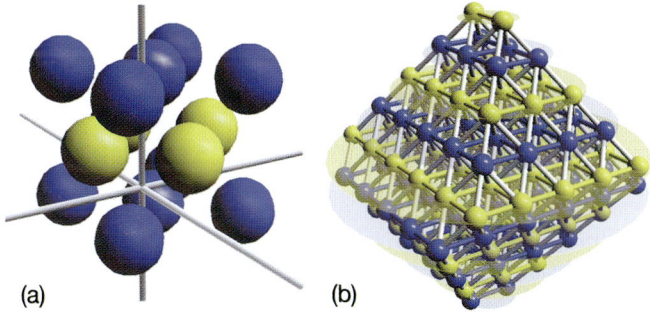

Fig. 9.11. Isospin layering along the z-axis is one means for keeping the proton charges away from each other

and intrinsic isospin of nucleons is also fixed in the model. The full description of nucleon quantum numbers can therefore be illustrated as in Fig. 9.12. The general significance of the quantum number assignments is that any nucleon (defined by n, j, m, s, i) has a unique position in three-dimensional space; moreover, the nucleon's lattice position defines a unique set of quantum values. When all of the possible combinations of quantum values are tabulated (Tables 9.1 and 9.2), it is found that the lattice model reproduces the known set of quantum values in the same sequence and with the same occupancy as indicated in the Schrödinger equation.

Parity

Parity can be defined in the lattice as the product of the signs of nucleon coordinate values:

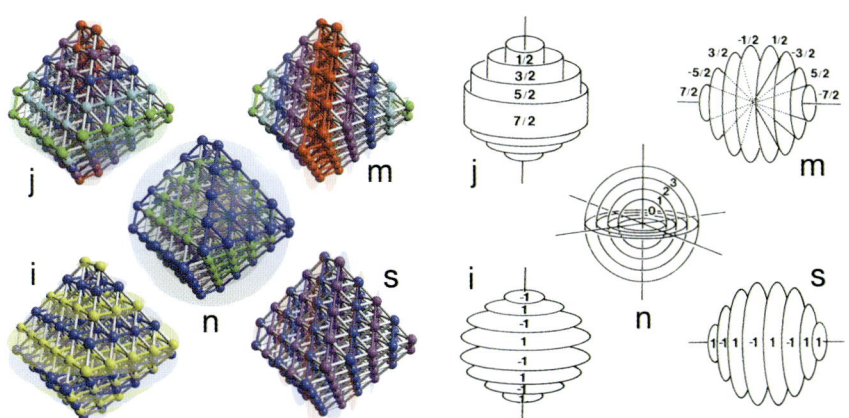

Fig. 9.12. The full labeling of quantum values in the fcc lattice for the ^{140}Yt isotope (on the *left*). Diagrams for the first 4 *n*-shells (^{80}Zr) are shown on the *right*

$$p = \text{sign}(x) * \text{sign}(y) * \text{sign}(z) \tag{9.14}$$

and the parity of the entire system is simply the product of all nucleon parities:

$$P_A = \Pi(\text{sign}(x_j) * \text{sign}(y_j) * \text{sign}(z_j)) \tag{9.15}$$

Parity in the lattice therefore alternates between positive and negative in successive n-shells.

Using the basic equations of the lattice model (9.6 through 9.13) every nucleon in the lattice is given a unique set of n, j, m, s and i values. It is important to realize that, for a given quantum number, only certain combinations of other quantum numbers are possible. For example, if a nucleon in the lattice model has an m-value of 7/2, then its $|x|$-coordinate must be 7 (9.11) and, as a consequence, it must have an n-value equal to $(7+|y|+|z|-3)/2$ (9.7). Because the absolute values of y and z must sum to at least 2 $\{(7+1+1-3)/2\}$, any nucleon with an m-value of 7/2 must have an n-value of at least 3. That is precisely what would be deduced from the independent-particle model.

Moreover, it is also possible to calculate how many nucleons with a given n-value could have a certain m- or j-value (or vice versa). For example, for an m-value of 3/2, $|x|$ must be 3 and, for an n-value of 1, the sum of $|x|$, $|y|$ and $|z|$ must be 5, so the sum of $|y|$ and $|z|$ must be 2. Possible permutations of the absolute values of odd-integers which will give a sum of 2 are of course (1 1), (1 −1), (−1 1) and (−1 −1). In other words, in the lattice model there can be up to four nucleons in n-shell = 1 with m-values = $|3/2|$, as predicted using the independent-particle model.

The relationship between quantum numbers and the lattice system can be summarized as:

$$n = (|x| + |y| + |z| - 3)/2 \tag{9.7}$$
$$j = |l + s| = (|x| + |y| - 1)/2 \tag{9.9}$$
$$m = (|x|/2)(-1)^{(x-1)} \tag{9.11}$$
$$s = (-1)^{(x-1)}/2 \tag{9.12}$$
$$i = (-1)^{(z-1)}/2 \tag{9.13}$$

Conversely, knowing the quantal state of a nucleon, its Cartesian coordinates can be calculated as:

$$x = |2m|(-1)(m + 1/2) \tag{9.16}$$
$$y = (2j + 1 - |x|)(-1)^{(i+j+m+1/2)} \tag{9.17}$$
$$z = (2n + 3 - |x| - |y|)(-1)^{(i+n-j-1)} \tag{9.18}$$

9.3 Symmetries of the Unit Cube of the fcc Lattice

The reason why an antiferromagnetic face-centered cubic lattice with alternating isospin layers is the only serious solid-phase candidate structure for

9.3 Symmetries of the Unit Cube of the fcc Lattice

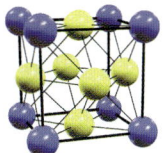
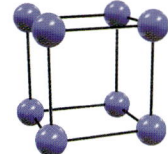
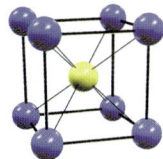

Fig. 9.13. On the left is shown the face-centered cube. The six nucleons on the six faces of the cube are shown in yellow. The overall cubic shape is emphasized by the thick lines, connecting second-nearest-neighbor blue nucleons. In the middle is the simple cubic packing lattice and on the right is the body-centered cubic lattice. Clearly, the different cubic lattices imply very different nearest-neighbor particle interactions (see Table 9.3)

nuclear structure theory is that it embodies the quantal symmetries implied by the Schrödinger wave-equation. If, therefore, we have reason to pursue a lattice model of the nucleus, the fcc lattice will necessarily be the place to start. What, then, are the known characteristics of this type of lattice? Let us examine the fcc lattice from a crystallographic perspective.

The fcc lattice has a unit structure with particles at each of the eight corners of a cube and on each of its six faces (Fig. 9.13). (The phrase "unit cell" is normally used in crystallography to refer to the minimal structure that contains the symmetries of the lattice type. Since the symmetries of the fcc lattice can be defined on the basis of only six lattice sites, the fundamental unit is not cubic. Since, however, the reduced "unit cell" introduces certain confusions of its own, the 14-site "unit cube" will be discussed here.) The fcc lattice is of course a well-known structure in crystallography and is one of the three basic types of cubic lattice – the other two being simple cubic packing (scp) and body-centered-cubic (bcc) packing, as shown in Fig. 9.13 (see Table 9.3 for basic lattice properties).

As a crystal structure, the fcc configuration is of particular interest because it is one of only two ways in which spheres can be "close-packed" at a maximal density of particles per unit volume (74.048%). The other is hexagonal-close-packing (hcp). The fact that only these two lattices are close-packed is not obvious unless the spheres are enlarged until they are contiguous (Fig. 9.14). The efficiency of the packing of nucleons, the size of nucleons relative to the average distance between them, and the density of the nucleus are all fundamental issues in nuclear theory and must be discussed in relation to any model that is claimed to give a realistic texture to the nuclear interior. Basic properties of common lattices are summarized in Table 9.3.

Note that the fcc unit cube can be depicted as two neutron layers, with a layer of protons sandwiched between them or as two proton layers with neutrons in-between (Fig. 9.15). In either case, the configuration of 14 nucleons does *not* correspond to a stable nucleus, but is simply a small portion of a nucleon lattice that extends indefinitely in all directions with alternating proton and neutron layers. (Nuclei that contain 14 nucleons in a cubic configuration,

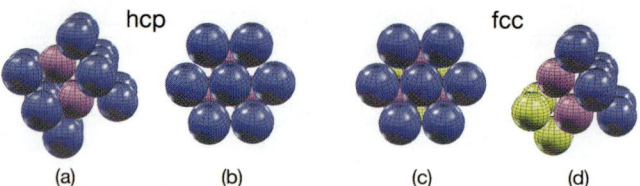

Fig. 9.14. The two kinds of close-packing of spheres, either of which leaves a minimum of free space among the spheres. All layers are hexagonal close-packed in 2D, but the layers are stacked differently in hcp and fcc. On the *left* (**a** and **b**) is shown hexagonal close-packing (hcp), in which the top layer (*blue*) repeats every other layer. On the *right* (**c** and **d**) is face-centered-cubic (fcc) close-packing, in which the top layer (*blue*) repeats every third layer (d), after the purple and yellow layers

Table 9.3. Fundamental lattice parameters (Cook & Hayashi, 1997)

Lattice type	fcc/hcp	bcc	scp	Diamond
Number of nearest neighbors	12	8	6	4
Nearest-neighbor distance (fm)[a]	2.0262	1.9697	1.8052	1.7896
Number of second nearest neighbors	6	6	8	8
Second neighbor distance	2.8655	2.2744	2.5529	2.9223
Coulomb force between protons[b]	0.7107	0.6331	0.7977	0.4927

[a] Nearest-neighbor distances are those that imply a lattice density equivalent to that of the nuclear core (0.17 n/fm^3).

[b] Assuming structures with alternating isospin layers, the Coulomb force is that at the nearest-neighbor distance in the fcc/hcp and scp lattices, with a larger Coulomb effect in the scp lattice because of the shorter nearest-neighbor distance. In the bcc and diamond lattices, the Coulomb force is that between second-nearest-neighbors (which is as close as protons come to one another in these lattices), with a stronger effect in the more compact bcc lattice.

e.g., ^{14}C and ^{14}N, can, of course, be constructed in the fcc model, but they are not optimized to give the minimal Coulomb repulsion or the maximal number of nearest-neighbor bonds among the nucleons, and are thus inherently unstable.)

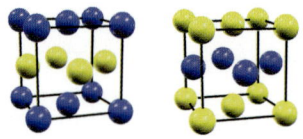

Fig. 9.15. Isospin layering in the fcc lattice. Within any large nucleus, cubic units with a neutron-proton-neutron layering (*left*) or proton-neutron-proton layering (*right*) can be found. Because of the large excess of neutrons or protons neither of these nuclei, ^{14}Be and ^{14}Ne, are stable in these cubic configurations

9.3 Symmetries of the Unit Cube of the fcc Lattice   197

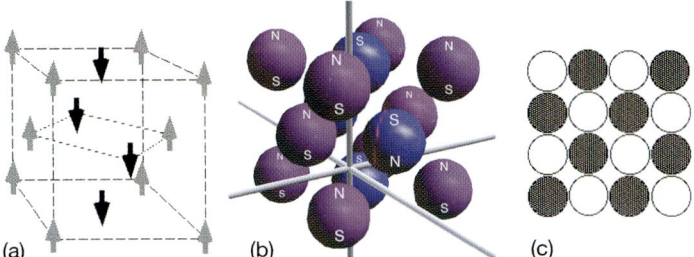

Fig. 9.16. The antiferromagnetic arrangement in the fcc lattice. All nearest neighbors within each layer have opposite spins, producing a checkerboard pattern (c) in each layer. The *dark circles* indicate north poles, and the *white circles* indicate south poles. Note that the structure within each layer of fixed isospin is not close-packed, giving each nucleon four nearest neighbors, all of which have opposite spin

The antiferromagnetic ordering of nucleons in the lattice means that all four nearest neighbors *within each horizontal layer* have their magnetic axes pointing in opposite directions (Fig. 9.16).

The exchange of charged pions among the nucleons of course greatly complicates the static picture of the fcc lattice with alternating isospin layers. To conserve charge, the transformation of a neutron into a proton (or vice versa) must occur together with the reverse transformation elsewhere within stable nuclei. Moreover, it is known that charged pion exchange leads to a reversal of the intrinsic spins of both nucleons. The coordinated exchange of both spin and isospin will therefore lead to a complex dynamic process in any multi-nucleon system. These dynamics undoubtedly play an important role in determining the electromagnetic characteristics of nuclei.

The existence of magnetic dipole moments in protons and neutrons is likely to be a result of the movement of their intrinsic charges within the confines of the particle. Depending on the net direction of that movement and the charges involved, the magnetic poles will be either north-up or south-up. In the case of the electron, the north pole is defined as the pole where magnetic flux lines leave when the flow of electric charge is counter-clockwise. For the proton, since the current consists of the flow of positive charge, the magnetic south pole is "up" and the north pole is "down". Although the total charge of the neutron is zero, it is known to have a positive charge located primarily in its interior and a negative charge located in its periphery (Fig. 9.17, left). The detailed substructure of both protons and neutrons remains uncertain (see Chap. 10 for further discussion), but the neutron can be usefully considered as having magnetic properties resulting from the intrinsic rotation of its peripheral negative charge.

The magnetic dipole interaction is necessarily one of the forces acting between neighboring nucleons. As a consequence of the checkerboard arrangement of nucleons in each layer of the fcc lattice (Fig. 9.16), the sum of the magnetic moments of the nucleons does not produce a large total magnetic

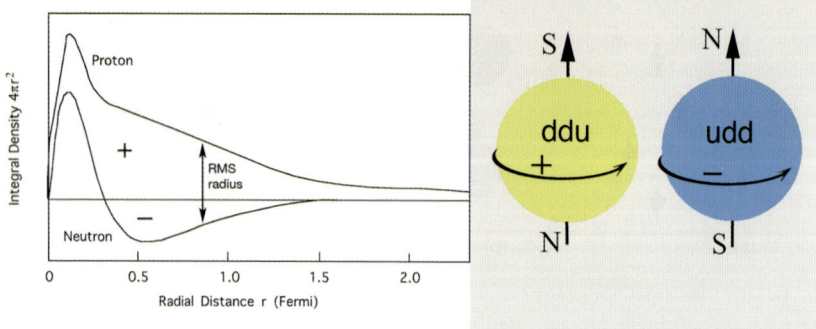

Fig. 9.17. The experimental distribution of charge in the proton and neutron (Littauer et al., 1961) is shown on the *left*. A cartoon of the structure of the nucleon is shown on the *right* (RMS radius ∼0.86 fm). Because the peripheral charges in the proton and neutron differ, their magnetic poles also differ when intrinsic spin is the same (*right*)

force field (which is the case in a ferromagnet), but rather the individual magnetic moments cancel one another out in pairs. Nuclei containing an even number of protons and an even number of neutrons will therefore have no net magnetic moment. Only the odd proton and/or neutron that does not have a corresponding oppositely-oriented neighbor with the same isospin will contribute to the total magnetic moment of the nucleus. In general, this means that the total nuclear magnetic moment will be due to one unpaired proton and/or one unpaired neutron (the Schmidt lines, Chap. 4). Clearly, the antiferromagnetic structure is important in order that the lattice model can produce magnetic moments that are of the right order-of-magnitude (that is, zero for even-Z, even-N nuclei, and approximately equal to the magnetic moments of the proton or neutron for odd-Z or odd-N nuclei).

The basic symmetries of the nucleons in the fcc unit cell can be shown as in Fig. 9.18a, where (i) the fcc structure, (ii) the antiferromagnetic alignment within each layer, and (iii) the alternating layers of protons and neutrons, are illustrated. The nearest-neighbor binding characteristics can be summarized qualitatively as in Fig. 9.18b.

Since the 1970s, many theoretical studies have examined the possible condensation of nucleons in the context of neutron star research (e.g., Tamagaki, 1979; Takahashi, 1991), but most have focused on pure neutron matter. Estimates of the density at which an assembly of protons and neutrons ($N = Z$) solidify – that is, the density at which a nuclear lattice is inherently more stable than a liquid or a gas – vary by more than a factor of ten, ranging from known nuclear densities to ten times that figure (Canuto, 1975). Quantitative justification of this particular fcc lattice structure on the basis of first principles is therefore still missing, but the only study that has examined the solidification densities of various crystal lattices of nuclear matter showed the

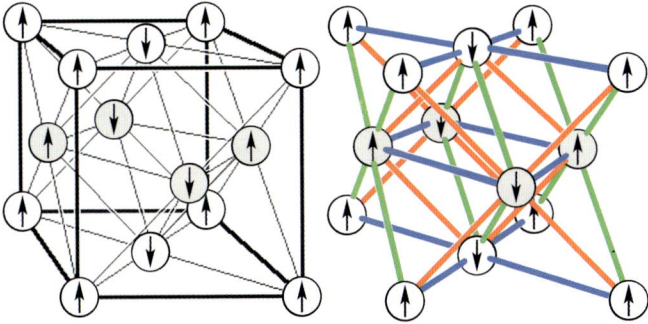

Fig. 9.18. The antiferromagnetic fcc lattice with alternating proton and neutron layers. Spins are indicated by the arrows on the nucleons. Protons are gray, neutrons are white (or vice versa). All nearest-neighbor (opposite-spin/same-isospin) effects *within* each horizontal layer are attractive (*blue*), whereas the interactions between same-spin/opposite-isospin nucleons in different layers are attractive (*green*) and those between opposite-spin/opposite-isospin nucleons (*red*) are repulsive

lowest energy configuration to be the antiferromagnetic fcc lattice with alternating isospin layers (Canuto & Chitre, 1974). Related results comparing the solidification densities of skyrmions in fcc, scp and bcc lattices have been reported by Castillejo et al. (1989) (Fig. 9.19). Final answers are not available, but it is likely that, if there are grounds for considering a lattice of nucleons as a nuclear model, then the antiferromagnetic fcc lattice with alternating isospin layers is the lowest energy lattice structure.

Taking both spin and isospin into account, there are 4 types of nucleon (conventionally described in terms of their quark constituents, unconventionally described in terms of their electromagnetic properties) and therefore 6 types of nucleon interactions (Fig. 9.20). Two attractive interactions (PP and

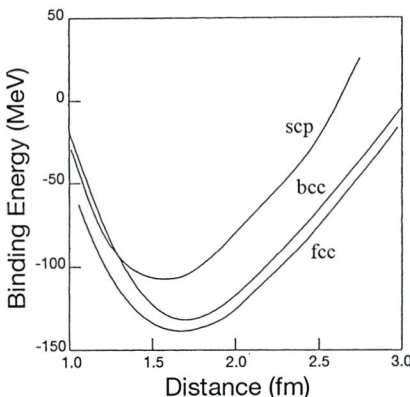

Fig. 9.19. A comparison of the condensation densities of skyrmions in various lattice types (Castillejo et al., 1989)

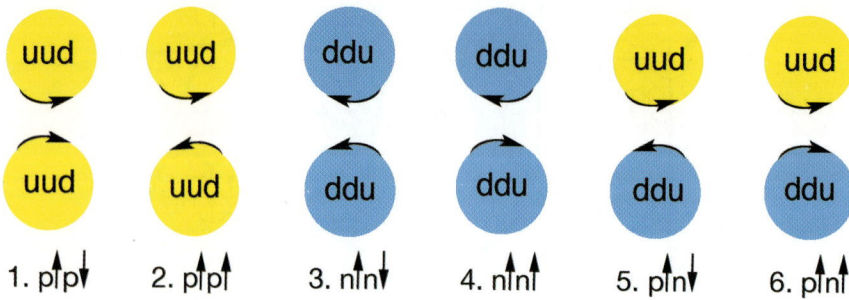

Fig. 9.20. The six possible interactions among protons (*yellow*) and neutrons (*blue*). Interactions 1, 3 and 6 are attractive and important for the binding of the fcc lattice. A classical description emphasizes electromagnetic properties, while a QCD description emphasizes quark properties, but there are, in either case, just these 6 permutations

NN) are microscopic analogs of the Biot-Savart law of electromagnetics (i.e., the parallel flow of like-charges at the point of nearest approach between the particles) and are also described as meson exchanges due to the pairing of quarks and anti-quarks (1 and 3 in Fig. 9.20). The third attractive interaction (6) can be understood as an analogous effect caused by the anti-parallel flow of opposite charges. Non-binding effects (2, 4, 5) arise from the anti-parallel flow of like-charges or the parallel flow of unlike-charges. Note that there are no nearest-neighbor effects of types 2 and 4 in the fcc lattice due to the anti-ferromagnetic structure.

Particularly during the 1970s, there was considerable theoretical interest in the structure of neutron stars and, within that context, many studies were carried out to determine the nature of nucleon interactions in liquid-phase and various solid-phase "condensates" of protons and neutrons. There is a consensus that greater stability is achieved when neutron stars contain mostly or only neutrons, but the most interesting theoretical results for issues of nuclear structure theory are those concerning the stability of lattices with equal numbers of protons and neutrons – as is roughly the case in normal nuclei. The interaction of nucleons can also be studied in various states with reduced dimensionality and, in fact, most published studies have examined the energetics of one- or two-dimensional condensation. A one-dimensional condensate means simply that, along one dimension of an aggregate of nucleons, there is a spatial regularity that is not seen in other dimensions. Similarly, a two-dimensional condensate means that there is a spatial regularity in two-dimensional planes, but that the planes are not aligned in a regular manner along the third dimension. There are mathematical advantages of studying systems with artificially-reduced dimensionality, but there is a danger of ignoring important effects that the second and third dimensions may bring. In

9.3 Symmetries of the Unit Cube of the fcc Lattice

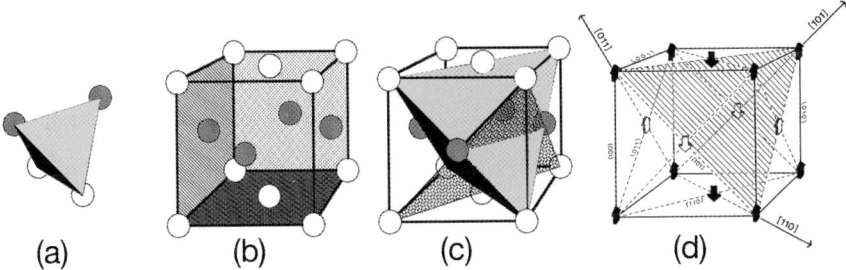

Fig. 9.21. The most important crystallographic planes in the fcc lattice include the four faces of the tetrahedron (**a**) and the three faces of the unit cube (**b**). The [100] plane is striped; the [010] plane is *light grey*, and the [001] plane is *dark grey*. The [100] planes contain nearest neighbor bonds only between nucleons of different isospin, but same spin; the between-layer magnetic effect is therefore attractive. The [010] planes contain nearest neighbor bonds only between nucleons of different isospin and different spin; the between-layer magnetic effect is therefore repulsive. The [001] planes contain nearest neighbor bonds only between nucleons of the same isospin, but different spin; the magnetic effect is attractive. In the middle cube, three of the four important oblique planes in the fcc lattice (001) (010) (100) (111) are shown. As seen also on the right, these planes correspond to the faces of the central tetrahedron in the lattice. Unlike the planes that run parallel with the x-, y- and z-axes of the lattice, all four of these planes contain all combinations of spin-up and spin-down bonds between nucleons of different isospin. Within-layer magnetic effects are always attractive, but between-layer effects are always mixed (Cook & Dallacasa, 1987)

the context of the fcc model, the antiferromagnetic arrangement is a 2D condensate within isospin layers, whereas the parallel-spin coupling of protons and neutrons is an orthogonal 2D condensate. The j–j coupling of like-isospin nucleons also occurs in 2D isospin layers.

As is true of any crystal structure, the antiferromagnetic fcc lattice has various planes and directions of symmetry – each with characteristic properties depending upon spin and isospin features and the packing of nucleons in the relevant one or two dimensions. Some of the most important planes of the fcc structure are illustrated in Fig. 9.21.

Another way of viewing the antiferromagnetic fcc lattice with alternating isospin layers is as four cubes, each of which has a unique combination of spin ($\pm 1/2$) and isospin ($\pm 1/2$) (Fig. 9.22).

In addition to the quantal symmetries in the fcc lattice, there is inevitably further substructure in the local grouping of particles. The substructure can be seen by emphasizing the tetrahedral grouping of nucleons in the lattice (Fig. 9.23). Because of the alternating isospin layers along the z-axis and the alternating spin layers along the x-axis, each tetrahedron corresponds to an alpha particle with 2-protons and 2-neutrons, and is the fundamental unit in the cluster models of nuclear structure.

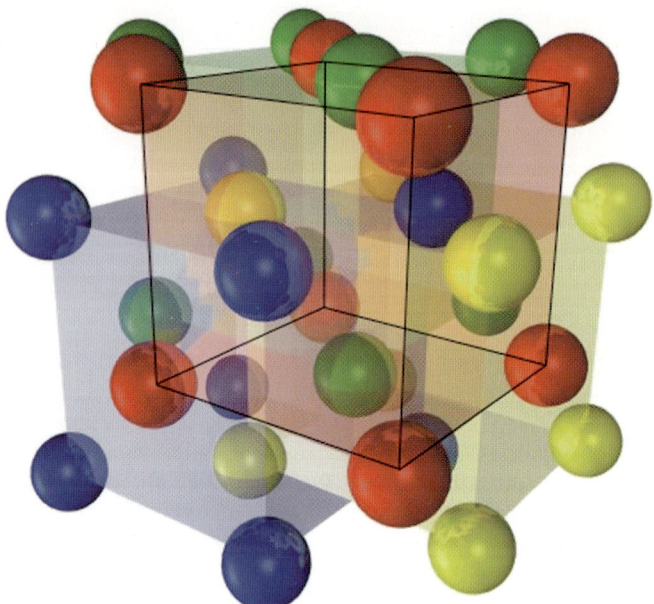

Fig. 9.22. The fcc lattice seen as four interpenetrating cubes. Each of the four cubes (*red, green, blue* and *yellow*) has a unique combination of spin and isospin. Assignment is arbitrary: every nucleon imbedded in the lattice has 12 nearest-neighbors, 4 of each spin/isospin combination that the nucleon itself is not!

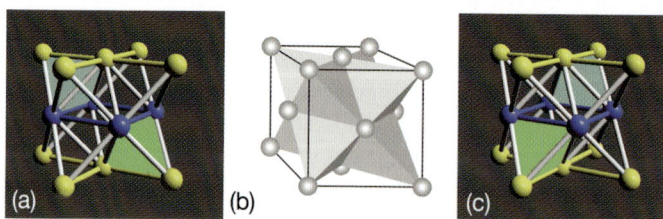

Fig. 9.23. The tetrahedra inherent to the fcc lattice. Any nucleon is a vertex simultaneously in 8 tetrahedra (**b**). If participation is restricted to only one tetrahedron, then only two distinct tetrahedrons are contained in the 14-nucleon unit structure (**a** and **c**)

The lattice model therefore implies that any nucleus can be considered as an array of alpha particles, often with some number of left-over protons and neutrons that are not a part of independent alpha particles. Certain of the large nuclei are known to be spontaneous alpha-emitters, and are generally thought to contain preformed alphas in the nuclear skin region. In the lattice model, these large nuclei are seen to have a surface virtually covered with alpha particles. This topic will again be addressed in Chap. 10.

Finally, the pairing of 1/2-spin fermions to give 0-spin (or other unit-spin) bosons is the central idea underlying the boson model of nuclear structure (Iachello & Arima, 1986). The permutations of possible pairings and their geometrical configurations within the fcc lattice are complex and depend critically on the nature of the spin/isospin pairing that is under consideration. The relationship between the boson description of nuclei and the fcc symmetries remains to be explored (see the *NVS* software for further details).

Summary of the fcc Lattice

Despite the many successes of the liquid-drop and cluster models and their continued use today, the shell model and its modern variants have dominated nuclear theory for the last 50 years – primarily because the shell model is based on quantum mechanics using a central potential-well analogous to the Coulombic central potential-well used in atomic physics. The fcc lattice model does not require the shell model's controversial potential-well, but reproduces – comprehensively and explicitly – all of the quantum mechanical symmetries, subdivisions and occupancies that are predicted by the shell model, while exhibiting a nuclear texture that is similar to the liquid-drop and cluster models. The fact that the lattice reproduces the properties of the other main models in nuclear theory is alone not formal demonstration that all of nuclear theory can be reworked in the framework of the lattice, but it does provide motivation to examine this lattice more closely.

And, indeed, lattices of nucleons continue to be studied in various contexts, primarily because of their computational simplicity. Representative are the "nucleon lattice" group at CalTech (Seki, 2004; Mueller, 2000; Lee, 2004; Abe, 2004), the "nuclear lattice" group at Michigan State University (Bauer et al., 1985, 1986, 1989, 2003), the "lattice-gas" theorists at McGill University (DasGupta, 1996, 1997, 1998; Pan, 1995, 1998; and colleagues), nuclear matter theorists in Kyoto (Tamagaki, 1979; Takahashi, 1991; and others), and heavy-ion multifragmentation groups at Grenoble (Cole, 2000; Desesquelles, 1993; et al.), Strasbourg (Richert and Wagner, 2001), and Paris (Campi and Krivine, 1986, 1988, 1997). What is striking about nearly all of the theoretical calculations and computer simulations of these groups is that they have explored *exclusively* the simple-cubic packing (scp) lattice (Chao and Chung, 1991, and Santiago et al., 1993, are exceptions). The conceptual and computational simplicity of the scp lattice is of course self-evident, but different lattices show remarkably different properties – notably, in the number of nearest-neighbors in the lattice and the nearest-neighbor distance at the same particle density (Table 9.3) and consequently differences in multifragmentation simulations (Cook & Hayashi, 1997). Clearly, in order to draw valid conclusions about the physics of a lattice of nucleons, comparisons among different lattice types need to be made.

From an orthodox shell model perspective, the reproduction of independent-particle model characteristics in a lattice might be dismissed as lucky numerology: by mere chance, it might be said, a certain geometrical structure contains symmetries that mimic the symmetries of the harmonic oscillator! But the converse situation is worth considering. Since these abstract numerical relationships are the basis for describing nucleon states in both models, could it not be the case that the harmonic oscillator is aping the lattice, rather than vice versa?

Rather than argue the case for the fcc model on the basis of theoretical considerations of the nuclear force (that today remain inconclusive), it is of interest to compare the various nuclear models in terms of what they imply about those nuclear properties that have been most problematic over the course of some 70 years of theorizing. The relatively concrete issues of lattice properties will therefore be discussed in Chap. 10. But, before the fit between theory and experiment is addressed, there is one variant of the fcc model that is worth examining in some detail: the lattice-gas model.

9.4 The Lattice-Gas Model

The advantages of the lattice model, as described thus far, are tied closely to the isomorphism between the fcc lattice and the independent-particle model description of nucleon states. In recreating the quantal description of nucleons within a lattice, the strengths of the shell model are retained, while the weak and long-range "effective" nuclear force has been entirely avoided. Furthermore, by employing a short-range nuclear force, the lattice model reproduces the essential nuclear texture of both the liquid-drop and the cluster models. It can therefore be concluded that, as counter-intuitive as a solid-phase model of the nucleus might initially seem to be, the unconventional lattice model appears to mimic the properties of all three of the firmly established, conventional models.

These merits of the lattice model should now be evident, but it is also clear that the insistence on a rigid geometry in the lattice model is unrealistic insofar as it implies a pre-ordained structure into which nucleons must fit. On the contrary, a fully convincing theory of nuclear structure would need to be derived from the properties of nucleons themselves. Given certain properties of the nuclear force, in principle it should be possible to predict how the system as a whole will self-organize into the lattice configuration described above – and there should be no need to "impose" the lattice symmetries as a preconceived notion. For this reason, in order to proceed from the curious isomorphism between the fcc lattice and the quantum mechanics of the nucleus, it is essential to show that the lattice of nucleons is consistent with what is known about the bulk properties of nuclear matter. The present section is therefore concerned with moving from the static billiard-ball geometry, described above, to a physically-realistic lattice model of nuclear matter. The theoretical argument gets somewhat more complex, but it will be seen that the fcc geometry remains the backbone of the more realistic model.

9.4 The Lattice-Gas Model   205

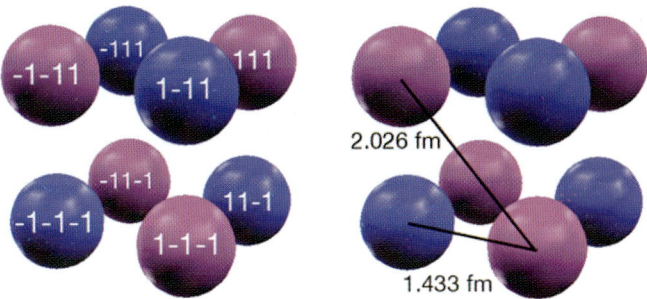

Fig. 9.24. The 8 nucleons at the center of the scp "lattice-gas". The purple nucleons constitute the central tetrahedron in the fcc lattice; the blue nucleons are from an overlapping fcc lattice twisted 90 degrees. Together, they form an scp lattice

First of all, it can be shown that increased fluidity of the lattice can be achieved while maintaining the isomorphism with the independent-particle model if the restriction on the (±) sign of available lattice sites is eliminated. Thus far, only those sites that correspond to the fcc lattice have been used, and that has meant allowing nucleons to occupy only lattice sites with odd-integer coordinates with certain restrictions on the sign of those odd-integers. As illustrated below, if the sign restriction that results in an fcc lattice is removed, the number of allowed sites for each and every shell and subshell in the lattice is doubled. In effect, this changes the lattice from fcc (with a nearest-neighbor distance of 2.0262 fm) to scp (with a nearest-neighbor distance of 1.4327 fm). For example, instead of the four tetrahedral sites (111, −1−11, 1−1−1, −11−1) in the first n-shell of the fcc lattice, we now get eight sites (111, −1−1−1, −1−11, 11−1, 1−1−1, −111, −11−1, 1−11) forming a cube in an scp lattice (where a cube encloses the origin of the coordinate system, with again the origin itself not corresponding to a lattice site) (Fig. 9.24).

The scp lattice can in fact be viewed as two overlapping fcc lattices (Fig. 9.25), and this fact implies that the quantal symmetries of the fcc lattice are simply doubled in the scp lattice. (This fact is easily verified using the *NVS* program on CD. By building any nucleus in the fcc model, the n-, j-, m-, s- and i-value symmetries can be displayed, and the isomorphism with the independent-particle symmetries confirmed. By then changing the visualization mode to the scp model, the 50%-occupancy of the scp lattice will mean some movement of nucleons from fcc sites to scp sites, but the overall symmetries of the nucleus will remain unchanged. See the Appendix for details.)

The definitions of quantum values in the lattice (9.4 through 9.17) remain the same and imply the same geometrical symmetries (spheres, cylinders, cones and orthogonal planes) when nuclei are built in the fcc or the scp lattice. But now, in the scp lattice, the occupancy of all shells and subshells is exactly twice that of the fcc lattice and twice what is known empirically. What the

Fig. 9.25. The melding of two fcc lattices gives a simple-cubic-packing (scp) lattice. The fcc lattice on the *left* is the 14-nucleon unit cube. The object in the *middle* is a 13-nucleon portion of the fcc lattice, with nucleons lying in-between those of the unit cube. When made to overlap, they give 8 unit cubes of the scp lattice (*right*)

increased occupancy means is that we are now in a position to "loosen up" the lattice by allowing any of the scp lattice sites to be occupied, but only up to the number of nucleons that is empirically known and that is allowed by the exclusion principle. Essentially, this means a 50% occupancy of an scp lattice.

What is implied by this change in the lattice model?

DasGupta's Lattice-Gas

As mentioned in Chap. 4, several lattice models have been developed as computational techniques to explain the multifragmentation data obtained from heavy-ion experiments. In such models, normally, a simple cubic packing (scp) lattice is used to represent the spatial positions of nucleons, and then a mechanism is invoked to induce the break-up of the system over time. These mechanisms include a "bond percolation" technique, where the adjustable parameter is the probability of breaking nearest-neighbor bonds within the lattice or a "site percolation" technique, where the probability of particle removal from the lattice is the adjustable parameter, but in either case a fully occupied lattice is the initial state. In contrast, DasGupta and colleagues have been able to reproduce the results of those models using a "lattice-gas" technique, where the sparse, random occupancy of the lattice is the parameter responsible for the fragmentation (Pan et al., 1995, 1998; DasGupta et al., 1996, 1997, 1998). In other words, a rather large lattice space, n^3, is randomly populated with a smaller number of nucleons, $\sim n^3/2$, and the local structure determines what fragments will survive.

Lattice Occupancy

If 100% of scp lattice sites were occupied by nucleons, each shell and subshell would have an occupancy which is exactly twice that of the independent-particle model, indicating that a mechanism for restricting occupancy to 50% is needed to obtain a correspondence with the independent-particle model.

Whether that mechanism might be justified theoretically on the basis of the exclusion principle or the repulsive core of the nuclear force is uncertain, but DasGupta has shown that the lattice-gas model predicts experimental multifragmentation results *only* if the lattice is occupied at somewhat less than 50% of full occupancy. In fact, the occupancy of the lattice that gives the best agreement with experimental multifragmentation data is 39% (Pan & DasGupta, 1995; DasGupta et al., 1998). The question therefore arises: since the lattice-gas model corresponds to the fcc model when the lattice-gas is occupied at exactly 50%, why would a value below 50% in the lattice-gas give the best simulation results?

The answer to that question comes not from the lattice-gas model itself, but from recent studies on the occupation of states in the independent-particle model. Contrary to conventional assumptions concerning the nucleon build-up procedure in all of nuclear theory since the emergence of the shell model in 1949, Pandharipande et al. (1997) have demonstrated that the nucleons in a ground-state nucleus occupy independent-particle model states only 65–75% ($\sim$70%) of the time. For about 1/4 of the time, nucleons are transiently in excited or in so-called "correlated" states (Fig. 9.26). Strictly within the framework of the independent-particle model, this re-evaluation of the nuclear texture in light of electron-scattering data (Pandharipande et al., 1997; Sick & deWitt Huberts, 1991) and microscopic calculations (Fantoni & Pandharipande, 1984; Schiavilla et al., 1986; Benhar et al., 1990) marks a major change in thinking about the nuclear texture. Instead of a fixed sequence of states for all nucleons, the $\sim$70% occupancy value indicates considerable local interactions and movement of nucleons into and out of low-lying excited states. An occupancy rate of less than 100% does not, of course, alter the essentially "independent-particle" nature of nuclear structure, but it does indicate a larger variability in individual nucleon states than what has been assumed throughout most of the developments of nuclear structure theory.

What does this mean with regard to the fcc model that had so clearly reproduced all of the quantal states of the independent-particle model? And what does it mean for the scp lattice-gas model that we have found to be equivalent to the fcc model when the scp lattice is half-filled?

Because there are potentially twice too many nucleons in any energy state in the scp lattice-gas model (all states being defined in a manner identical to the fcc model), the *experimentally-known* $\sim$70% occupancy of the independent-particle model (or fcc model) corresponds to $\sim$35% filling of the scp lattice. In other words, in order to reproduce the empirical $\sim$70% occupancy value of the independent-particle (fcc) model (Fig. 9.26), an occupancy of half that value in the scp lattice must be assumed. Starting with a fully-occupied scp lattice for their theoretical work, DasGupta et al. examined a range of occupancies in their model and found that it fits the experimental data on multifragmentation when the occupancy of scp states was 39%. The closeness of their best-fit occupancy of 39% to the 35% value required in the doubling of states in the fcc model is striking.

208   9 The Lattice Model: Theoretical Issues

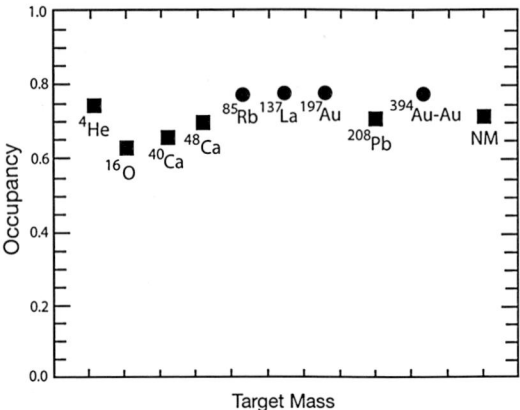

Fig. 9.26. The occupancy of independent-particle states. Regardless of the size of the nucleus, only 65–75% of nucleons are found to be in independent-particle model ground-states, with the remaining 25–35% of nucleons at any given time being transiently in excited or correlated states. The squares indicate values taken from a review by Pandharipande et al. (1997). The circles are twice the occupancy values in the lattice gas model of Das Gupta and colleagues. NM refers to nuclear matter calculations

In other words, a 39% occupancy of the scp lattice is equivalent to a 78% occupancy of the fcc lattice – and 78% occupancy is roughly equivalent to the slightly under-occupied state of the independent-particle model that Pandharipande and colleagues maintain is the normal state of most nuclei. Stated within the framework of the fcc model, nucleons are in fcc/scp lattice positions for 3/4 of the time, but are in transient in-between non-lattice positions the remaining 1/4 of the time.

A Unified Model?

The lattice-gas model has been advanced strictly as a computational technique (a "quick tool" according to Pan & DasGupta, 1998) for dealing with the otherwise intractable many-body problem presented in heavy-ion reactions. Nevertheless, DasGupta and colleagues have emphasized the advantages of the low-occupancy scp lattice-gas model over previous lattice models precisely in the fact that it reproduces various *realistic* physical quantities of the nucleus (in addition to fragmentation results). They have obtained: (i) a realistic equation of state for describing bulk nuclear matter, (ii) a nuclear temperature that can be related to levels of nuclear excitation, (iii) realistic Coulomb effects, and (iv) realistic kinetic energy values in the lattice-gas – features that are not obtained in other lattice models. While the equations that define the quantum numbers in relation to the lattice geometry (9.5 ∼ 9.18) explicitly relate the lattice-gas model to the independent-particle description of nuclei, the above physical quantities relate the lattice-gas model to the collective model of the nucleus and to the huge body of theoretical work on nuclear matter. Together, these properties of the lattice-gas suggest that a unification of the strong-interaction and independent-particle conceptions of nuclear structure can be achieved within the lattice-gas (Cook, 1999).

To reiterate, the lattice-gas model (a sparsely-populated scp lattice) has advantages over other lattice models specifically due to the fact that it is physically realistic – and is *more than* simply a "computational technique". By virtue of the correspondence with the fcc lattice model described above, it also shows a remarkable link with the traditional independent-particle model of nuclear structure theory. For these reasons, there are grounds for arguing that this particular form of a lattice-gas is a potentially unifying model of nuclear structure theory – capable of explaining multifragmentation data, while embodying the principal strengths of the independent-particle model, as well as the strengths of the strong-interaction collective (liquid-drop) model. Cluster effects are thrown in for free.

On first exposure, the occupancy issue in the lattice-gas model may appear to be an unwanted confusion added onto the geometrical complexity of a 3D lattice, but in the end it does not alter any of the geometrical symmetries of the fcc lattice and its correspondence with the independent-particle model, while providing a mechanism for dynamic effects within the lattice. For calculating nuclear properties using a lattice, either the 50%-occupied scp lattice or the 100%-occupied fcc lattice will produce equivalent results, but for consideration of the texture of nuclear matter, the more fluid scp lattice-gas may be an important improvement.

9.5 Conclusions

The starting point of the independent-particle model is a nuclear potential-well that attracts nucleons toward the center of the nuclear system. That assumption is clearly a convenient fiction, insofar as: (i) there is no centrally-located body to which nucleons are attracted, (ii) the nuclear force does not act over distances of $5 \sim 10$ fm, and (iii) nucleons are too large and (iv) the nuclear force too strong to allow intranuclear orbiting. Nevertheless, it is a historical fact that an extremely capable model of the nucleus, i.e., the shell model, followed directly from that fiction. The fcc lattice model, in contrast, starts with a realistic, strong local force acting solely between nearest-neighbor nucleons (~ 2 fm), but arrives at the same independent-particle-like description of nucleon states based on the relative positions of the nucleons in the lattice. From the perspective of conventional nuclear theory, the fcc symmetries are a totally unexpected result, but they clearly suggest that a high-density lattice approach to nuclear structure theory may be as valid as the low-density, orbiting nucleon approach.

Any close-packed lattice implies a dense nuclear texture and contains tetra-hedral (alpha-particle-like) substructures throughout the nuclear interior and on the nuclear surface. Moreover, the fcc lattice with a tetrahedron of particles at the center of the system uniquely shows a one-to-one correspondence with the quantal values implied by the shell model. The ability to reproduce the major properties of the shell, liquid-drop and alpha-particle models is a necessary, but not sufficient condition for taking the lattice model seriously. The next theoretical step is to reproduce the gross properties of so-called nuclear matter ($N = Z$) within the lattice. Precisely such work has been accomplished by DasGupta and colleagues. Although their focus has been on a ~50% occupied scp lattice, it is easily shown that the removal of every other nucleon in each layer of an scp lattice produces an fcc lattice. In that respect, the nuclear matter arguments that DasGupta has marshaled in favor of the 50%-occupied scp lattice apply identically to the 100%-occupied fcc lattice. A slight reduction in occupancy from 50% to 39% in the scp lattice, or equivalently from 100% to 78% in the fcc lattice seems to be required to obtain an acceptable fit between theory and experiment.

The correspondence between the independent-particle model and the fcc lattice provides a vital connection between the lattice-gas and mainstream nuclear structure theory. Lattice models are inherently "strong-interaction" models because nearest-neighbor nucleon effects play the dominant role in nuclear binding. The lattice approach (where local nuclear force effects are predominant) is nonetheless related to the shell-model approach, since it reproduces the independent-particle model symmetries. The lattice-gas model may therefore provide the basis for a truly unified theory of nuclear structure that exhibits the principal strengths of both approaches. In the next chapter, the nuclear properties implied by the lattice will be examined more closely.

10
The Lattice Model: Experimental Issues

This chapter shows how the lattice model can explain some of the most important experimental data on nuclei. Discussion is restricted to only those issues that have already been reported in the physics literature, but covers a broad enough range of topics that the strengths of the lattice approach will be evident. Most of the results were obtained from calculations using the fcc lattice, but some have been obtained using the $\sim$50%-occupied scp lattice. In either case, the comparisons between theory and experiment reveal only the first-order plausibility of the lattice model, and no attempt is made to massage the model into a better fit with the experimental data by appending parameters, even when that might be theoretically justifiable. The reason for taking an ultra-conservative approach is that the history of nuclear theory has abundantly demonstrated that any model (gas, liquid, solid or cluster!) can be twisted into compatibility with the experimental data by means of the proliferation of adjustable parameters. Ultimately, parameters and their fine-tuning are of course essential to determine what remains unexplained in any given model, but those are issues for research. Here the general plausibility of the lattice as a model of nuclear structure is the matter of primary concern.

What the fcc model is *not* is a theoretical explanation of elementary particle structure. The underlying physics of space-time and how matter arises from it are truly fundamental, but largely unresolved issues. I expect that theories not dependent on incomprehensible higher dimensions, alternative universes, and non-local, acausal, indeterministic mechanisms will eventually emerge (e.g., Palazzi, 2003, 2004, 2005; Tamari, 2005), but crucial tests of self-consistency and predictive power must be passed to achieve general acceptance. Today, the predominant paradigm at the particle level is the so-called standard model of quantum chromodynamics. In that model, nucleons contain quarks with partial charges – a view that flies in the face of all of classical and quantum physics. Although it is uncertain how the quark model may eventually evolve into a theory consistent with the basic facts of nuclear structure physics, it is already certain (1) that the nucleons themselves have substructure and (2) that the properties of nucleons are accurately described

212   10 The Lattice Model: Experimental Issues

on the basis of −1/3 and +2/3 charge characteristics. In other words, there is little doubt that the quark *description* of nucleon states is useful, although the existence of quarks (as discrete entities) remains uncertain.

The present chapter is concerned with how the fcc model can be used to predict specifically the properties of nuclear structure, but it is important to ask how the lower-level quark description of nucleons might be consistent with the nucleon lattice. I remain agnostic regarding the conceptual or heuristic value of visualizing nucleons as containing three discrete quarks (Fig. 10.1), but it is worth noting that a certain "parton" substructure of the nucleons themselves is implied by the fact that each nucleon is embedded in an fcc environment and necessarily has an intrinsic character related to its existence in a many-nucleon system.*

From what is known about the charge density of the nucleon (Littauer et al., 1961; Fig. 6.2) and theoretical estimates concerning the distance over which quarks interact with one another, the nucleon can be depicted as shown in Fig. 10.1. Given those dimensions, each nucleon embedded in an antiferromagnetic fcc lattice with isospin layering would therefore have potentially 12 nearest-neighbors of known characteristics. For a neutron, 1/3 of its 12 neighbors are opposite-spin neutrons (N↓N↑ bonds), 1/3 are same-spin protons (P↑N↑ bonds) and 1/3 are opposite-spin protons (P↓N↑ bonds) (Fig. 10.2). Similarly, for a proton, 1/3 of its 12 neighbors are opposite-spin protons (P↓P↑ bonds), 1/3 are same-spin neutrons (P↑N↑ bonds) and 1/3 are opposite-spin neutrons (P↓N↑ bonds). Stated more succinctly, in the fcc lattice any nucleon (with definite spin and isospin) interacts with three kinds of nearest neighbors that have spin and/or isospin properties different from itself.

These 1/3 ratios for the local interactions of all nucleons hold true regardless of the detailed structure of any given nucleus, and can be restated in the terminology of QCD. That is, the 1/3 charge character attributed to quarks might be a consequence of these regularities of local nucleon-nucleon binding (Fig. 10.2). The two +2/3 constituent quarks of the proton are involved in binding with their eight neutron neighbors and the −1/3 constituent is involved in binding with the four neighboring protons. For neutrons, the −1/3 quarks are involved in binding with eight protons and the +2/3 quark is involved in binding with four neutrons. Seen in this light, the quark properties of nucleons are less indicative of "extractable" partial charges residing inside of nucleons than of the color-symmetrical multiplicity of local interactions: 4/12

---

* Most nucleons in the universe are not bound within nuclear systems, but the expanding-universe of modern cosmology implies that all nucleons in the early moments following the big bang were once imbedded in matter far denser than nuclear matter. If there was a time at which the density of matter was similar to that of the nuclear core, then all nucleons would necessarily have gone through a stage of close-packing. Canuto & Chitre's (1972) finding that an antiferromagnetic fcc lattice with alternating isospin layers is the lowest energy state of a nuclear lattice then becomes relevant in suggesting that the properties common to all nucleons may be related to the structure of the high-density lattice.

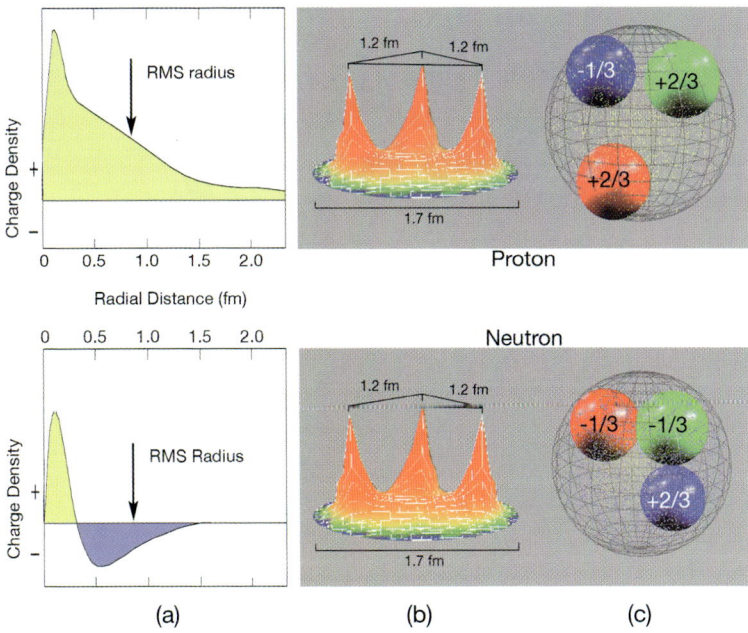

Fig. 10.1. Nucleon structure. (**a**) The experimental charge distributions of the neutron and proton (e.g., Littauer et al., 1961), (**b**) the so-called action-density of constituent quarks implies an inter-quark distance of 1.2 fm within each nucleon of diameter = 1.7 fm (after Barnes, 2004 and Bali, 2000); (**c**) the nucleons depicted as containing three discrete quarks. There is little prospect that quarks as discrete particles will ever be isolated, but the +2/3 and –1/3 characterization of nucleons is consistent with the properties of all known baryons and mesons

with nucleons of the same isospin/different spin, 4/12 with nucleons of different isospin/same spin and 4/12 with nucleons of different isospin/different spin. In other words, the 12-valence positions of each nucleon reflect three sets of possible nucleon-nucleon interactions, each of which entails a different kind of quark. It is noteworthy that the systematics of nucleon-nucleon bonding in the fcc model are entirely consistent with the quark description of mesons. Each nucleon is a set of three types of quarks: uud and ddu (u and d denoting the up/down quarks). In the antiferromagnetic fcc lattice with alternating isospin layers, there are four (and only four) types of quark-quark meson interactions between nucleons: $u\bar{u}$ between neutrons (the π^0 meson), $d\bar{d}$ between protons (the η meson), the $u\bar{d}$ between proton and proton (the π^+ meson), and the $d\bar{u}$ between neutron and proton (the π^- meson) (see Fig. 10.2).

More detailed accounts of quark models that lead specifically to a close-packed lattice model of the nucleus can be found in: Goldman et al. (1988); Goldman (1991); Maltman et al. (1994); Benesh et al. (2003); Musulmanbekov (2002, 2003, 2004). However such puzzles may eventually be resolved, suffice

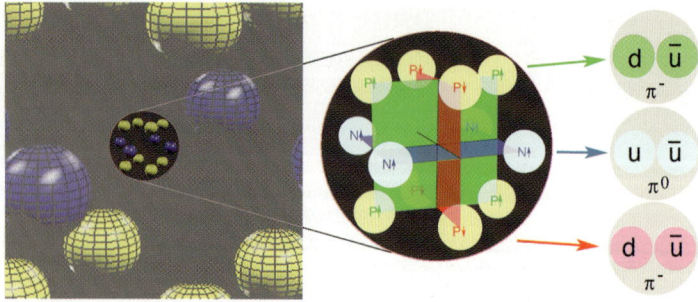

Fig. 10.2. Given the surrounding environment for any nucleon within the fcc lattice, its local interactions are 1/3 each with the other 3 types of spin/isospin nucleons. On the left, a spin-down neutron is depicted imbedded in the fcc lattice. In the middle are shown its 12 possible interactions with its neighbors: 4 exchanges (those in the horizontal *blue* plane) with the quarks in neighboring spin-up neutrons (uu mesons, π^0), 4 exchanges (those in the vertical *red* plane) with the quarks in neighboring spin-down protons (du mesons, π^-), and 4 exchanges (those in the vertical *green* plane) with the quarks in neighboring spin-up protons (du mesons, π^-)

it to say here that the fcc model and the known spatial dimensions of quarks, nucleons and nuclei imply local (nearest-neighbor nucleon) effects at distances up to ~2 fm, approximately as shown in Figs. 10.1 and 10.2. So, starting with this 12-valence structure for the nucleon, what kinds of fcc nuclei are implied?

10.1 Nuclear Size and Shape

Because any lattice can be considered as a "frozen" liquid-drop, the fcc lattice model deals with questions concerning the nuclear size, density and shape in a manner similar to the liquid-drop model, but there are a few interesting differences where the constraints of the specifically fcc lattice allow for predictions not implied by a liquid-drop.

Nuclear Density

From electron scattering experiments in the 1950s, the density of *protons* in the nucleus was determined to be about 0.085 protons/fm^3– almost regardless of the size of the nucleus (Fig. 10.3). On the basis of results from uncharged pion scattering experiments done in the 1970s, the distribution of *neutrons* in the nuclear interior was determined to be similar to the proton distribution, implying that the nuclear core contains 0.17 nucleons/fm^3. This density figure will be used in the following discussion.

The density curves shown in Fig. 10.3 indicated that: (i) the nuclear radius increases with the number of nucleons (this was *not* the case for atomic size, as discussed in Chap. 2); (ii) the maximal charge (mass) density remains

10.1 Nuclear Size and Shape

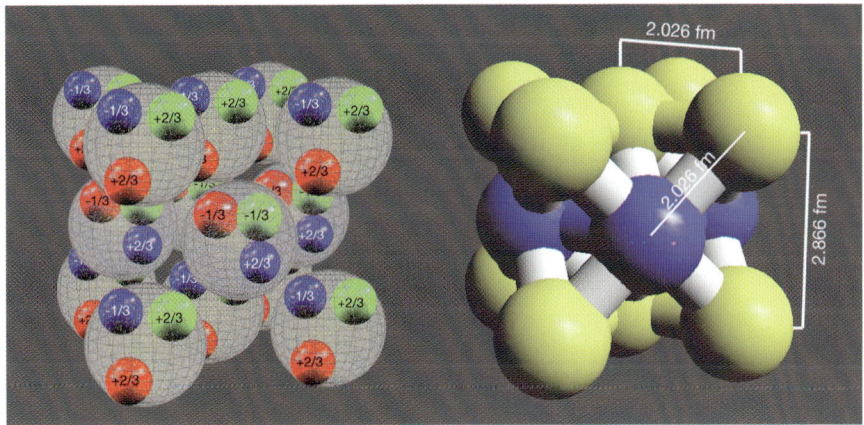

Fig. 10.3. Dimensions of the nucleons and quarks in the fcc model. The cube edge is 2.866 fm and the center-to-center internucleon distance is 2.026 fm, giving the known nuclear core density. The relative size of the lattice, the nucleons ($r = 0.86$ fm) and the quarks ($r = 0.4$ fm) are drawn to scale

roughly constant for all nuclei (with the notable exception of Helium); and (iii) the nuclear density falls to zero over a constant distance (the nuclear skin), regardless of the size of the nucleus. To account for these spatial features of the nucleus on the basis of a lattice model, certain assumptions must be made about the dimensions of the unit cell of the lattice. The most basic is the edge length of the face-centered-cube (2.8655 fm), which implies a nearest neighbor distance of 2.0262 fm and therefore the known density of 0.17 nucleons/fm^3 (Fig. 10.3).

In contrast to a liquid-drop conception, what is unusual about a close-packed lattice is that it has both dense tetrahedral configurations and less-dense octahedral configurations within the unit cube. As shown in Fig. 10.4, the high-density tetrahedral cells (~ 0.35 nucleons/fm^3) and the low-density octahedral cells ($\sim .09$ nucleons/fm^3) sum to a core nuclear density of 0.17 nucleons/fm^3. Specifically, 8 tetrahedral cells and 4 octahedral cells (one at the center of the unit cube and fully 12 quarter-octahedrons along the edges) are contained in the cube. Large nuclei built in accordance with the fcc lattice must have a mean density identical to that of the unit cell, but deviations from the mean are possible for the smallest nuclei containing too few nucleons to establish an unambiguous fcc texture. The unusually high density of the ^{4}He nucleus is the most interesting example and has a simple explanation in terms of the lattice: It is *not* a small chunk of generic "nuclear matter" typical of all nuclear cores, and it is *not* such a small nucleus that it is all low-density "skin". From geometrical considerations alone, the lattice model suggests that the first four nucleons of ^{4}He will exhibit a tetrahedral structure that has a high density (~ 0.35 n/fm^3) approximately 2-fold that of the nuclear

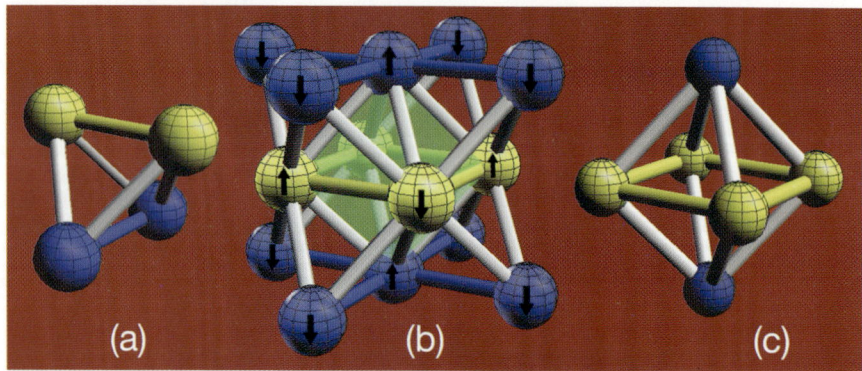

Fig. 10.4. The face-centered-cube (**b**) with a mean density of 0.17 nucleons/fm^3 contains high-density tetrahedral substructures (0.35 nucleons/fm^3) (**a**) and low-density octahedral substructures (0.09 nucleons/fm^3) (**c**) within it. The size of the nucleons is reduced here and in the following figures to facilitate an understanding of the lattice geometry

core ($\sim$0.17 n/fm^3) – in rough agreement with the results of Hofstadter, and shown in Fig. 10.3.

Other small nuclei of interest are ^{6}Li and ^{9}Be. Unlike ^{4}He, these nuclei have a large ratio of octahedral-to-tetrahedral substructures (Fig. 10.17 and 10.18), and therefore are expected to have relatively large radii and low densities (<0.17 n/fm^3). A quantitative estimate would demand further assumptions concerning the dimensions of the core and skin regions, but clearly the density predictions of the fcc model for these nuclei must lie between 0.09 and 0.17 n/fm^3, because of the abundance of octahedral (or portions of octahedral) subunits relative to tetrahedral subunits in the lattice structures for these nuclei. Other of the smallest nuclei exhibit relatively more octahedral structures than tetrahedral structures, in comparison with the 2:1 ratio in the fcc unit structure and should therefore show relatively lower densities – again solely on the basis of geometrical considerations.

The Nuclear Skin

Since most nuclei are approximately spherical, theoretically there should be no low-density skin region whatsoever in the naïve liquid-drop model. Even a prolate or oblate deformation of a droplet would give a constant density spherical core with a rapid drop from the core toward the poles of the ellipsoid shape. In other words, the known, relatively mild deformations of the majority of the non-spherical nuclei are such that, on average, there will be a skin region whose density falls from 100% to 0% over a distance of about 1.0 fm (see Chap. 7).

Because the shape and thickness of the nuclear skin (as distinct from the nuclear interior) is accurately known from electron-scattering experiments,

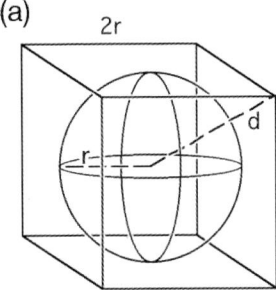 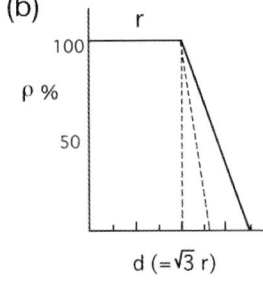

Fig. 10.5. The radial density curve of a solid cube. (**a**) A cube with edge length, 2r, shows a constant density core over the volume occupied by a sphere of radius r contained within. (**b**) The density falls gradually from its maximum core value at r to zero when the corners of the cube at distance d are reached. The *dotted lines* in (**b**) show the skin densities for a spherical liquid-drop and a liquid-drop with oblate deformation. Note that the density of the cube itself is not actually "diffuse" in its surface region except in terms of an average radial density value

the 2.4 ± 0.3 fm skin remains essentially unexplained in a liquid-drop formulation. The lattice model, in contrast, exhibits both a constant density core and a low density skin region as properties *inherent to the lattice build-up*, for the medium to large nuclei. In the lattice, the polyhedral build-up of nuclei inevitably implies a skin region into which the angular corners of the polyhedrons project. Although the density of the polyhedral shapes themselves (the lattice of nucleons) is constant, the average *radial* density will necessarily fall gradually from the core to the most peripheral nucleon.

Regardless of the exact structure of such polyhedrons, the thickness of the nuclear "skin" region is considerable. For example, consider the shape of the density curve for a cubic volume (Fig. 10.5). If the cube is homogenously packed with particles, the cube itself will of course have a constant density throughout its volume. However, as measured *radially* from the center of the cube, the constant-density region will extend only as far as the center of each face of the cube (r). Beyond that distance, the corners of the cube will protrude, so that the average radial density will gradually decrease until it reaches zero at the radial distance of the corners (d).

It is for this reason that the polyhedral shapes (generally approximating octahedrons) of the lattice model reproduce quite reasonable nuclear skin thicknesses. Portions of the lattice for any given nucleus extend beyond the spherical core region. In contrast to models which require further parameters to account for the diffuse nuclear surface, a skin region of gradually decreasing density is thus an intrinsic part of all lattice models, in general, and the fcc model, in particular (Fig. 10.6).

As illustrated in Fig. 10.7 for the ^{59}Co, ^{115}In and ^{197}Au nuclei, the fcc model unambiguously shows a nuclear surface region that is qualitatively

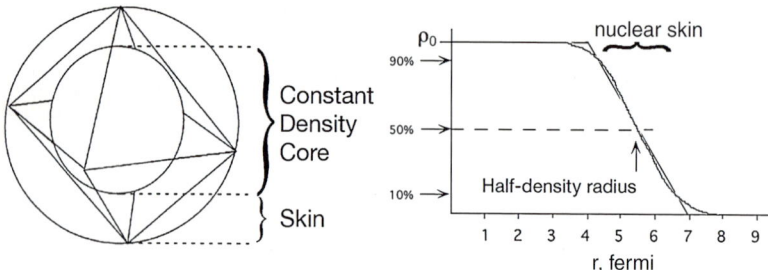

Fig. 10.6. The octahedral symmetry of the fcc build-up procedure produces both approximately spherical nuclear shapes ($x = y = z$) and thick skins ($2 \sim 3\,\text{fm}$) for all nuclei

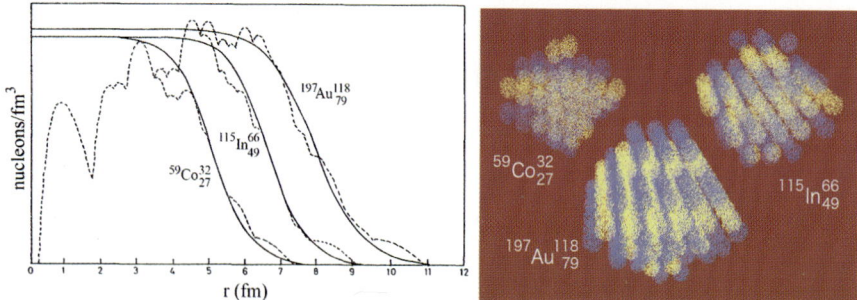

Fig. 10.7. The nuclear density of medium and large nuclei (*solid curves* from Hofstadter, 1956). The *dashed lines* are the densities measured from the center of the fcc lattice with a nucleon radius of 0.86 fm. Despite the fact that the nuclear core has a constant density, low densities at small r are obtained because of the computational technique, in which the nuclear density is calculated from the nuclear center. The density of the nuclear skin region in the fcc model reproduces the Fermi curves (Cook & Dallacasa, 1987)

similar to that which is empirically known, without the use of a "surface thickness" parameter.

Moreover, due to the requirements of the antiferromagnetic fcc lattice with alternating isospin layers, all excess neutrons (N>Z) must go to the surface of the lattice, implying that the nuclear core will have similar neutron and proton densities and the nuclear skin will be neutron-rich. This does not mean, however, that the nuclear skin is entirely made up of neutrons. The fcc structure even for N:Z ratios of 3:2 has protons at sites well into the skin region, but there is inevitably an abundance of neutrons in the skin, as shown in Fig. 10.8. In experiments using uncharged pions to measure mass radii, comparisons of isotopes of the same element have shown small excesses of neutrons in the skin region. A particularly good example is seen in the comparison of ($N = Z$) ^{40}Ca and (N > Z) ^{48}Ca. Egger et al. (1977) measured the elastic scattering of 130 MeV pions from these nuclei and found ^{48}Ca to be 0.3 fm larger. In the

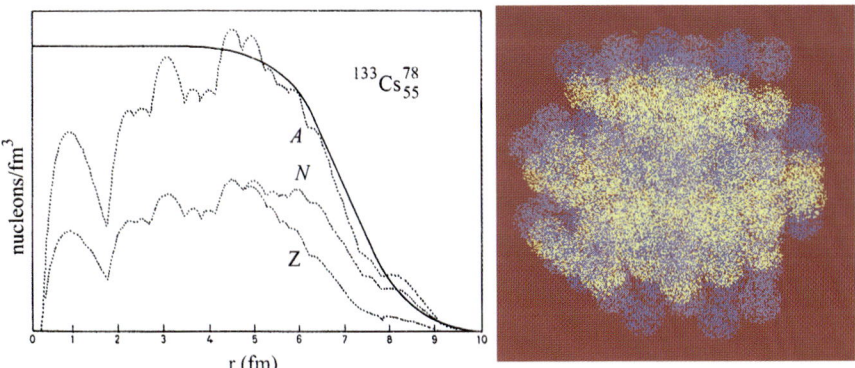

Fig. 10.8. The density of protons and neutrons in the fcc lattice for ^{133}Cs. The interior densities for both N and Z are identical, but there are excess neutrons on the nuclear surface (Cook & Dallacasa, 1987)

fcc model, the difference is also 0.3 fm. The density of protons and neutrons for the ^{133}Cs isotope is illustrated in Fig. 10.8. Note that the core densities for Z and N are identical, but the skin region shows an excess of neutrons. (This figure illustrates the static fcc structure; if the effects of charged pion exchanges were also considered, then all surface neutrons would have some proton-like properties that would lessen the high density of neutrons in the skin region.)

The difficulty for conventional models in explaining the nuclear surface arises from the fact that the Coulomb force among protons implies that freely-moving charged particles in a gaseous or liquid-phase equilibrate at positions where the repulsion is minimized. If indeed nucleons are free to move about in response to local forces, a lower energy configuration of protons and neutrons could be achieved if, all things being equal, the protons drifted to the outer surface. This is empirically not the case, so that theoretical mechanisms fundamentally at odds with liquid- or gaseous-phase models must be invoked in such models to achieve the approximately equal density of protons and neutrons in the nuclear interior. In the fcc model, the mechanism is the lattice structure itself.

MacGregor (1975) has calculated that the penalty that a large nucleus such as Uranium pays for not allowing protons to drift to the surface of the nucleus is in excess of 150 MeV. Yang and Hamilton (1996, p. 381) give an estimate of 100 MeV, but, in either case, the fact that there is no proton excess on the nuclear surface indicates that the protons are not free to move about within the nucleus. MacGregor argues that protons are bound within clusters in the nuclear interior, but another way in which protons could be trapped within the nuclear core would be if nuclei are stable only when they are arranged in a lattice.

Another interesting shape property of nuclei is that they tend to have an equatorial bulge (Powers et al., 1976). That is, the filling of nucleon orbitals proceeds from the equator and works toward both poles. The theoretical structures of the fcc lattice reproduce this feature, since – within any given n-shell – the nucleons with highest j-values have equatorial positions and are generally filled prior to lower j-value subshells (7/2 spin nucleons prior to 5/2 nucleons prior to 3/2 nucleons, and so on) within the same n-shell.

Charge and Mass Radii

The most accurately known measure of nuclear size is the root-mean-square (RMS) charge radius, for which experimental data are available for 621 isotopes. As discussed in Chap. 6, a formula that accurately ($R^2 > 0.99$) expresses the dependence of the nuclear radius on the number of nucleons (A) present is:

$$\text{RMS}(Z, N) = r_Z(2Z)^{1/3} + r_N(A - 2Z)^{1/3} \qquad (10.1)$$

where r_Z and r_N are constants indicating the size of the nucleons in the nuclear core and that of excess neutrons in the nuclear skin, respectively. This formula and its two constants ($r_Z = 1.002$ fm, $r_N = 0.010$ fm, when all known nuclear radii are used in the multiple regression and slightly different values when the smallest nuclei are excluded) accurately reflect the character of the experimental data, but it is important to note that there is nonetheless an implicit model in the use of these two constants. That is, the radius of the core nucleons, r_Z, is obtained under the assumption that the core contains equal numbers of protons and neutrons that contribute equally to the core density. The charge radius of the neutrons that lie beyond the $Z = N$ core, r_N, is assumed to be different. There are empirical grounds for making both of these assumptions and the high correlation between the experimental data and the calculated radii support the idea that should be treated differently, but this is nonetheless a model that assumes there is a $N = Z$ core and a neutron-rich skin. By using the values obtained for the constants in (10.1), the lattice model does not introduce any new parameters for calculating radii. The lattice structures themselves specify the relative positions of the nucleons in nuclear space and the empirical constants determine how great the contribution of each proton and neutron is to the overall radial measure.

The results from the lattice model are shown in Fig. 10.9. It is seen that, with the exception of a few of the smallest nuclei, the error is generally within ±0.2 fm (mean and SD of 0.08 ± 0.08). This fit between the fcc model and the experimental data is only marginally worse than the fit between (10.1) and the data (mean and SD of 0.06 ± 0.08) (Chap. 6). In other words, the lattice model reproduces nuclear size data extremely well *without adding any adjustable parameters*, and shows quantitatively how similar the lattice and liquid-drop conceptions of the nucleus are.

Of particular interest is the fact that, by using the default nucleon build-up sequence to obtain radial values (rather than manipulating the build-up

10.1 Nuclear Size and Shape   221

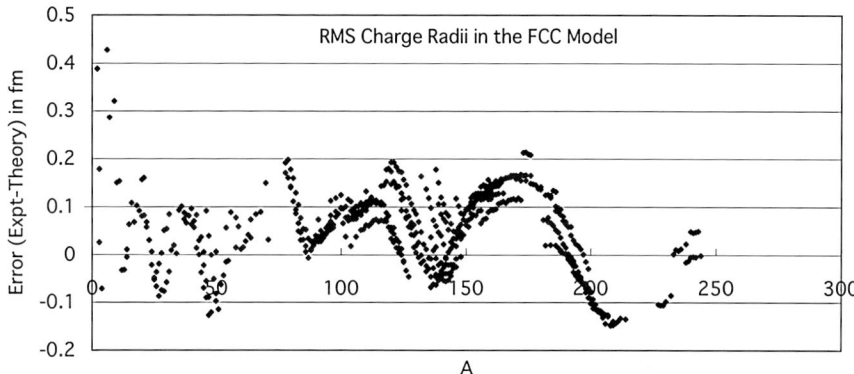

Fig. 10.9. The difference between experimental radial values and those obtained with the fcc model

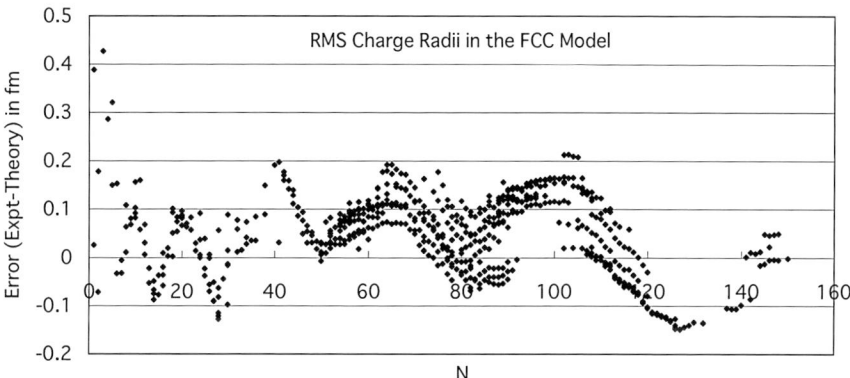

Fig. 10.10. A plot of the mean error (experiment-theory) versus neutron number

sequence to obtain better agreement between theory and experiment), the difference between experiment and theory shows signs of shell structure. These effects reflect the J-subshell structure inherent to the lattice. As shown in Fig. 10.10, when the error is plotted against the neutron number, N, the model is found to over-estimate the size of nuclei at magic numbers 28, 50, 82 and 126. When plotted against the proton number (Fig. 10.11), magic effects are seen at 82, with some indication of subshell effects around 14, 20, 50 and 54.

Because the high-angular momentum j-values in the fcc model correspond to equatorial lattice positions, they are sometimes loosely bound (having a relatively small number of nearest-neighbors). Optimization of the structure of all nuclei in the lattice model would mean moving high-j nucleons to lower j-value positions. This would imply a small increase in the number of two-body bonds and therefore greater nuclear binding energy; it would also simultaneously bring small reductions in the RMS charge radii. These effects imply that

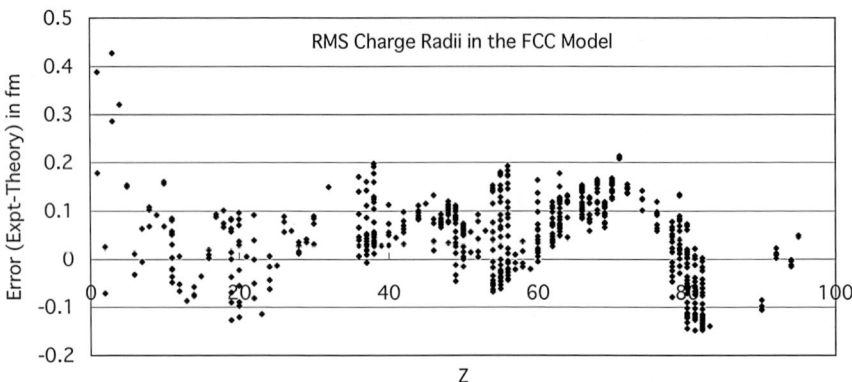

Fig. 10.11. A plot of the mean error (experiment-theory) versus proton number

a hand-crafted, *post hoc* manipulation of the fcc structures would reduce the spurious indications of magic shells (for which there is in fact little empirical indication, as shown in Figs. 2.26 and 2.27 and improve the fit between theory and experiment.

It remains to be seen how close a fit between theory and experiment for radial measures might be obtained by individually constructing fcc nuclei in light of J-values and binding energies. Nevertheless, the results shown in Figs. 10.9 through 10.11 already show that fcc structures reproduce experimental RMS charge radii reasonably well without appending parameters that are essentially foreign to the basic model assumptions (i.e., the fcc build-up procedure and the differential contribution of protons and neutrons to the charge radius).

10.2 The Alpha-Particle Texture of Nuclei

One of the original motivations behind the alpha-particle model in the 1930s was the fact that, among the naturally radioactive nuclei, the alpha particle was known to be one of the principal emissions. Since such radioactivity is conceptualized as the evaporation of alpha-particles from the nuclear surface, the high rate of alpha-particle production suggested that alphas might exist, at least transiently, as bound systems on the nuclear surface.

Subsequent to the early interest in the alpha-particle models, the shell and liquid-drop models became dominant and the cluster models were applied almost exclusively to problems involving small numbers of nucleons, generally 40 or less, where the configuration of 10 or fewer alphas remained a tractable problem in solid geometry. In the early 1970s, however, there was renewed interest in nuclear clustering in light of results from quasi-fission and multifragmentation experiments. The basic experimental finding has been that, when medium and large nuclei are bombarded with relatively high-energy

10.2 The Alpha-Particle Texture of Nuclei

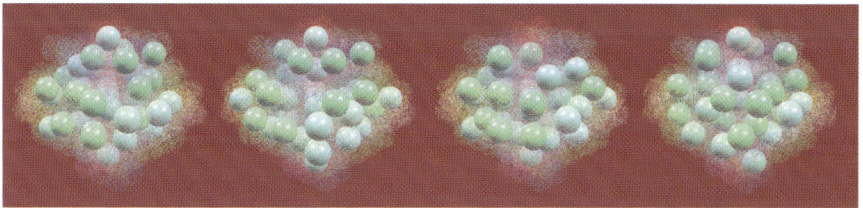

Fig. 10.12. Changing configurations of alphas (*solid spheres*) in the interior and on the surface of the fcc structure for a large, alpha-emitting nucleus, $^{246}\text{Cm}^{148}$. Nucleons are depicted as probability clouds ($r = 1.5\,\text{fm}$)

particles – not merely enough to strip the nucleus of one or a few nucleons, but enough to shatter it into small fragments, there is an unexpectedly large number of alpha particles and multiples of alpha particles among the breakup fragments. Such results are strong indication that there is alpha clustering *throughout the interior* of *all* nuclei – small, medium and large (MacGregor, 1976).

Qualitatively, the fcc model can of account for such clustering simply from the fact that a close-packed lattice of nucleons can be viewed as a regular array of tetrahedrons. All nuclei ($Z > 2$) contain tetrahedral alphas within the lattice structures and, as seen in Fig. 10.12, any large nucleus with excess ($N > Z$) neutrons necessarily has tetrahedral alphas on the surface. Quantitative prediction of which nuclei will eject such clusters from the lattice would require a dynamical theory of the movement of charges within the lattice (that has not yet been achieved), but the qualitative effect of having alpha tetrahedrons on the nuclear surface is consistent with the phenomena of alpha decay.

Undoubtedly the most interesting example of alpha clustering among the small nuclei in the fcc model is ^{40}Ca. The *only* geometrical configuration of the ground-state of the 10 alphas for ^{40}Ca that has been reported in the physics literature is a structure described as a "tetrahedron of alphas lying inside of an octahedron of alphas" (e.g., Hauge et al., 1971; Inopin et al., 1979) (Fig. 10.13b). That configuration has been used to explain many of the low-lying excited states (Fig. 10.13a), as well as the electron form factor for this nucleus (Fig. 10.13c). Interestingly, an *identical* alpha-particle geometry is found in the default build-up of the fcc lattice (see Figs. 10.13d and 10.14).

As is evident in Fig. 10.14, the fcc structure for ^{40}Ca is simultaneously: (i) a (frozen) liquid-drop with all nucleons interacting only with nearest-neighbors, (ii) three symmetrical closed shells containing 4, 16 and 40 nucleons – corresponding to the doubly-magic ^{4}He, ^{16}O and ^{40}Ca nuclei, and (iii) a conglomerate of 10 alpha particles – 4 forming a tetrahedron on the inside and 6 forming an octahedron on the outside. Could there be a better example of how all three of the conventional models of nuclear structure theory accurately describe one and the same nucleus in completely different ways? And all three are reproduced in the fcc lattice.

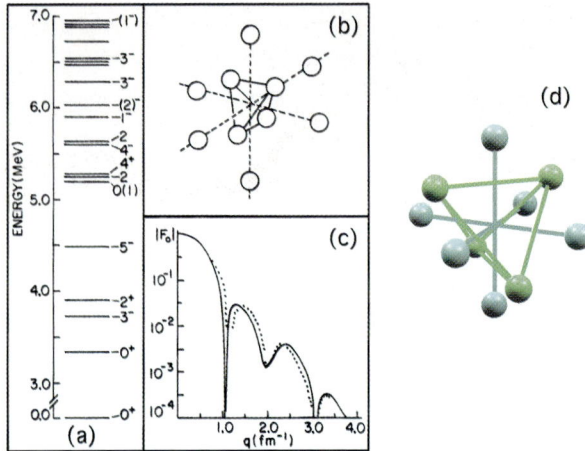

Fig. 10.13. The conventional alpha configuration for the 10 alphas of ^{40}Ca. The excited states in (**a**) and form factor in (**c**) are implied by the alpha structure in (**b**) (from Hauge et al., 1971; Inopin et al., 1979). The alpha configuration implied by the fcc model is shown in (**d**) (compare with Fig. 10.14)

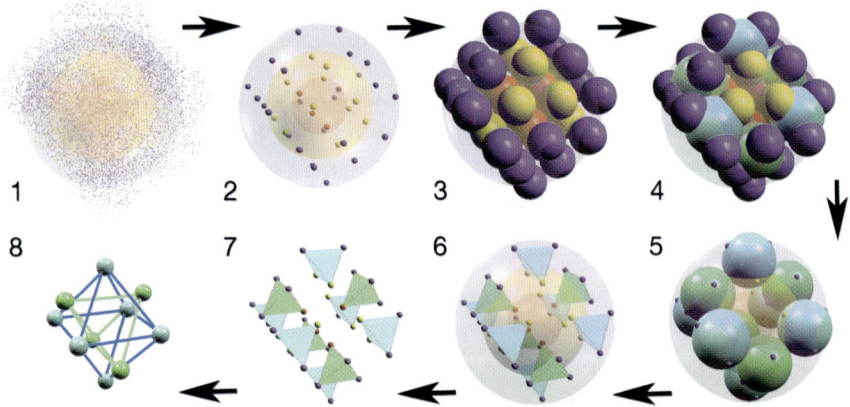

Fig. 10.14. The ^{40}Ca nucleus in the fcc model. (**1**) shows the fcc configuration for ^{40}Ca with nucleons depicted as diffuse probability clouds in relation to the three first n-shells. (**2**) shows the same 40 nucleons as point particles located at fcc lattice sites. (**3**) shows the same nucleus with nucleons depicted with realistic radii (0.86 fm). (**4**) shows the same nucleus with 10 alpha particle spheres joining neighboring nucleons. (**5**) shows the same alpha configuration with nucleons again reduced to points. (**6**) shows the same alpha configuration with alphas depicted as tetrahedra. (**7**) shows the same 10 alpha particles with the three translucent n-shells removed. And finally (**8**) depicts the alphas as a tetrahedron of green alphas inside of an octahedron of blue alphas

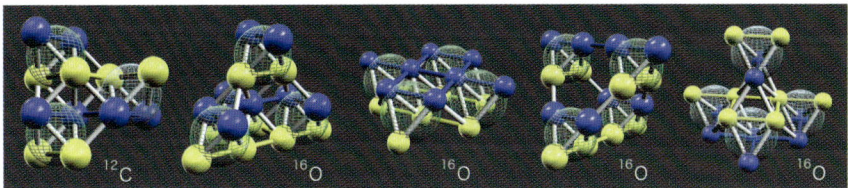

Fig. 10.15. A few of the many possible alpha-cluster configurations for ^{12}C and ^{16}O in the fcc model. Unlike the ^{40}Ca alpha-cluster model in Fig. 10.14, none of these structures are likely to be ground-states since the number of 2-body bonds is not maximized

Because of the inherent tetrahedral packing in the fcc lattice, there are inevitably many possible cluster configurations for the small $Z = N$ nuclei. A few examples are shown in Fig. 10.15. Judging from the total electrostatic repulsion and the total number of nearest-neighbor bonds in such structures, it is unlikely that these configurations are ground-states, but a more detailed examination of binding energies would be needed to draw firm conclusions.

10.3 Nuclear Spin in the Lattice Model

One of the main strengths of the shell model is its ability to account for the total angular momentum (spin) values of a wide range of nuclei. As reviewed in Chap. 2, the shell model's predictions concerning the magic numbers are less impressive and raise some difficult questions about the meaning of magicness, but the spin predictions are unambiguous and agree well with experimental data. Given the starting assumptions of the independent-particle model (the central potential-well of the nuclear force, spin-orbit coupling and the independent-particle description of nucleon states), the build-up procedure correctly predicts many nuclear spin values and the remaining J-values can be plausibly accounted for on the basis of configuration-mixing. In other words, even when the shell model does not predict the correct experimental spin from a "default" nucleon build-up procedure, the correct spin values can be obtained simply by assuming that there is some variation in the sequence of nucleons (likely due to local interactions among nucleons). The explanation of nuclear spins is a genuine strength of the shell model and, as a matter of historical fact, was one of the principal reasons why the manifestly unrealistic assumptions of the "orbiting-nucleon" independent-particle model were taken seriously in the 1950s.

How does the lattice model compare with the shell model successes here? In Chap. 9, it was shown how the j-value of an individual nucleon is defined in terms of its coordinate values within the lattice and that the occupancy (the number of nucleons per n-shell with a given j-value) is identical to the results of the shell model. If it is assumed that the nuclear J-value is simply the summation of nucleon j-values – as postulated in the shell model, then the lattice model will necessarily predict nuclear spins in the same way as the shell model.

What distinguishes the lattice model from the shell model is that there are geometrical constraints on the binding configurations in the fcc model. Specifically, the geometry of any configuration of Z and N will have a total Coulomb repulsion that is calculable from the configuration of protons and a total number of nucleon-nucleon nearest-neighbor bonds that is also directly computable from the fcc geometry. These two factors will determine the optimal fcc structure for any given combination of Z and N. As a consequence, it can be said that there are three assumptions in the lattice model that are required to make predictions about nuclear spins. The first is identical to that of the shell model, whereas the latter two are unique to the lattice model:

(i) The total spin of a nucleus is due to the summation of all nucleon spins, with up/down spins of like-isospin pairs of the same j-value canceling each other out. This means that, in general, the nuclear spin will be zero for the ground-state even-even nuclei, equal to the spin of any unpaired proton or unpaired neutron for the odd-even or even-odd nuclei, and equal to the sum or difference of the odd nucleons for the odd-odd nuclei.

(ii) In first approximation, the number of nearest-neighbor nucleon-nucleon bonds must be maximized in the fcc lattice.

(iii) Given equivalent numbers of nearest-neighbor bonds, the configuration which minimizes the electrostatic repulsion among the protons will be favored.

If fcc structures are *not* optimized using assumptions (ii) and (iii), then the fcc model is identical to the unadorned shell model, i.e., the harmonic oscillator plus spin-orbit coupling, with many equivalent configurations of nucleons showing the same J-values. Using a default build-up sequence in the lattice – essentially adding nucleons incrementally to lattice positions around a central tetrahedron – an approximate correspondence with the shell model and with experimental J-values is in fact obtained. The default sequence, however, ignores relevant and, in the lattice model, calculable effects of the local structure of the lattice – i.e., the placement of nucleons within the various subshells and the local correlations (bonds) with other nearby nucleons. For that reason, a default (=shell model) sequence does not always provide the best possible lattice structures, and the structure implied by maximization of two-body bonds and minimization of Coulomb repulsion should be used instead.

Given these constraints on the nuclear build-up, the configuration for given Z and N that gives maximal nuclear binding can be found and a total nuclear spin is necessarily implied by the coordinate positions of any unpaired proton and/or neutron. Further manipulation of the lattice structures, not unlike the manipulations within the shell model, can of course be made on the basis of a more sophisticated treatment of the nuclear force (the effects of nucleon spin

10.3 Nuclear Spin in the Lattice Model

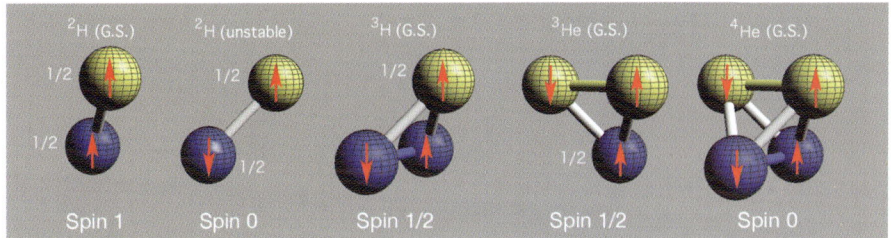

Fig. 10.16. The configurations of nucleons in the smallest nuclei. The up/down spins of all nucleons are indicated by the red arrows. Protons are yellow, neutrons are blue. The j-values of unpaired nucleons are as calculated in the fcc model, and are shown near to any unpaired nucleon(s) in the nucleus. The total spin is shown below. On the far left is the ground-state (GS) of the deuteron, with a total spin of 1. If a proton-neutron pair with opposite spins is chosen, the spins sum to zero and give an unbounded state. In ^{3}H and ^{3}He, the spins of the paired neutrons and protons, respectively, cancel one another, so that the nucleus as a whole has a spin dependent on the last, unpaired nucleon (1/2). On the right, again the spins of the paired nucleons cancel, giving a net spin of 0 for the ground-state of ^{4}He

and isospin, etc.), the weaker effects of second-nearest-neighbors, and so on. Such manipulations are of interest for obtaining the best fit between theory and experiment, but the present analysis will remain at a more basic level without addressing problems concerning the nuclear force.

Let us examine the implications of the nucleon build-up procedure in the lattice model for nuclear J-values, starting with the smallest nuclei (Fig. 10.16). For clarity in the description of the underlying geometry, only the billiard-ball depictions of nucleons will be used, but the reality is undoubtedly more complex. The configurations of the smallest nuclei are rather trivial, but the process of summation of nucleon spins, as illustrated in Fig. 10.16, is the starting point for demonstrating the relationship between the fcc lattice and experimental J-values. It is noteworthy that there is evidence for a slightly elongated prolate shape for the deuteron, an oblate shape for the triton and helion, and a tetrahedral structure for the alpha particle, with a low density central region. Beyond ^{4}He, the number of possible fcc (or shell model) configurations increases rapidly, so that the combinations of nucleon j-values in relation to the number of nucleon bonds and total Coulomb effects must be examined.

Figure 10.17 shows the ground-state and first three excited states of ^{6}Li$_3$ in the fcc model. It is important to note that *the spin states of all of the nucleons are identical to the states as specified in the independent-particle model*. What the fcc model adds to the independent-particle model description is specific (sometimes unique, but more often a small number of possible) geometrical configurations of nucleons, implying a specific set of nucleon-nucleon interactions.

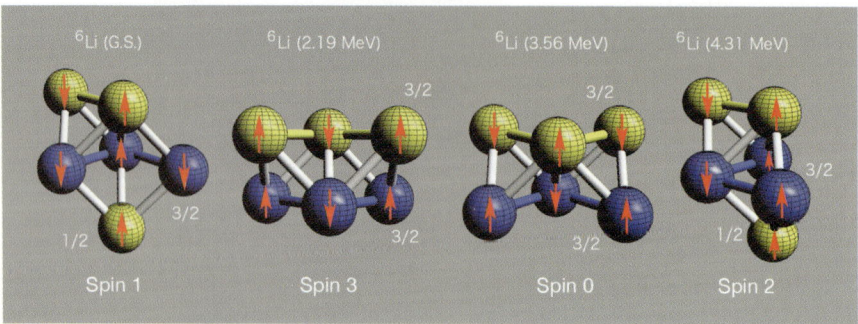

Fig. 10.17. The configurations and spins of ^{6}Li. On the left is shown the ground-state with a spin of 1 (the summation of spin 3/2 and spin 1/2 nucleons with opposite orientations). In the *middle* are shown two excited states, whose unpaired nucleons sum to 3 or 0, depending on the orientations of the 3/2-spin proton and neutron. On the *right* is shown an excited state that differs from the ground-state solely in the location of the unpaired 1/2-spin neutron; the odd-proton and odd-neutron sum to a spin of 2. Protons are yellow, neutrons are blue. Note that all nearest-neighbor protons (neutrons) within a layer always have opposite spins (the antiferromagnetic arrangement)

It is a remarkable empirical fact of nuclear structure that, with the exception of 12 alpha-emitting isotopes, there is a stable combination of protons and neutrons for every value of A ($Z + N$) from 1 (Hydrogen) to 209 (Bismuth) – and all nuclei remain in that stable configuration essentially forever unless subjected to strong external excitation. For any combination of Z (>2) and N (>2) that exists in a stable state, there are various possible excited states, but nuclei remain in such excited states for extremely short periods before decaying to the one stable configuration of Z and N. In contrast, there are another 1800+ isotopes with unfavorable (unstable) Z:N ratios that nonetheless have significantly long half-lives. Although these radioactive isotopes eventually emit radiation and settle into a different ratio of protons and neutrons on their way toward stability, they are of great importance for understanding nuclear stability. In comparison with the excited states of stable combinations of Z and N, the half-lives of most of the isotopes with unstable combinations of Z and N are *not* short by the time-scale of nuclear events. This is easily understood if we compare virtually any of the radioactive isotopes with the lifetimes of their excited states. For example, the ^{6}Li isotope is stable in its ground-state, but the first three low-lying excited states exist for less than a femtosecond (10^{-12} sec). The significance of this fact for nuclear structure theory is that, for any given combination of Z and N, there is one, and normally only one (discounting a small number of isomers) configuration that is stable or semi-stable. In contrast, the vast majority of excited states decay to the ground-state in a few femtoseconds. Once having settled into that ground-state, however, either the ground-state is stable or the transition to a different

10.3 Nuclear Spin in the Lattice Model

isotope with a more favorable Z:N ratio occurs quite slowly – half-lives that can be millions of years long and are always many orders of magnitude greater than the half-lives of excited states.

Therefore, the display of possible excited states in the fcc model (Figs. 10.17 through 10.19) is of interest in demonstrating that there are theoretical fcc (or, equivalently, shell model) states that correspond to the measured spin states of excited nuclei. However, since all such excited states are extremely short-lived, their significance for deciphering the riddle of nuclear binding energies is limited. In this respect, the ground-states of the stable and radioactive isotopes are more important than these excited-states, insofar as only the ground-states provide information on nuclear configurations that have some degree of longevity.

As illustrated in Figs. 10.18 and 10.19, isotopes of ^{7}Li through ^{16}O show ground- and excited-states with total J-spins that are the sum of all nucleon j-spins. The spin 7/2 excited state of ^{7}Li is an exotic configuration with only one unpaired 7/2 spin proton. Other configurations are possible, and the more complex coupling LS-coupling schemes used in the shell model make for a much larger number of possible combinations. Whether they are needed in the fcc model remains unclear, and must be determined from more fundamental considerations of the nuclear force. The ground- and low-lying excited states for Beryllium, Boron, Carbon, Nitrogen and Oxygen nuclei are also shown. The configurations of protons and neutrons in these figures are plausible structures on the basis of the number of two-body bonds, electrostatic repulsion among protons and the implied nuclear spin values, but other structures with equivalent J-values are also possible. A rigorous, deductive method for determining fcc structures, spins, binding energies, etc. remains to be devised, but what these figures illustrate is that it is virtually always possible to construct fcc structures with the experimentally known spins (Table 10.1), simply because the nuclear build-up procedure is identical to that of the independent-particle model.

The build-up *procedure* in the fcc model is similar for all nuclei – requiring maximization of the number of two-body bonds and minimization of Coulomb repulsion among all protons. In general, this leads to compact structures, although the reduction of proton repulsion and the tendency to favor higher over lower j-value positions tend to make nuclei less compact, while maximization of two-body bonds tend to make nuclei more compact.

Because the fcc lattice has the exact same set of individual nucleon states as specified in the independent-particle model, there is ambiguity in both models regarding which of several equivalent sites (in terms of j-values) are to be filled first. Configuration-mixing and intruder states are an empirical reality, but the remarkable strength of the independent-particle model has been that it has always been possible – often on a *post hoc*, but not arbitrary basis – to justify why a particular isotope has one or another total spin value, simply as the summation of nucleon j-values. In this regard, the fcc model has the same strengths as the independent-particle model, but the fcc model has

230   10 The Lattice Model: Experimental Issues

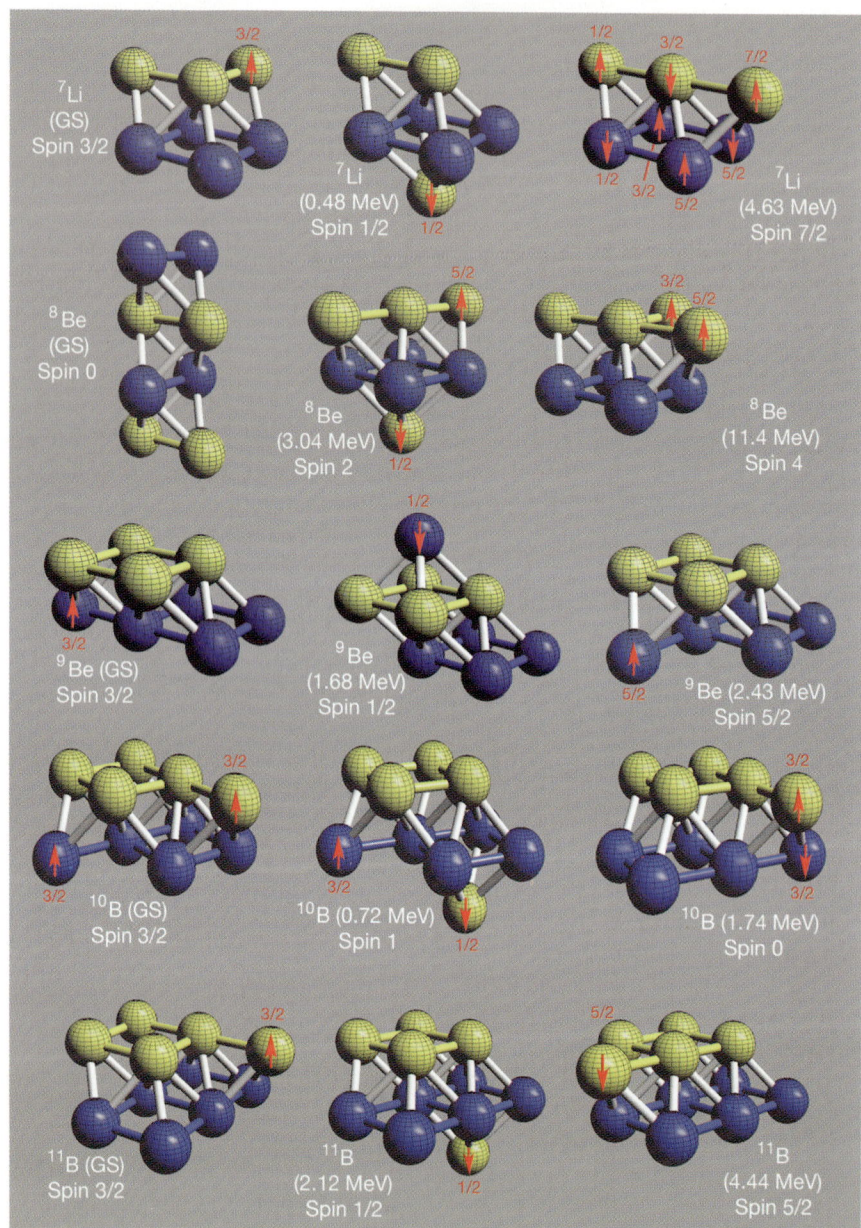

Fig. 10.18. The ground-state and low-lying excited states of ^{7}Li, ^{8}Be, ^{9}Be, ^{10}B and ^{11}B. Only the spin orientations (*red arrows*) and *j*-values of the unpaired nucleons are shown. The spins of the ground- and excited-states of these nuclei are determined by the spin of the last unpaired neutron/proton (noted by the *red arrows*)

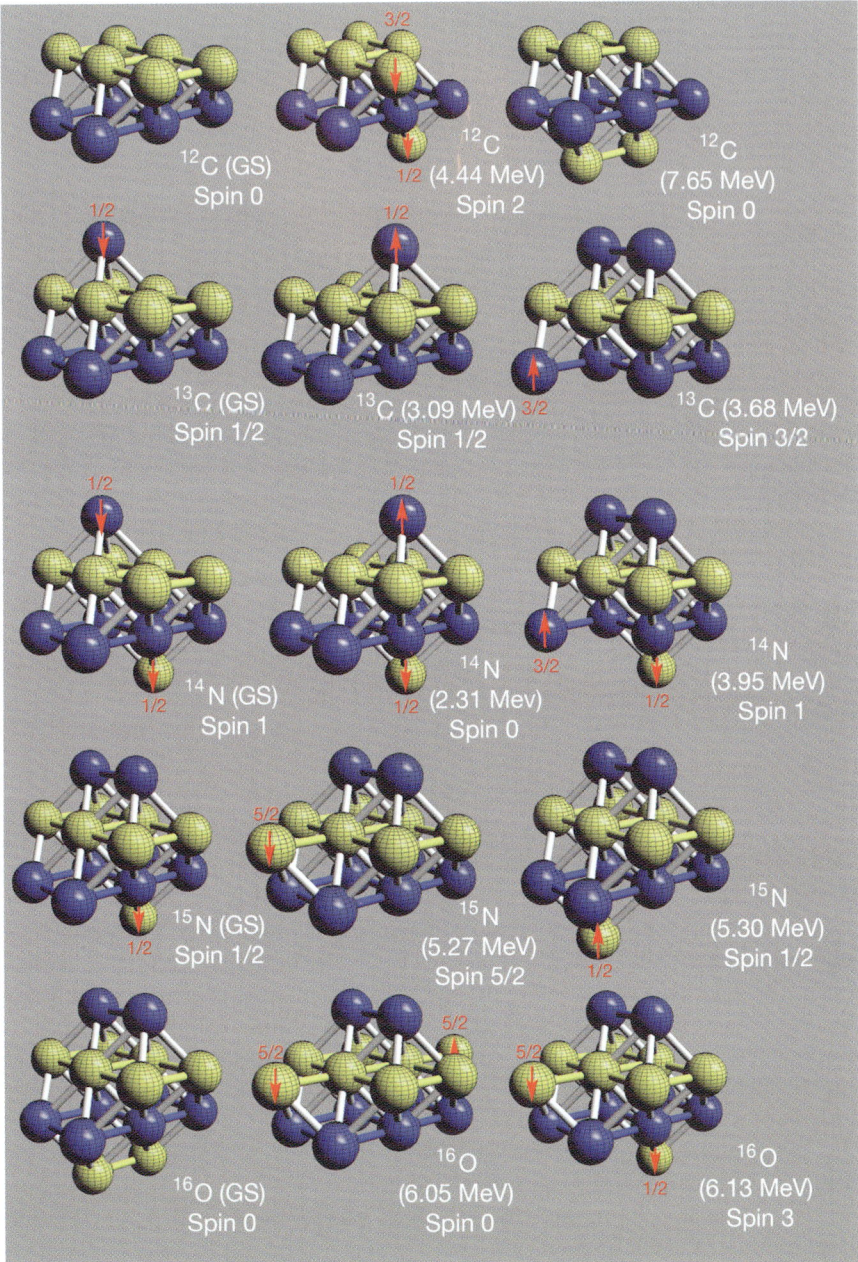

Fig. 10.19. The ground-state and possible excited states of ^{12}C, ^{13}C, ^{14}N, ^{15}N, and ^{16}O

Table 10.1. Spins in the fcc model for $Z < 21$

Isotopes	No. of spins predicted by the default fcc build-up	No. of spins explained by fcc configuration-mixing	Unexplained spins
neutron	1	0	–
$^{1\sim 6}H_1$	5	1	–
$^{3\sim 8}He_2$	4	2	–
$^{5\sim 11}Li_3$	5	2	–
$^{7\sim 15}Be_4$	9	0	–
$^{8\sim 19}B_5$	9	3	–
$^{10\sim 21}C_6$	11	1	^{21}C (1/2)
$^{10\sim 20}N_7$	10	1	–
$^{11\sim 24}O_8$	14	0	^{23}O (1/2)
$^{14\sim 28}F_9$	9	6	–
$^{15\sim 31}Ne_{10}$	14	3	–
$^{17\sim 31}Na_{11}$	5	10	–
$^{19\sim 32}Mg_{12}$	12	2	–
$^{21\sim 31}Al_{13}$	5	6	–
$^{22\sim 36}Si_{14}$	13	2	–
$^{24\sim 33}P_{15}$	0	10	–
$^{26\sim 44}S_{16}$	17	2	–
$^{28\sim 49}Cl_{17}$	9	13	–
$^{30\sim 45}A_{18}$	15	1	–
$^{32\sim 52}K_{19}$	8	13	–
$^{34\sim 52}Ca_{20}$	14	5	–
Totals	189	83	2

additional geometrical constraints that are totally absent in the independent-particle model.

The nuclear binding problem is far from trivial in the fcc model (see Sect. 10.5, below), and has not in fact been solved at the level of first-principles concerning the nuclear force. At present, what is clear is that there is a direct correspondence between the fcc lattice and nucleon quantum numbers, so that any independent-particle model configuration of protons and neutrons has an equivalent fcc model configuration with a spin value that can be calculated from the fcc nucleon coordinates. In ground-states, the spin is always zero for even-even nuclei and otherwise the sum or difference of the spins of the odd nucleons. By such logic, it is possible to construct a small number of equivalent nuclei in the fcc model that have the correct j-value and, at the same time, minimal Coulomb repulsion and maximal two-body bonds.

Without a microscopic theory of the nuclear force that takes all quantum characteristics into account to calculate the binding energies of ground- and excited-states, it cannot be said that fcc nuclei are built on the basis of first principles, but it is worth pointing out that the process of nucleon build-up in the lattice is fundamentally a tractable problem. Unlike the shell model approach, that suffers from an unmanageable combinatorial explosion (e.g., 10^{18} states for medium-size nuclei), the lattice model has geometrical constraints that are absent in a gas and greatly reduces the permutations of possible states.

To illustrate the capabilities of the fcc model with regard to nuclear J-values, all 274 isotopes of the first 20 elements have been constructed using the *NVS* software (see the Appendix), and the J-value results are summarized in Table 10.1. It is seen that the default build-up accounts for most (69%) of the J-values, and all but two very short-lived isotopes (for which there is some uncertainty concerning the spin value) can be accounted for by means of configuration-mixing. Such results are virtually the same as those obtainable in the shell model and thus not a particular strength of the fcc model, but, again, that is precisely the point. Using either the default build-up procedure or reasonable possibilities for configuration-mixing, both models can account for the spins of nearly all known isotopes.

10.4 The Coulomb Force and Super-Heavy Nuclei

There is a small set of nuclei, known as the mirror nuclei, for which the binding energies are experimentally known for both possible combinations of Z and N summing to A. These nuclei are of importance to theoretical physics because they present a relatively clear picture of the Coulomb force acting within nuclei. The simplest example is the $A = 3$ mirror nuclei, ^{3}H and ^{3}He. They have the same spin properties, so that it is likely that the second neutron in ^{3}H and the second proton in ^{3}He are in the same quantal state. The difference in binding energy is 0.764 MeV – approximately the value obtained from a classical calculation of the Coulomb effect for two charges separated by 2 fm – and therefore thought to be due primarily to the electrostatic repulsion of the two protons in ^{3}He.

For larger mirror nuclei that differ solely by the presence of one last proton or neutron, the Coulomb effect of the last proton grows significantly, since it feels the repulsion of many more protons than was the case for ^{3}He. In the liquid-drop model, the Coulomb term in the binding energy formula is proportional to A (and to the nuclear radius), and that term produces a good approximation of the Coulomb effect (Fig. 10.20). Not surprisingly, all of the various lattice models reproduce the experimental data reasonably well.

The slight overestimate in the liquid-drop model and the slight underestimate in the lattice models are consequences of model parameters, but what is significant for all models of the nucleus is the ever-increasing penalty that the nucleus must pay for containing positively-charged particles. Experimental data for the mirror nuclei extend only as far as $Z = 38$, where already there is a penalty of ~12 MeV for every proton added. In any of these models, the theoretical Coulomb cost for adding one last proton when Z approaches 100 is therefore more than 20 MeV. At the same time, the additional binding

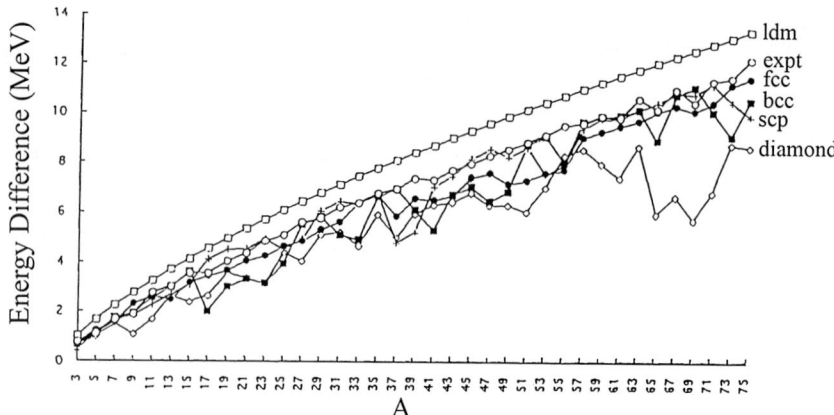

Fig. 10.20. The binding energy differences in the mirror nuclei. Assuming that the nucleons in mirror nuclei are in the same quantal states, the difference in binding energy is due solely to the electrostatic effect of the last proton

force per nucleon that is obtained by adding a proton or neutron is known not to increase beyond $Z = 28$ – suggesting that the indefinite growth of nuclei will not be possible. At some point, the increased electrostatic repulsion incurred by adding protons will outweigh the gain in binding energy – and that is where the periodic chart must come to an end.

In the conventional view of the nucleus, the only factor that might delay the end of the periodic table is the slight increase in binding obtained by the closure of further magic shells – and that effect has been the basis for several generations of calculations concerning the so-called super-heavy nuclei. In the fcc model, all variations in nuclear bonding – including "magic" stability – are a consequence of calculable local binding between nearest-neighbors, i.e., the relative compactness of lattice structures and Coulomb effects of various proton configurations. In other words, there is no "magic" other than nearest-neighbor effects, and therefore no possibility of constructing super-heavy nuclei once the addition of a proton implies more Coulomb repulsion than the total nearest-neighbor attraction generated by nearest-neighbor bonding.

Hypothetical super-heavy nuclei of any size can of course be built in the fcc model, but there are energetic limitations on nucleon build-up inherent to the lattice, as illustrated as in Fig. 10.21. Here a nucleus containing 112 protons and 168 neutrons is shown. Still more neutrons might be added to surface positions, giving even larger shelves of neutrons above and below every proton layer, but all nuclei will have the basic $N = Z$ core with alternating isospin layers and excess neutrons on the surface, as shown in the figure. As indicated from the mirror nuclei, the addition of one more proton to $^{280}Xx_{112}$ implies a Coulomb repulsion of about 20 MeV (with values calculated in the fcc model ranging between 17.57 and 24.59 MeV, depending on the proton's precise location). No matter where the proton is added, however, its contribution to

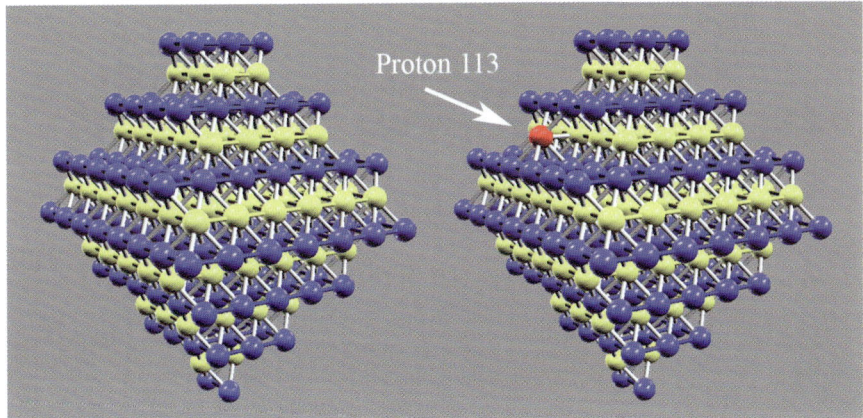

Fig. 10.21. Addition of a proton to one of many favorable positions on the surface of the ^{280}Xx$_{112}$ isotope adds only 7 nearest-neighbor bonds

nuclear *binding* cannot be more than the maximum produced by creating 12 new nearest-neighbor bonds. Moreover, the likely surface positions (Fig. 10.21) allow for only 7 new nearest-neighbor bonds. From the liquid-drop model, we know that the so-called volume term of the binding energy formula implies a binding force of only 16 MeV, so that 7/12 (on the surface) or 12/12 (in the nuclear interior) of 16 MeV can produce only 9 ∼ 16 MeV attractive binding to counteract the 17 ∼ 24 MeV of Coulomb repulsion (Cook, 1991).

These admittedly rough calculations clearly indicate that, no matter how great the neutron excess on the nuclear surface, the construction of stable super-heavy nuclei in the fcc model will be impossible insofar as nuclei have $Z = N$ nuclear cores and surface positions that can contribute only ∼16 MeV to nuclear binding. In the lattice model, "magic" stability beyond what local-binding provides is not possible, so that an island of stability at $Z > 114$ (super-heavy nuclei) is not predicted.

10.5 Nuclear Binding Energies

In the fcc model, nuclear binding is necessarily due to local (nearest-neighbor) effects, plus the Coulomb force acting among all protons. Since the semi-empirical mass formula requires the assumption of terms that reflect volume, surface, Coulomb, pairing, symmetry and shell effects, a lattice account of nuclear binding must also produce such effects. The easiest and most obvious are the volume and surface terms: any nucleon in the lattice will have 1 ∼ 12 nearest-neighbors, 12 for all core nucleons and 3 ∼ 9 for most surface nucleons, but both terms are merely the summation of nucleon-nucleon bonds. The Coulomb effect can be calculated exactly, so that the rough outlines of nuclear binding energies can be obtained simply by assuming a "net binding

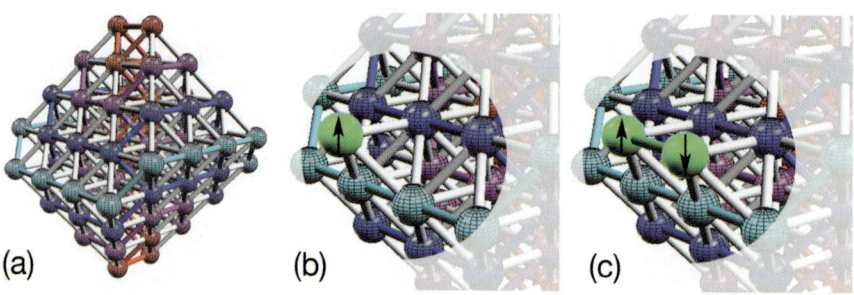

Fig. 10.22. Pairing effects in the fcc lattice. (**a**) depicts a closed n-shell nucleus (^{80}Zr^{40}). (**b**) shows the bonding of one added (*green*) nucleon to the core. (**c**) shows the pairing effect, i.e., three more bonds with the addition of a second (*green*) nucleon, plus a bond (*green*) between the two new nucleons

force" for the nucleon-nucleon bond (∼2.8 MeV) (regardless of spin/isospin properties) in the lattice. Somewhat better fit with the data is obtained by assigning bond energies that are dependent on relative spin and isospin, and excellent agreement is obtained by including a binding energy parameter that is dependent on the number of bonds per nucleon (see the NVS program for details).

It is worth pointing out that the correlation coefficient that expresses the fit between experimental data and theoretical values is high ($R^2 \sim 0.99$) *regardless* of the lattice type (fcc, hcp, scp, etc.) (Cook, 1994) – and reflects only the fact that nuclei are dense conglomerates of locally-interacting nucleons, not unlike a liquid-drop! As a consequence, any liquid- or solid-phase model (unlike a gas) necessarily has volume and surface effects, the magnitudes of which become adjustable parameters in the specific model. The more parameters used, the better the fit, but, simultaneously, the worse the model unless the parameters can be justified on the basis of the nuclear force.

Interestingly, because of the spin- and isospin-layering in the fcc model, there are, respectively, pairing and symmetry effects implicit to the lattice structure. With regard to the pairing effect in the model (Fig. 10.22), the sequential addition of same-isospin nucleons in each j-subshell begins with an unpaired nucleon; addition of a second nucleon with the same j-value and same isospin will most frequently lead to a nucleon located at a neighboring (same-layer, opposite-spin) site. The spins (and magnetic moments) thereby cancel out, but there is an increase in binding energy (BE/A) that reflects not only the addition of several nucleon-nucleus bonds, but also the nucleon-nucleon bond between opposite-spin pairs in the current j-subshell. In other words, the known pairing effect in the binding energy reflects the fact that, in the fcc model, an even-number of neutrons (protons) will normally have one more nucleon-nucleon bond (between the pairs), in addition to the bonds that each nucleon makes with the core lattice. Depending on the specific j-subshells and the local lattice structure, the strength of the pairing effect

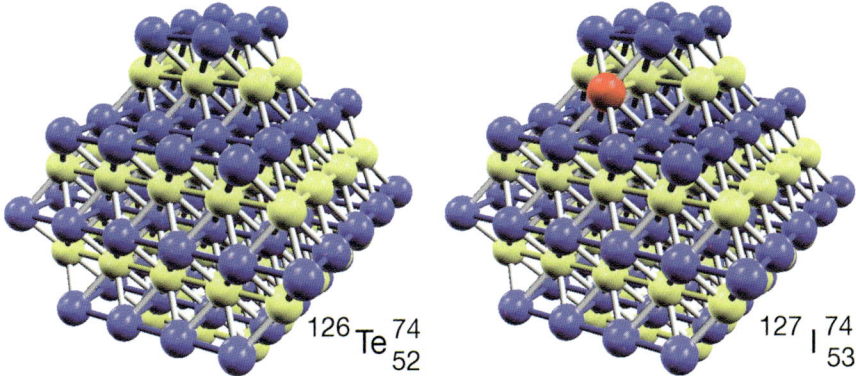

Fig. 10.23. The so-called symmetry effect on binding energies in the lattice. (**a**) shows the stable $^{126}\text{Te}^{74}$ nucleus, with characteristic neutron shelves extending beyond the $Z = N = 52$ core. (**b**) shows the 7 new bonds made by adding a proton onto an available 5/2 site, forming $^{127}\text{I}^{74}$

will differ somewhat, but, on average, the number of nuclear bonds in the lattice will be greater for even-N than odd-N, and greater for even-Z than odd-Z nuclei, as is empirically found.

The symmetry term in the liquid-drop model has a similar geometrical explanation in the lattice (Fig. 10.23). The isospin layering in the model implies that, in general, the local lattice geometry of adding nucleons to coordinate sites is identical for both protons and neutrons. However, the empirical fact that excess neutrons can always be added to an N = Z core (and excess neutrons are required for all N = Z > 20 cores) means that the addition of protons to the lattice will normally entail a greater increase in the number of 2-body bonds than the addition of neutrons. The effect is therefore small for small nuclei, where the neutron excess is zero or small, but increases with increasing A. It is for this reason that the symmetry effect (Fig. 4.3) is more accurately calculated in relation to the number of *excess* neutrons [(A-2Z)²/A], and, specifically, *not* in relation to N, Z or A (4.1).

Finally, the shell effects known from the semi-empirical mass formula can be explained in terms of the rearrangement of surface nucleons to maximize two-body binding. This effect in the lattice can be illustrated as in Fig. 10.24. The filling of j-subshells is known empirically to proceed from higher to lower values (e.g., 8 nucleons in 7/2 positions, followed by 6 nucleons in 5/2 positions, etc.) within any given n-shell – presumably due to an angular momentum factor inherent to the nuclear force. In the lattice, this corresponds to the completion of layered "shelves" of protons or neutrons. However, "corner" nucleons in each shell necessarily have one-fewer bond than comparable "edge" nucleons. Assuming only that maximization of 2-body bonds is the dominating principle of nuclear binding, the sequential filling of subshells will therefore be altered whenever a nucleon can form more bonds by shifting "cor-

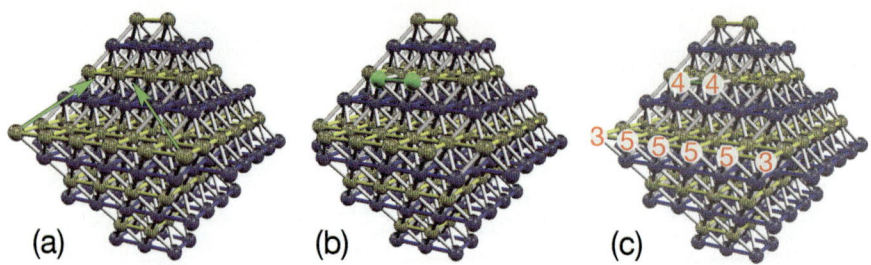

Fig. 10.24. Shell effects in the lattice. (**a**) shows a filled $j = 11/2$ proton shelf, the corner protons of which are weakly bound. (**b**) shows the corner nucleons now in $j = 7/2$ positions. (**c**) shows the number of bonds per proton

ner" nucleons to "edge" positions in a different j-subshell (suffering a slight decrease in binding energy due to the lower j-value, but gaining a net increase due to the increased number of bonds). In effect, this implies that nucleons will settle into lattice sites with maximal j-values, all else being equal, but will move to lower j-value sites when a greater number of bonds can be formed. It then becomes a (conceptually simple, if permutationally complex) matter of the local geometry that determines when there will be continuous increases in binding energy with the addition of nucleons, and when there will be small discontinuities reflecting the unavailability of maximal binding sites.

As shown in Fig. 10.24, depopulating corner positions (sites marked as "3") in otherwise filled j-subshells means the partial filling of lower j-value subshells (sites marked as "4"). This shift alone increases the number of 2-body bonds, and will *always* further increase the number of bonds whenever there are nearby valence nucleons of opposite-isospin, with which additional bonds can form (in Fig. 10.24b, bonds with any of the unpopulated neutron positions directly beneath the green protons). In the lattice model, this shifting of nucleons to maximize bonding is the source of the well-known influences of proton build-up on neutron magic numbers, and vice versa. As a result, within each n-shell for $Z > 20$, adding nucleons to fill higher j-subshells will, in general, allow for nucleon shifts that maximize 2-body bonds, whereas the subsequent filling of lower j-subshells eventually implies rather loose bonding on the nuclear surface (i.e., no subshell-closure [=magic] effects).

The fit between the fcc model and experimental data is similar to that obtained in the liquid-drop model. The lattice has the merit of explaining all effects on the basis of local geometry and, for that reason, the meanings of all model parameters are transparent. This can be examined in detail using the *NVS* program on CD. It must be said, however, that the results are still a phenomenological fit between a model and experimental data and do not constitute a rigorous theory of nuclear structure. Once a quantitative theory of the nuclear force becomes available, it should become possible to build individual nuclei in the fcc model that are optimized for binding energy, and then compare theoretical predictions concerning radii, nuclear moments and spins with experimental data.

10.6 Fission of a Lattice

The fcc model provides a natural way to account for the asymmetry of fission fragments simply due to the break-up of a large nucleus along its lattice planes. The possibility that a nuclear solid might split along crystal lattice planes was first suggested by Winans in 1947, but only the simple cubic packing of nucleons was considered and previous accounts of the fission of a lattice have remained qualitative. Without adding any new parameters to the fcc model, however, a few basic predictions can be made.

The technique used for simulating nuclear fission in the fcc model includes three basic steps: (i) determination of nucleon lattice positions, (ii) fission of the nucleus along all of its dominant lattice planes, and (iii) collection of fission fragment statistics. The most likely configuration of protons and neutrons for a fissionable nucleus must first be obtained from the default build-up sequence. For specification of the positions of *surface* nucleons, for which there are many possibilities, a randomization technique can be used. In this way, several hundred roughly similar structures for a given Z and N are obtained – all of which have identical cores ($N = Z = 70$) and variable surface occupancies. In general, the build-up procedure leads to roughly octahedral nuclei, but the large excess of neutrons implies mildly oblate structures for nuclei with $Z > 70$. The randomization process can be repeated many times and each nucleus individually split along lattice planes to obtain statistics on fragment size.

Once a core structure with randomly occupied surface lattice positions has been obtained, the lattice can be fractured. There are seven major planes [crystallographically, (001), (010), (100), (111), ($\underline{1}$11), (1$\underline{1}$1), (11$\underline{1}$)] that pass through the center of the fcc lattice, and another 14 which run parallel on either side of the seven central planes. To determine which of these 21 fission planes would be energetically favored, the number of nearest-neighbor internucleon bonds that cross each fission plane is counted, and the total electrostatic repulsion between the two fragments calculated. This procedure is illustrated in Fig. 10.25 for a small nucleus.

Using the total binding energies of first-neighbor bonds crossing the lattice planes and subtracting the electrostatic repulsion between the two fragments, the net attraction between the fragments is computed and taken as an inverse measure of the likelihood of fission along the lattice plane. That is, the probability of a nucleus fissioning along a particular lattice plane, P_{fission}, is defined as:

$$P_{\text{fission}}(Z, N) = 1 \bigg/ \left(\beta \sum_{m}^{A_{f1}} \sum_{n}^{A_{f2}} b_{m,n} - \sum_{j}^{Z_{f1}} \sum_{k}^{Z_{f2}} Q_{j,k} \right) \quad (10.2)$$

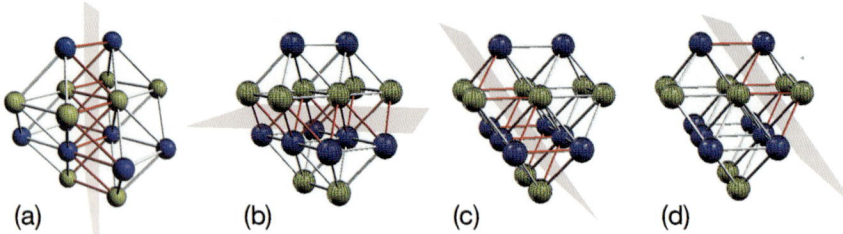

Fig. 10.25. The fission calculations done on a small nucleus, ^{16}O. The *red* bonds between the fragments are counted and the total electrostatic repulsion between the protons (*yellow*) in the separate fragments is calculated. The fewer the bonds crossing the given lattice plane, the more likely that the nucleus will fracture along the plane (**d**), but the electrostatic repulsion between the fragments will be higher when equal numbers of protons are in each (**a** and **c**)

where Z, N and A refer to the number of protons, neutrons and nucleons, respectively; subscripts f1 and f2 denote the two fragments; β denotes the coefficient for the binding force between first-neighbor nucleons; $b_{m,n}$ is the number of bonds crossing the fission plane; and $Q_{j,k}$ is the sum of all Coulomb effects between the two fragments.

To evaluate the probability of symmetric or asymmetric fission of Uranium and Plutonium isotopes, a total of 21 possible fission events was calculated for each randomly assembled nucleus and repeated for 1000 configurations with given Z and N, resulting in 42,000 fragments per isotope for statistical analysis. In effect, each nucleus showed several ($\sim$3 – 6) high probability fission modes along lattice planes with relatively few two-body bonds, and a larger number ($\sim$15 – 18) of low probability modes along planes with many bonds and/or low inter-fragment repulsion. The data accumulated from such simulations are summarized in Fig. 10.26.

The simulation data in Fig. 10.26 show several effects. First of all, before considering the likelihood of different types of fragmentation, it is clear that the lattice geometry implies five major fragment sizes. Approximately symmetrical cuts (with a peak at $\sim$118) are obtained whenever the lattice is cut vertically or horizontally through its center. Asymmetrical fragments (with peaks at $\sim$98 and $\sim$138 nucleons) are obtained by oblique cuts, and super-asymmetric fragments of $\sim$78 and $\sim$158 nucleons are obtained when fission of the lattice occurs off-center relative to the origin of the coordinate system. These numbers simply reflect the geometry of the lattice, and do not indicate which cuts are energetically favored.

To obtain the favored fragments, the relative effects of Coulomb repulsion between the fragments and inter-fragment binding must be calculated. The Coulomb effect between protons is therefore computed for every pair of protons in separate fragments, and is found to approach 160 MeV for symmetrical cuts through Uranium. Although the total number of protons is always 92, a

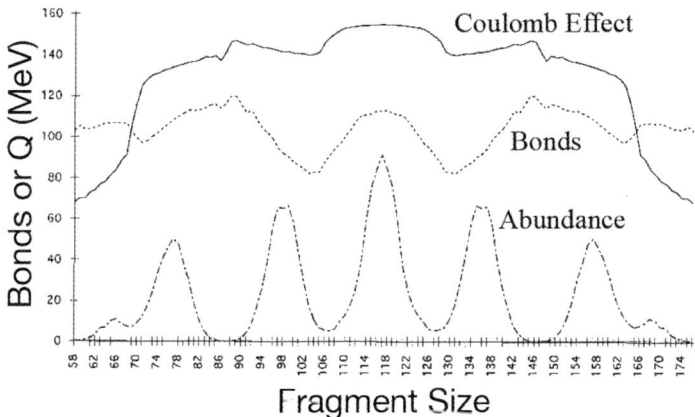

Fig. 10.26. The simulation data set on 21,000 fission events for ^{235}U + n$_{th}$. Given the randomization technique and the constraints of the lattice, the 21 fission planes produce five peaks of fragment size (the Abundance curve). The actual incidence of fission will, however, be influenced by the number of Bonds and the total Coulomb Effect between the fragments. Symmetric fragments show high mean electrostatic repulsion (the central plateau in the Coulmb Effect curve), whereas asymmetric fragments often have fewer nearest-neighbor bonds between the fragments (notably, the two dips in the Bonds curve at ∼106 and ∼130 nucleons). The likelihood of fission is a result of the relative balance between electrostatic repulsion and fragment bonding (Cook, 1999)

somewhat greater Coulomb effect is obtained for symmetrical fission (46 protons in each fragment) relative to asymmetrical fission (e.g., 40 and 52 protons in the fragments). It is of course this effect that leads the liquid-drop model to predict that symmetrical fission should always be favored energetically.

In the fcc model, however, the nuclear texture is not an amorphous liquid, but a lattice within which there are necessarily planes with greater or lesser numbers of bonds. A rough estimate of the nuclear force binding the two fragments together across each plane can be obtained by assuming that each nearest-neighbor bond in the lattice (disregarding the effects of spin and isospin in first approximation) has a mean binding force of several MeV (β).

When β is small, the contribution of the bonding between the fragments is theoretically small and therefore the Coulomb effect between fragments containing equal numbers of protons will favor symmetrical fission. With increases in the strength of nearest-neighbor nucleon binding toward a realistic value of 2 ∼ 4 MeV per bond, however, the number of internucleon bonds between the fragments becomes more important, relative to the importance of the electrostatic repulsion between the fragments. As a consequence, there is an increase in the probability of fission along lattice planes with relatively few two-body bonds and this effect favors asymmetrical fission (specifically, in the troughs of the Bonds curve shown in Fig. 10.26).

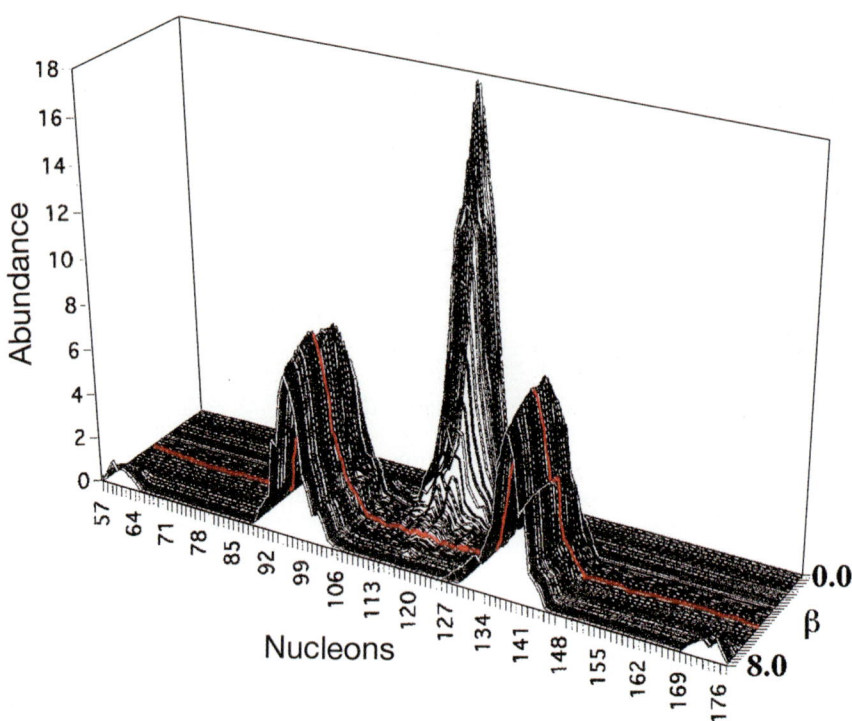

Fig. 10.27. The abundance of symmetric vs. asymmetric fission for ^{235}U + n$_{th}$ as a function of the strength of nearest-neighbor nucleon binding (β). For $\beta > 2.0$, the symmetric peak is suppressed and most fragments are in the regions of 92 $\sim$ 104 and 132 $\sim$ 144 nucleons. At very large values of β (>7.0 MeV), extremely asymmetric fragments (f$_1 \sim$ 62, f$_2 \sim$ 174) emerge. High energy fission of ^{235}U shows the gradual filling in of the symmetric region, corresponding in the lattice model to a weakening of the nearest-neighbor binding relative to the Coulomb repulsion, and thus the inclusion of symmetric fission events. The lattice model indicates that extremely asymmetric fragments (1:3) will appear at very high excitation energies – concerning which there is some experimental evidence (e.g., Barreau et al., 1985)

In order to understand the relative importance of electrostatic repulsion and the number of internucleon bonds, the total interfragment binding for the 21 possible fission events for ^{235}U + n$_{th}$ was calculated for each value of β (0.0–8.0 MeV). The abundance of the fragment masses was then plotted for the various β values (Fig. 10.27). It is seen that asymmetric fission becomes dominant when the net attractive binding force between nearest-neighbor nucleons is greater than about 2.0 MeV. This general pattern of fragment size – with peaks in the vicinity of 98 and 138 nucleons – is found until extremely asymmetrical fragments become favored at large β values (>8.0 MeV), i.e., when the Coulomb effect becomes small relative to the fragment binding.

Figure 10.27 clearly shows that the abundance of asymmetrical fragments for the actinide nuclei is strongly influenced by the adjustable nuclear force parameter (β), but it is noteworthy that only values near to 3.0 MeV give realistic total nuclear binding energies [BE(A) = $\beta \cdot b_{A-}$ $Q_A \sim 1800$ MeV for A $\sim$ 230–240]. Employing only this one nuclear force parameter, the lattice model thus indicates that neither symmetrical ($\beta < 2.0$ MeV) nor extremely asymmetrical ($\beta > 6.0$ MeV) fission will be favored for the actinide nuclei. This parameter is of course a gross simplification of the complexities of the nuclear force, but any constant-density lattice model will imply a mean value of nucleon-nucleon binding at a given nearest-neighbor lattice distance. Using default lattice structures (and ignoring all spin, isospin, etc. effects in first approximation), the known binding energies of all nuclei are reproduced using a β-value of 2.7 $\sim$ 2.8 MeV (e.g., 2.78 for ^{40}Ca, 2.77 for ^{90}Zr, 2.76 for ^{142}Nd, 2.76 for ^{174}Yb, 2.79 for ^{202}Hg) and specifically those of ^{233}U, ^{235}U and ^{239}Pu are reproduced with a β-value of 2.77 MeV. This indicates that a β-value near 2.8 MeV is the correct magnitude for the mean nearest-neighbor nuclear force effect at $\sim$2 fm.

As shown in Fig. 10.28, the width and shape of the theoretical fragment peaks are influenced strongly by assumptions concerning the randomization technique used to construct the fissioning nuclei. When only lattice structures with maximal two-body binding are used (i.e., default structures with no randomization of lattice occupation), fewer fragment sizes are possible ($\beta = 3$ MeV, dotted lines in Fig. 10.28), but these peaks broaden as random vacancies are allowed at various lattice positions. It can be concluded that, in the lattice model, the small number of interfragment bonds along certain planes leads to the predominance of asymmetrical fragments for all reasonable values for the strength of the nearest-neighbor internucleon effect (i.e., $\beta > 2.0$).

For $\beta \sim 3$ MeV, the overwhelming majority of fission events for ^{233}U+n$_{th}$, ^{235}U+n$_{th}$ and ^{239}Pu+n$_{th}$ are asymmetrical (Fig. 10.28). Of the seven planes that cut through the center of the default lattice structure for ^{233}U, three produce fragments whose masses lie in the symmetrical region (e.g., Fig. 10.29a), but such cuts inevitably entail breaking a large number of bonds in the fission plane and are therefore not energetically favored. Four oblique planes through the nuclear center (e.g., Fig. 10.29b) cut significantly fewer two-body bonds and result in asymmetrical fragments (93–100 vs. 133–140 nucleons). Many of the 14 cuts that do not pass through the nuclear center produce extremely asymmetrical fragments (e.g., 77 and 156 nucleon fragments). They are not, however, favored modes of fission, despite sometimes having a relatively low number of two-body bonds in the plane of fission, because of a correspondingly low level of electrostatic repulsion between the fragments (see Fig. 10.26). As a consequence, regardless of minor variations due to the randomization process, the dominant modes of low-energy fission of the actinides in the lattice model are moderately asymmetrical (2:3), oblique cuts through the nuclear center.

The fundamentals of fragment symmetry/asymmetry in the lattice are already apparent in the seven major fission planes of the default ^{235}U

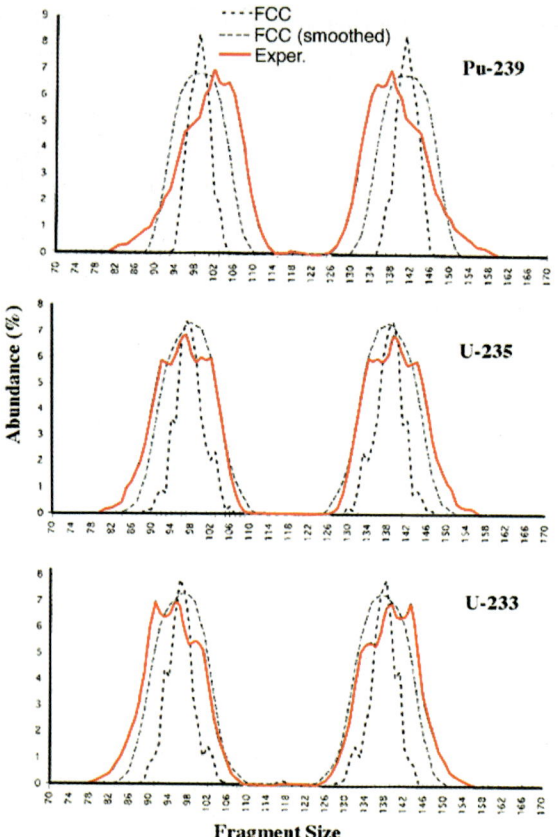

Fig. 10.28. A comparison of experimental (*red lines*) and theoretical (preneutron-release) fragment masses for the three most important fissionable nuclei, ^{233}U + n$_{th}$, ^{235}U+n$_{th}$, and ^{239}Pu+n$_{th}$. (data from Gönnenwein, 1991). The effects of the randomization technique can be seen from a comparison of the two theoretical curves in each graph. The narrow peak curves show the abundance using maximally close-packed default lattice structures without randomization of the occupancy of lattice sites, whereas the smoother curves are obtained by calculating a moving average of the narrow curves

structure (Fig. 10.29a and b). Moreover, the advantage of an oblique cut is easily understood from the symmetry/asymmetry of cuts through the central-lying ^{4}He tetrahedron (Fig. 10.30a and b), where symmetrical cuts demand breaking more bonds than asymmetrical cuts. This qualitative result finds confirmation in the statistical analysis of nuclei with randomized configurations (Fig. 10.28). The results presented here are based on the simplest possible assumptions concerning the nuclear force (i.e., a nearest-neighbor binding of 3 MeV regardless of spin and isospin effects). A more realistic treatment of the nuclear force in the study of fission remains to be studied.

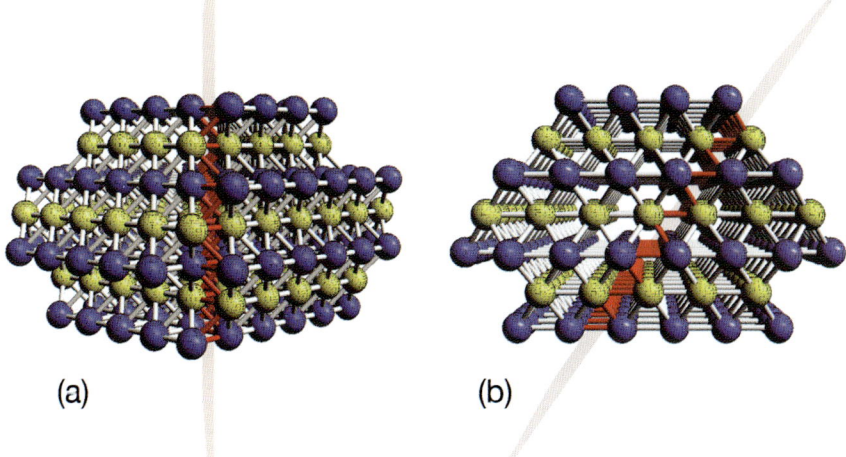

Fig. 10.29. Symmetrical and asymmetrical slices through a lattice structure for ^{236}U. The vertical slice in (**a**) cuts 120 nearest-neighbor bonds and produces symmetrical fragments, whereas the oblique slice in (**b**) cuts 86 bonds and results in asymmetrical fragments

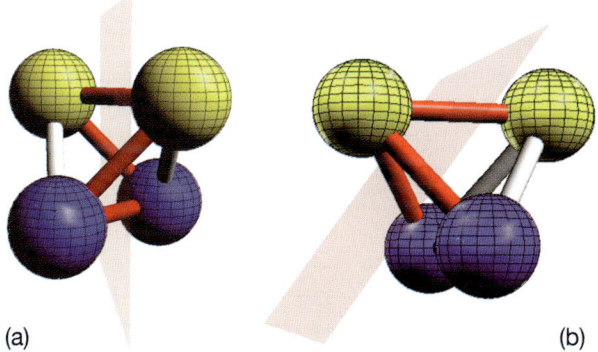

Fig. 10.30. Symmetrical (**a**) and asymmetrical (**b**) cuts through the centrally located ^{4}He nucleus require cutting 4 and 3 bonds, respectively (all broken bonds are shown in red). Already at this level, an oblique cut that breaks fewer bonds produces fragment asymmetry, whereas a vertical cut that requires breaking more bonds produces fragment symmetry

In comparison with other fission models – where parameters are chosen specifically to produce fragment asymmetries, the lattice model reproduces the broad outline of the known asymmetries of actinide fission using only one parameter that is chosen solely to reproduce the total nuclear binding energy, not the fragment asymmetry itself. Various fission phenomena remain to be explained within the lattice model, such as the competition among the different decay modes, the slight excess of neutrons released from the light fragment in thermal fission, and the return to symmetric fission at $A > 257$. Although the remarkable mass asymmetry of thermal fission fragments is one of the oldest, unsolved puzzles in nuclear physics, but it may have a rather easy explanation in terms of the fragmentation of a lattice of nucleons.

10.7 Conclusions

Ultimately, the fcc nuclear model is still a model with adjustable parameters and plausible assumptions, rather than a formal theory built on first principles concerning the nuclear force. This is a weakness not unlike those of the other nuclear models – and one that eventually must be addressed. The model does, however, have the unusual merit of allowing for a self-consistent understanding of the major themes of nuclear structure without bouncing back-and-forth between incompatible ideas about the nature of nucleon interactions. Given the unambiguous mapping between the symmetries of the Schrödinger wave-equation and the fcc lattice (Chap. 9), the configuration of any number of protons and neutrons in an fcc lattice is determined in a non-arbitrary manner, and nuclear properties can, in principle, be deduced directly from the lattice structures.

Much work remains to be done – particularly on the nature of the nuclear force, nuclear moments and binding energies, and the meaning of angular momentum within the lattice, but it can be said that the first-order calculations on nuclear sizes, shapes, spins, binding energies and fission phenomena have produced results that are consistent with what is known about the nucleus. Many properties that are usually calculated on the basis of the shell, liquid-drop or cluster models can also be calculated from the lattice structures – suggesting that a unified lattice model is possible. Moreover, the implications of a lattice model concerning the nuclear force are unusually clear insofar as the parameters of nearest-neighbor interactions must be precisely specified within the lattice to obtain the known density and radial measures of nuclei.

A

The "Nuclear Visualization Software"

This appendix is concerned with visualizing the invisible: the graphical display of the nucleus. Because of the extremely small size of all things atomic, the only possibility for viewing nuclear structure is through the tricks of computer graphics. Although direct observation on an external reality has traditionally been at the heart of classical science, quantum physics, and especially nuclear structure physics, have led us into a microworld where observation is indirect and the final "picture" is not a photograph, but rather a concept or an image generated by the mind. With the help of computer hardware and software, such mental images can be made explicit and consequently manipulated as if they were macroscopic objects.

Visualization of natural phenomena is worthwhile primarily because it facilitates understanding and literally lets us see relations and connections that are not obvious without the visual mode of thinking. This is not to say that other forms of understanding are unimportant or that a visual image can replace symbolic learning, but graphical display can provide insights distinct from those obtained through symbolic manipulation. In fact, computer-aided visualization of the otherwise-invisible structure of atoms and molecules is a well-established technique in chemistry and is widely used for both research and education. In other words, quantum mechanics and the unusual implications of the uncertainty principle are themselves not an obstacle to visualization (Ghirardi, 2005), but graphical display of the nuclear realm has not been thought to be practical or important and has consequently gone largely undeveloped. In nuclear physics, the subject matter is, visually, quite simply missing! This appendix, together with the related Nuclear Visualization Software, *NVS*, on CD, fills this gap.

There is full unanimity that quantum mechanics underlies all of nuclear physics, but the various models of nuclear structure are based upon assumptions that are seemingly quite incompatible with one another. For the specialist, details of the formal models will normally suffice for research purposes, but anyone trying to come to grips with the basic ideas in the nuclear realm there is an over-riding interest that comes prior to the technical details: What is the "overall picture" of nuclear structure? Is the nucleus a tiny gas? A liquid-drop? A quantum solid? Or something else? Unlike the situation in chemistry or genetics or even brain science, it is uncertain what constitutes a more-or-less accurate, first-order understanding of nuclear structure. Fortunately, advances in computer science have made it possible to represent and manipulate complex data sets and to display theoretical models in ways that were once only vaguely imaginable. Using the tricks of computer graphics, a more visual approach to theoretical nuclear structure physics has become possible and is discussed here.

By placing nucleons at the coordinate sites of a lattice, it can be shown that the known energy shells, sub-shells and symmetries inherent to the quantum mechanics of the nucleus have a simple, entirely-intuitive 3D geometry. The basic structures are unambiguous and, if perhaps "difficult" the way all problems in solid geometry are difficult, the argument is nonetheless fundamentally schoolboy geometry. It does not demand adventures into the murky realms of the various interpretations of the uncertainty principle, ideas about time reversal or an "n-dimensional" universe, but it does demand attention to the 3D structure of many-body systems.

What can be seen using the software is that small changes in the lattice can lead directly to the other models of nuclear structure theory. By allowing the lattice to "melt" into a liquid phase, "evaporate" into a gaseous phase, or "coalesce" into a molecule-like solid, the diverse models can be easily simulated, visually displayed and manipulated at will. Above all else, such visualization shows that the seemingly contradictory models of nuclear structure are in fact each describing the same underlying reality from slightly different perspectives. By adjusting the relative motion of the nucleons, the main features of the traditional models emerge essentially intact within the lattice. In short, there are relevant pictures of the nuclear realm – pictures that are, to be sure, simplifications of a more complex reality, but, importantly, pictures that (i) are based on the known quantum physics of the nucleus, (ii) are consistent with a wealth of experimental data, and (iii) reproduce the basic properties of the established models of nuclear structure.

A.1 A Brief User's Manual

The basic idea underlying the software is that the major properties of the traditional models of nuclear structure theory (Sect. A.6) can be visualized within an fcc lattice. By adding "thermal motion" or some degree of random movement of each nucleon around its lattice site, the lattice can be "melted" into a liquid or "vaporized" into a gas, while maintaining certain types of internal structure. Below is presented a discussion of how such changes can be brought about within the *NVS* program for an intuitive understanding of nuclear structure theory.

A.1 A Brief User's Manual   249

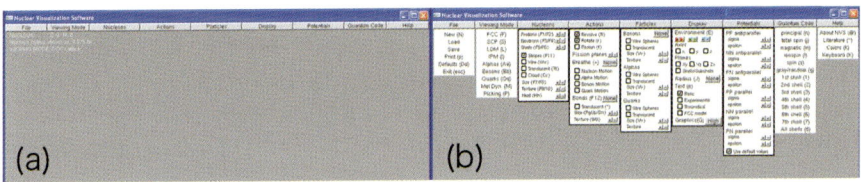

Fig. A.1. (a) The blank start-up screen. A menu bar appears at the top of the screen and current information on the nucleus and the viewing parameters are listed at the upper left-hand corner. (b) The blank screen when the nine menu bar items have been unscrolled

Start up the software by double-clicking the *NVS* icon or typing **NVS** in command mode. A blank screen will appear, as shown in Fig. A.1a. On the top of the screen is an unopened menu bar, from which various options can be selected using the mouse. At the upper left within the blank screen is shown information concerning the current nucleus, the current viewing mode and relevant experimental data. By clicking on any of the items in the menu bar, they will unscroll, revealing various options (Fig. A.1b).

Most options are available both through the menu items on screen and through keyboard shortcuts, but the shortcuts are generally a more efficient way of modifying the nuclear structures. The specifications for the keyboard shortcuts are intended to be intuitive in the sense of being triggered by a relevant letter of the alphabet (e.g., **A** = alphas, **B** = bosons, **C** = clouds, **D** = defaults, etc.; see the listing in Sect. A.4). The most common commands for building and displaying nuclei will soon become familiar, and can always be reviewed by directly checking the keyboard items on screen (from the menu bar, "Help/Keyboard," or from the keyboard **K**). The major exceptions to the "intuitiveness" of the commands are the effects produced by the Function Keys, so that keyboard templates on which the Function Key effects are printed are provided in Sect. A.5.

File Functions: `File`

The `File` item is located on the far left-hand side of the menu bar, under which the most common filing commands are to be found, i.e., Save and Load, for saving specific nuclear structures for a given number of protons and neutrons, and later reloading them. Other functions found here are `New` (**N**) to erase the current display; `Defaults` to restore the default buildup sequence of nucleons (**D**), and the default graphics settings (**d**); Print (**p**) to print relevant information on the current nucleus to an output file; and `Exit` (**Esc**) to quit the program.

A The "Nuclear Visualization Software"

The Basic Viewing Modes: `Viewing Mode`

There are nine viewing modes, corresponding to seven important classes of nuclear model and two additional modes of theoretical interest. The choice of mode can be made in two different ways: selecting the `Viewing Mode` option from the menu bar using the mouse, or directly through keyboard shortcuts. The items under `Viewing Mode` (and the keyboard short-cuts) are as follows:

```
Viewing Mode
    FCC          (F)     the solid-phase Face-centered-cubic lattice model
    SCP          (S)     the solid-phase Simple-cubic-packing lattice model
    LDM          (L)     the liquid-phase Liquid-drop model
    IPM          (I)     the gaseous-phase Independent-particle model
    Alphas       (Aa)    the Alpha-cluster model
    Bosons       (Bb)    the Boson model
    Quarks       (Qq)    the Quark model
    Molec. Dyn.  (M)     the ''Molecular dynamics'' simulation mode
    Picking      (P)     the Picking mode
```

Clicking on these items in the `Viewing Mode` panel will change the display of the current nucleus to: (F) the face-centered-cubic (fcc) lattice model, (S) the simple-cubic-packing (scp) lattice model, (L) the liquid-drop model (ldm), (I) the independent-particle model (ipm), (A) the alpha-cluster model, (B) the boson model, (Q) the quark model, (M) the "molecular dynamics" mode for simulating nuclear dynamics, or (P) the picking mode for constructing specific nuclei using the mouse.

Note that by changing the viewing mode via the menu bar panel, all graphics parameters are automatically set to default values that facilitate visualization of the chosen nuclear model (for example, nearest-neighbor bond settings and nucleon sizes). In contrast, by selecting the viewing modes via the keyboard commands (**FSLIABQ**), the various visualization parameters are left unchanged while the nuclear model is altered. These two methods of switching between models thus facilitate (1) optimal visualization of each of the models (via the `Viewing Mode` options of the menu bar), or (2) direct visual comparison of the models with all visualization parameters held constant (via the keyboard).

Also listed under the `Viewing Mode` item are three related options that can be selected through the keyboard (or through options under `Particles`). The **a** option changes the current alpha cluster display for a given nucleus to a different cluster geometry. Note that **A** toggles alpha clusters on-and-off, whereas **a** displays different alpha configurations. This latter function is necessary since there are inevitably a large number of possible groupings of 4-nucleon clusters. Similar functions for boson configurations can be triggered by **B** and **b** from the keyboard. **B** toggles among several boson model variations, whereas **b** displays different boson configurations for the current boson model.

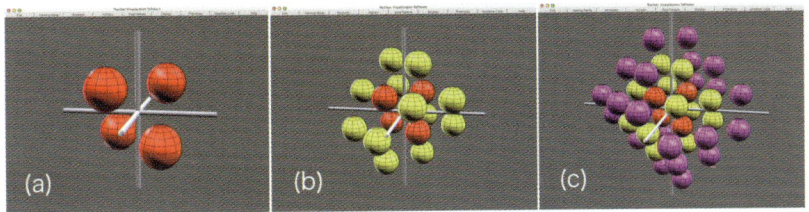

Fig. A.2. The first three n-quantum number shells containing 4, 16 and 40 nucleons, as seen in the fcc lattice model. Each n-shell can be found embedded within the outer shells. No distinction is made here between protons and neutrons

The **Q** option also works to toggle the quark display itself on-and-off, while the **q** option results in the display of an alternative quark configuration within each nucleon. Further options for alphas, bosons and quarks can be found under the Actions and Particles items in the menu bar.

In order to make use of any of these display modes, it is of course necessary to first specify a nucleus.

Building a Nucleus: Nucleons

To construct a specific nucleus, three nuclear build-up techniques are available. The first and easiest to use is **Function Key 6** to add consecutive harmonic-oscillator shells of protons and neutrons. Press **F6** once and the first shell containing four ($n = 0$) nucleons will appear (Fig. A.2a). Press it again and again, and the second ($n = 1$) and third ($n = 2$) shells containing 16 and 40 nucleons appear on the screen (Fig. A.2b, c). Shells can also be added by clicking on the Shells (+) radio button under Nucleons.

The nucleons are displayed by default as small spheres with the colors corresponding to the chosen quantum number – initially, the principal quantum number n-values ($n = 0$, red; $n = 1$, yellow; $n = 2$, purple; etc.). The default display mode is the fcc lattice, so that the consecutive shells are regular (tri-axially symmetric) structures within the lattice. No distinction is made between protons and neutrons in this display, since the default coloration is for the principal energy level, regardless of isospin value. Note that the numbers of nucleons in the n-shells in the fcc lattice correspond precisely with the occupancy of nucleons in the harmonic oscillator n-shells for $n = 0 \sim 7$. As is known from the independent-particle model, these numbers are identical to the doubly-magic nuclei for $n = 0$, 1 and 2. The larger magic nuclei correspond to the filling of j-subshells due to the fact that all stable nuclei with $A > 40$ have an excess of neutrons ($N > Z$).

For color-coding on the basis of other quantum values, the items under Quantum Code in the menu bar can be used or the related characters typed on the keyboard: **n** (principal quantum number), **j** (total angular momentum quantum number), **m** (magnetic or azimuthal quantum number),

i (isospin quantum number), **s** (spin quantum number), and **g** (gray/random coloration). Note that help on the coloring scheme can be obtained from the menu bar Help/Colors item or by typing **k**.

Shells of nucleons can also be deleted by clicking the Shells radio button (−) under Nucleons or by pressing **F5**. After three presses, the display will return to the original blank screen.

The second technique for building nuclei is to add nucleons one-by-one. To add protons or neutrons individually, press **F2** or **F4**, respectively. By pressing each of these keys twice, 8 times or 20 times, the exact same shells of 4, 16 and 40 nucleons can be displayed as were displayed using the shell command, **F6**. To delete these nucleons, either **F5** can be used to remove entire shells, or protons and neutrons can be individually deleted with **F1** and **F3**. Corresponding (+) and (−) radio buttons can be found under Nucleons in the menu bar, and give the same results.

The third technique for building nuclei is with the mouse via the picking mode (see **Picking and Choosing**, below).

Under Nucleons in the menu bar, there are seven options for setting the general appearance of protons and neutrons. They are: (1) display with or without latitude/longitude stripes (keyboard shortcut **F11**), (2) solid-sphere or wire-frame nucleons (shortcuts **Ww**), (3) opaque or translucent nucleons (**Tt**), (4) solid sphere-like or cloud-like nucleons (**Cc**), (5) various size (nucleon radius) settings (**F7/F8**), (6) various texture (nucleon smoothness) settings (**F9/F10**), and (7) various degrees of zero-point thermal movement of the nucleons around their lattice sites due to applied heat (**Hh**). None of these settings except the heat setting has any physical significance, but they are included to facilitate visualization. For example, the nucleon RMS radius is experimentally known to be ∼0.86 fm, but the display of nuclei using a slightly smaller radius will often allow for a better understanding of the 3D geometry of the nucleus.

Using these three techniques for adding and subtracting nucleons, any nucleus with any combination of protons and neutrons ($Z < 169, N < 169$) can be constructed. Note that many unstable or experimentally unknown nuclei can be constructed in this way, but if a nucleus with the specified numbers of protons and neutrons has been experimentally detected, then the available empirical data on the nuclear ground state are displayed in the upper left-hand corner of the screen.

More on Viewing Modes

Having built a nucleus by specifying the desired numbers of protons and neutrons, it is of interest to see how the nucleus appears within the context of the various nuclear models. Let us examine the viewing modes by again constructing the ^{40}Ca nucleus (press **F6** three times). By default, the ^{40}Ca nucleus will be shown initially as an fcc lattice structure. The other viewing modes can

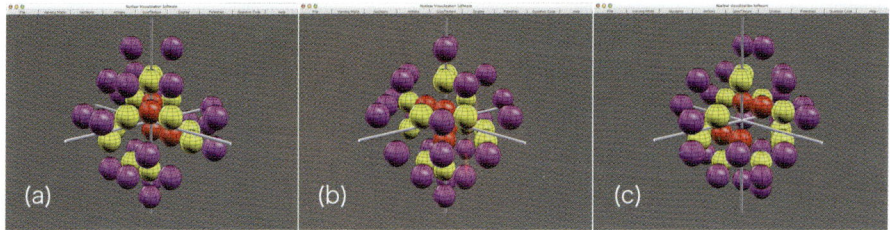

Fig. A.3. Three depictions of the n-shells of ^{40}Ca in the scp lattice mode. The randomization procedure fills one half of all scp sites to obtain the known occupancy of quantum states, but with slightly different configurations (a, b, c) each time the scp mode is selected

be chosen from the menu bar or from the keyboard, and the FCC mode can always be recovered from the keyboard by typing **F**. In the FCC mode, nucleons are depicted as spheres located at coordinate sites in an fcc lattice. By clicking SCP under the Viewing Mode item or by typing **S** the display is switched to an scp lattice.

As discussed in Chap. 9, the scp lattice contains all of the coordinate sites of the fcc lattice as well as an equal number of in-between sites. (The scp lattice can be thought of as two overlapping fcc lattices.) For this reason, the scp lattice for ^{40}Ca shown in Fig. A.3 is populated randomly at 50% occupancy to give the known numbers of nucleons in the various shells and subshells. Repeated selection of **S** will produce different scp structures with somewhat different occupancy configurations. It is of importance for an understanding of the lattice models, however, to see that, whatever the detailed occupancy of the scp or fcc lattices, the numbers of nucleons with any given quantum number are identical to the numbers known from the independent-particle model (see **Quantum Shells and Subshells**, below).

To change to the liquid-drop model for viewing the ^{40}Ca nucleus, click the LDM item under Viewing Mode, or type **L** on the keyboard. The nucleus changes to a liquid phase, in which the nucleons move slowly at random from their initial lattice positions, interacting locally with nearest neighbors (Fig. A.4a). The independent-particle model can be selected by clicking the IPM item under Viewing Mode, or typing **I** (Fig. A.4b). The liquid and gas modes differ predominantly with regard to the movement of the nucleons relative to one another.

The next three display modes, alpha clusters (**A**), bosons (**B**) and quarks (**Q**), are compatible with simultaneous display of the nucleus in the lattice-, liquid- or gaseous-phase modes. As a consequence, choosing alphas and/or bosons and/or quarks will result in the visual display of 4-nucleon alpha clusters and/or 2-nucleon bosons and/or intranucleon quarks together with nucleons themselves displayed as a lattice, a liquid-drop or a gas (for examples, see Fig. A.5). The idea underlying these diverse viewing modes is that, although the phase-state of the nucleus itself remains controversial, useful nuclear

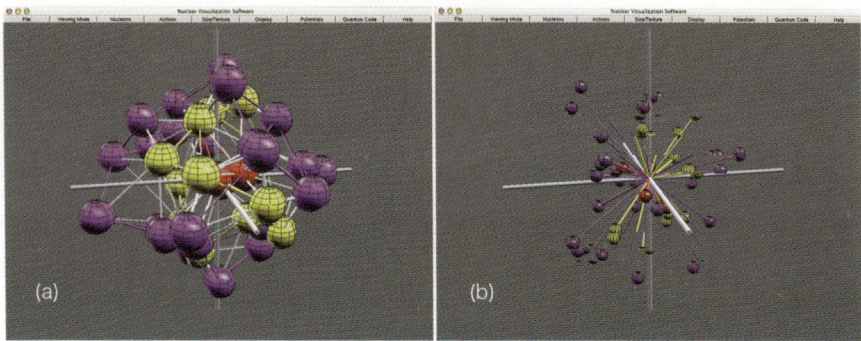

Fig. A.4. The liquid-drop model (**a**) and independent-particle model (**b**) depictions of the ^{40}Ca nucleus. Depending on the bond display setting (**F11**), nearest-neighbor bonds are shown for the LDM, whereas central "bonds" due to the nuclear potential well are shown for the IPM. Note that the numbers of $n = 0, 1, 2$ color-coded nucleons remain the same in both models – and identical to those in the fcc and scp models

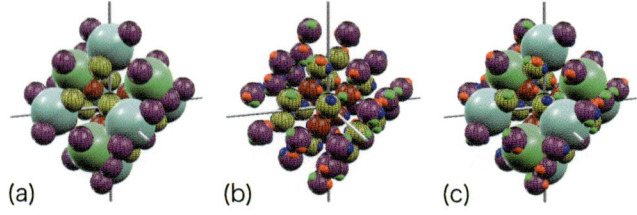

Fig. A.5. Various depictions of ^{40}Ca within the fcc lattice mode: (**a**) nucleons and alpha clusters, (**b**) nucleons and quarks, and (**c**) nucleons, quarks and alphas

models have been developed by assuming certain kinds of nucleon grouping within the nuclear interior, as displayed here.

Usage of the full set of viewing modes for the phase-state of the nucleons, plus the grouping of nucleons into alpha and boson structures with quark substructure, can produce visual displays of bewildering complexity. The usefulness of such complex displays is perhaps dubious, but it bears emphasis that it is precisely such complexity that is the topic of study in nuclear physics. By necessity, one or a few of the properties of the nucleus are normally singled out for experimental or theoretical study, but the nucleus itself is a multi-faceted, dynamic, many-body problem that includes all such groupings of nucleons. For the sake of simplicity, only one or two features should normally be displayed, but no single model, to the exclusion of all others, can explain all nuclear properties.

The final display mode is the molecular dynamics (**M**) option. Unlike the other modes, the molecular dynamics mode contains realistic nuclear forces acting between all pairs of nucleons, so that the display will vary depending on

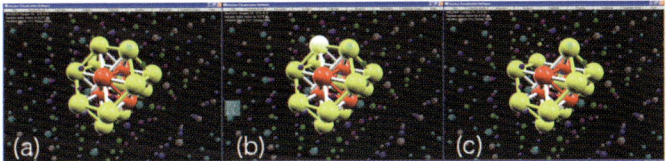

Fig. A.6. The ^{16}O nucleus: (**a**) Oxygen as displayed in the Picking mode with all available unoccupied fcc lattice sites displayed as small translucent spheres. (**b**) A single click on a nucleon will highlight it and display relevant information. (**c**) A second click *on* the same nucleon will remove it, whereas a second click *off* the nucleon will leave it unchanged

the parameter settings of the force (see **Molecular Dynamics:** `Potentials`). As a default setting, the molecular dynamics mode uses an fcc lattice at the onset of the simulation simply as a means of beginning the dynamics with homogenous matter-, spin- and isospin-densities, but the system evolves due to small randomization effects and the nuclear force settings, unrelated to the configuration at the time of initialization. Most of the visualization options are turned off for the molecular dynamics simulation, but are reinstituted when a different viewing mode is selected.

Picking and Choosing: `Picking`

The **P** option toggles the picking mode on-and-off. When picking is on, individual nucleons (displayed, by default, in the fcc mode) can be added to the nucleus by double clicking on visible lattice sites with the mouse. In contrast, existing nucleons will be removed (and displayed as small lattice sites) when double-clicked (see Fig. A.6). In this way, the `Picking` mode allows one to select any configuration of nucleons (any set of nucleons with unique quantum numbers), and then display them in any of the display modes, once the `Picking` mode has been exited.

The picking function is the most convenient means for constructing a nucleus with a set of nucleons with specific quantum numbers. For the vast majority of semi-stable nuclei, the total nuclear spin value is known, so that the individual nucleons should be selected such that their spins sum to the experimental value. Within that constraint, however, there are often many possible nucleon configurations that give the same sum. By selecting different combinations of nucleons that sum to the experimentally-known J-value, a variety of structures with different binding energies, magnetic moments, quadrupole moments, and radial properties can be constructed and compared with experimental data.

Once a specific nucleus has been constructed and displayed, it is often of interest to move the last proton or neutron individually from one energy-state to another to find the configuration that best reproduces experimental values.

This can be done most easily using keyboard commands. The following four keys move protons/neutrons forward/backward to the next available lattice site. Specifically, the position of the last proton can be moved forward and back along the default sequence using **O** and **o** and that of the last neutron using **U** and **u**, respectively. For every step forward in the sequence, the nucleon will move to a different position in the lattice – and that change in the lattice site corresponds to movement of the nucleon in nuclear "quantum space" (a change in the nucleon's principal n-shell, s-spin, m-value, and/or j-subshell).

The default sequence for the specified number of protons and neutrons can always be imposed from the bar menu item, `File/Defaults`, or by typing **D** at the keyboard. To initialize the build-up sequence, return all graphics options to default values, and clear the screen, type **N**.

Dynamic Aspects of the Display: `Actions`

The *NVS* program is designed primarily to illustrate the fact that many of the *known* properties of the nucleus can be depicted in a highly-intuitive, geometrical fashion. Real nuclei, however, are far from static, so that several dynamic features of the nucleus have also been implemented in the software. Similar to static nuclear features, dynamic aspects can be selected using keyboard shortcuts or by clicking on menu items under `Actions` in the menu bar.

Both `Revolve` (**R**) and `Rotate` (**r**) are switches that turn nuclear revolutions and nucleon rotations off-and-on. By default they are both on, but can be turned off from the keyboard or by clicking on the menu check boxes. The speed of revolution/rotation can be increased or decreased using keyboard commands. The left and right arrow keys ($\Leftarrow$) ($\Rightarrow$) control nuclear revolution, whereas faster/slower nucleon rotations can be chosen using the up/down arrow keys ($\Uparrow$) ($\Downarrow$).

The third and fourth options under `Actions` entail the display of fission planes in the chosen nucleus. (For details, see **Asymmetric Fission Fragments: Fission**).

The fifth dynamic option is the "breathing" of the nucleus as a whole. There are three distinct breathing modes that have been studied in nuclear theory. The first can be selected by clicking once on the `Breathe` button; this results in the expansion and contraction along one axis (the vertical z-axis). The second mode involves breathing along two axes (the x- and y-axes) simultaneously, and the third mode involves expansion and contraction along all three axes. These options can also be accessed from the keyboard using the plus (**+**) key.

The next four options under Actions allow for continuous changes in particle positions: Nucleon Motion, Alpha Motion, Boson Motion and Quark Motion. When these options are turned on, they provide for a more realistic display of particle dynamics and interactions and the continual regrouping of nucleons within the nuclear interior. All of these options, however, also make the visual display considerably more chaotic. Again, this reflects the reality of nuclear structure theory, but selective use of these options is recommended to obtain visual displays that facilitate understanding.

The final options under Actions concern the internucleon bonds.

Internucleon Bonds

It is known that nucleons interact via the so-called "strong" nuclear force only when they are separated by less than 3.0 fermi, but the various nuclear models make strikingly differing assumptions about the nature of that force. To view the different kinds of nucleon-nucleon "bonds" implied by the nuclear models in any of the display modes (except the molecular dynamics mode), type **F12** on the keyboard, or click the bonds button in the panel under Actions in the menu bar.

The bonds option has several settings that differ depending on the nuclear model in use. In the fcc or scp mode, the first setting (B1) displays only the bonds that connect nearest neighbor nucleons with the same (currently-displayed) quantum number (same nucleon color-coding) within the lattice. This option is useful for visualizing the geometry of the various shells and subshells. By typing **F12** again, all nearest neighbor bonds (B2) are displayed, regardless of quantum values. If the nucleons have the same color-coding, then the bonds will be the same color as the nucleons, whereas the bonds will be white if the color-coding of the nucleons differs. Such color-coding options allow one to view simultaneously the constant-density texture of the nucleus and its shell or subshell structure (Fig. A.7).

Other bond properties can be altered from the Actions menu bar item or directly from the keyboard. These include increasing (**Page Up** Function Key) or decreasing (**Page Down** Function Key) the thickness of the bonds, and decreasing or increasing the smoothness of the bond cylinders (**9/0**). Note that the bond thickness cannot exceed the current value of the nucleon

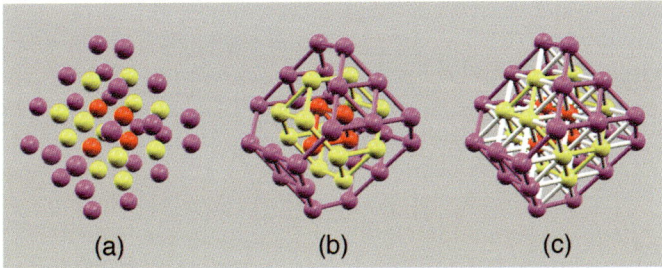

Fig. A.7. The various bond options for the ^{40}Ca nucleus depicted in the fcc lattice mode. (**a**) No bonds. (**b**) Bonds between nearest-neighbors with similar n-quantum numbers (revealing the triaxially-symmetrical n-shells). (**c**) Bonds between all nearest-neighbors (illustrating the frozen liquid-drop, constant-density nature of the same nucleus)

radius. Finally, the bonds can be toggled between solid and translucent using the asterisk key (*).

In the scp mode, the first and second bond settings (B1 and B2) are similar to those of the fcc model, but the nearest-neighbor distance in the scp lattice is shorter than that in the fcc lattice. As a consequence, a different set of bonds is drawn in the fcc and scp modes. The third bond setting (B3) in the scp lattice shows bonds between second-nearest neighbors (corresponding to first-nearest neighbors in the fcc lattice); the relationship between the scp and fcc lattices thus becomes visible with this bond option selected.

In the liquid-drop mode, the displayed bonds are identical to those in the fcc lattice, except that the nucleon locations are fluid and continually changing – leading to bonds sometimes being drawn between very near or distant neighbors.

In the gaseous mode, only the bonds to the center of the nucleus (the central potential well postulated in the IPM) are displayed in the first setting (B1). By pressing **F12** again, the local interaction bonds among nucleons that approach within 2.0 fm of each other are displayed. All bonds can be hidden by pressing **F12** twice more.

Finally, in the quark mode, visualization of current ideas about the nature of nuclear binding due to quark effects can be obtained. Here, with the quark radius enlarged, nucleons set to wire-frame or translucent display and the nucleon radius set to 0.86 fm, it is seen that the quarks in neighboring nucleons overlap, thus implying a strong quark-quark interaction that may account for the nuclear force and the binding of nucleons to form stable nuclei.

Fine-Tuning the Display: Particles

The basic options for altering the size and texture of nucleons were discussed under **Building a Nucleus: Nucleons**, but similar adjustments can be made to the other types of particles: alphas, bosons and quarks. The four available controls are: (1) solid-objects versus wire-objects, (2) opaque versus translucent objects, (3) increased or decreased size (radius) of the particles, and (4) various degrees of particle smoothness. From the keyboard, the size of alphas, bosons and quarks can be increased (decreased) simultaneously with **V** (**v**). All other particle adjustments must be made through the menu bar, rather than through keyboard shortcuts.

The (+) and (−) radio buttons for adjusting the size and texture of particles are similar to those for adjusting nucleon size. The default sizes are the empirically known values (alpha radius = 1.67 fm, boson [∼deuteron] radius = 2.1 fm, quark radius = 0.5 fm), but larger or smaller radii often facilitate visualization. Figure A.8a shows a nucleus with the nucleons depicted as unrealistically small "points" ($r = 0.14$ fm); Fig. A.8b shows the same nucleus with nucleons depicted with their experimentally known radii (RMS charge radius of 0.86 fm); Fig. A.8c shows the same nucleus with nucleons depicted as large, overlapping probability clouds ($r = 4.3$ fm). Settings for rougher or

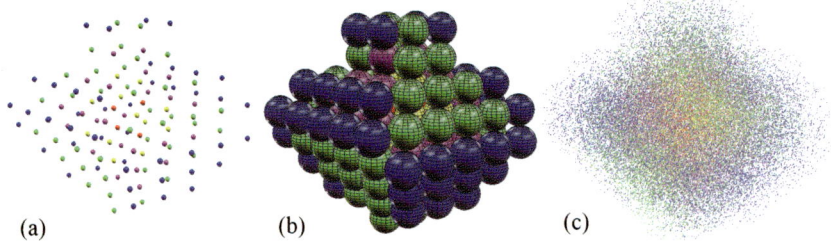

Fig. A.8. Display of the $^{126}\text{Ce}_{58}$ nucleus with nucleons illustrated as (**a**) "points" free to orbit independently within the nuclear interior; (**b**) solid spheres with the experimentally known radial value (0.86 fm), and (**c**) diffuse probability clouds ($r = 2.8$ fermi)

smoother particles refer to the number of facets drawn per particle. The larger the value, the smoother (more spherical) is the object. Smaller values allow for more rapid animation.

The **F11** command changes the appearance of the nucleons by drawing longitude and latitude lines on the colored spheres. The default setting shows thin lines on the nucleons, but typing **F11** once will make them thicker, and typing **F11** twice will remove them entirely.

There are a great many combinations of size and texture for the four types of particles. For illustration of any particular feature of the nucleus (alpha geometry, nucleon shell structure, etc.), it is often best to make other features less evident by making some particles wire-framed, translucent and/or smaller. The translucency of the particles can be changed from the menu bar or from the keyboard. Consecutive typing of **T** will produce, in sequence, every combination of opaque and translucent nucleons, alphas, bosons and quarks. A similar series of wire-frame objects can be selected with consecutive typing of **W**, and countless combinations of translucent/opaque and wire/solid structures can be created using **T** and **W** together.

In contrast, the lower-case **t** and **w** commands are means by which the current settings for translucency and wire-frame objects can be reversed simultaneously for all types of particles. Pressing **t** or **w** again will return the display to its initial translucency/wire-frame settings. All of these options can be used in various combinations, and often facilitate the visualization of complex structures. Some examples are shown in Fig. A.9. If and when the visual display becomes too complex and chaotic, it is always possible to return to the simplest default graphics settings by typing **d**.

Miscellaneous Display Options: `Display`

The environmental (background) color can be altered with **E** or by clicking on the red/green/ blue radio buttons below the `Display` item in the menu bar.

260   A The "Nuclear Visualization Software"

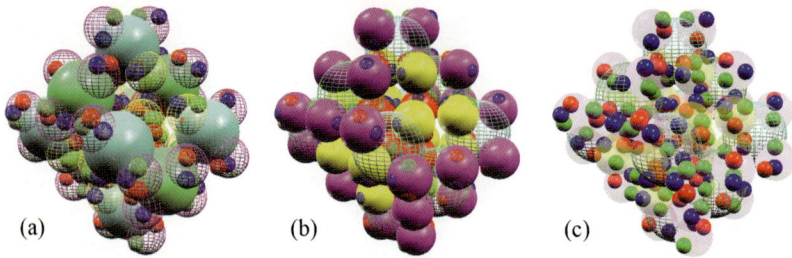

Fig. A.9. Selective use of translucency and wire-framed objects facilitates the visualization of complex structures. (**a**) ^{40}Ca emphasizing the alpha structure by depicting the alphas as opaque spheres, and the nucleons and quarks as wire-frame spheres. (**b**) The same nucleus with the nucleons shown as opaque spheres, but the alpha clusters and quarks as wire-frames. (**c**) A quark model of the same nucleus obtained by depicting the nucleons as translucent, the alphas as wire-frame spheres and the quarks as solid

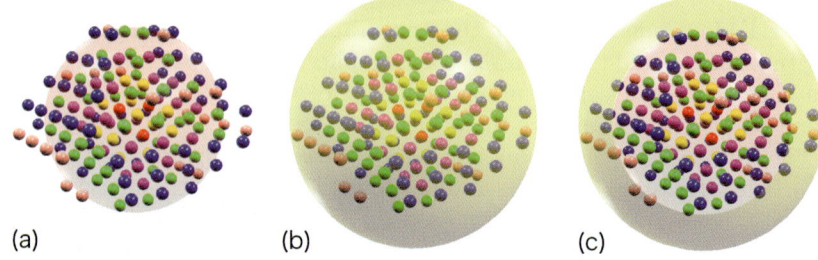

Fig. A.10. The three possible radial displays for the ^{152}Gd^{88} nucleus: (**a**) the constant-density core radius, (**b**) the nuclear surface radius, and (**c**) the core and surface radii together, revealing the nuclear "skin". Nucleons are depicted here with 0.36 fm radii

Repeated typing of **E** changes the environment to 12 different preset colors; fine-tuning can be made with the radio buttons.

Display of the coordinate axes can be chosen from the menu or the keyboard (**x**, **y** and **z**). Similarly, planes that slice the nucleus along the XY-plane, YZ-plane and ZX-plane can be specified with **X**, **Y** and **Z**. A checkbox is also available for displaying the shell/subshell structure. All of these displays can be useful for illustrating the various symmetries of the nucleus.

The nuclear radius can be displayed by clicking the radius button in the `Display` panel, or by typing **J**. There are three varieties of radial display. The first corresponds to the RMS mass radius (Fig. A.10a) and the second to the nuclear surface radius (Fig. A.10b). The third radial display shows both the RMS mass radius and the nuclear surface radius – corresponding roughly to the so-called skin region (Fig. A.10c). The nuclear skin in the lattice models is of particular interest because a considerable skin thickness is implied due solely to the lattice buildup sequence. Because of the triaxial symmetry of the lattice shells, there is necessarily a surface region where the nuclear density falls gradually from the interior value to zero. No arbitrary "skin thickness parameter" is needed to reproduce this well-known feature of nuclei.

Various kinds of explanatory text information can be displayed using the Explanatory Text checkboxes or the e command, which toggles among several levels: (i) basic information on the current nucleus, (ii) experimental data and (iii) theoretical calculations.

Finally, three levels of graphics settings can be chosen from the keyboard by repeatedly typing **G**. The three levels correspond to (i) default settings (mid-level resolution), (ii) fine-grained graphics, (iii) and coarse-grained graphics. The speed of the animation at these levels also differs, with the fine-grained graphics being the slowest and the coarse-grained graphics being the fastest.

Quantum Shells and Subshells: Quantum Code

Every nucleon has a unique set of quantum numbers, together which determine its energy state as described in the Schrödinger wave-equation. The relationships among the quantum numbers and the occupancy of various energy shells and subshells are known and can be formally stated as in the independent-particle model (the central paradigm of nuclear structure theory since ~1950). Such regularities also have a straightforward graphical representation in the fcc and scp lattices. It is therefore of interest to view the various quantum numbers of the nucleons separately within the framework of the chosen nuclear model. The quantum number color-coding of the nucleons can be selected in two ways. The first is from the Quantum Code option in the menu bar, and the second is directly from the keyboard. The menu panel appears as:

```
Quantum Code
   principal    (n)  color-coding according to the principal quantum number, n
   total spin   (j)  color-coding according to the nucleon's total
                     angular momentum, j
   magnetic     (m)  color-coding according to the azimuthal quantum number, m
   isospin      (i)  color-coding according to the nucleon's isospin value, i
   spin         (s)  color-coding according to the nucleon's spin value, s
   gray/random  (g)  gray or random coloring of nucleons
   1st shell    (1)  emphasis drawn to nucleons in the first shell (subshell)
   2nd shell    (2)  emphasis drawn to nucleons in the second shell (subshell)
   3rd shell    (3)  emphasis drawn to nucleons in the third shell (subshell)
   4th shell    (4)  emphasis drawn to nucleons in the fourth shell (subshell)
   5th shell    (5)  emphasis drawn to nucleons in the fifth shell (subshell)
   6th shell    (6)  emphasis drawn to nucleons in the sixth shell (subshell)
   7th shell    (7)  emphasis drawn to nucleons in the seventh shell (subshell)
   All shells   (8)  toggle on-and-off substructure of all shells/subshells
```

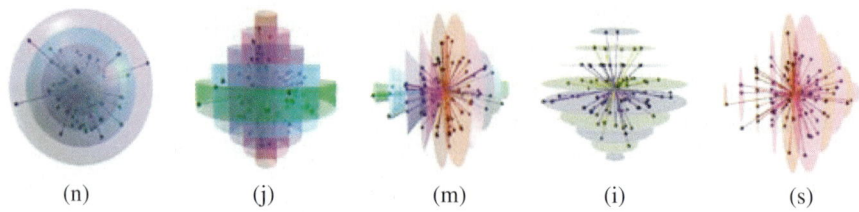

(n)　　　　(j)　　　　(m)　　　　(i)　　　　(s)

Fig. A.11. The coloring scheme for nucleons in Yt^{140} in the gaseous-phase independent-particle model. The translucent substructure option has been turned on to emphasize the geometry of the various shells and subshells. The background color has been changed to white in order to show the subshells more clearly

For viewing the intrinsic geometry of the quantum numbers in any nucleus, the nucleons can be color-coded by selecting in turn the different quantum conditions. They are: **n** (the principal energy level), **j** (the total angular momentum value), **m** (the magnetic or azimuthal quantum number), **i** (the isospin quantum number), and **s** (the spin quantum number). These five coloring schemes with the subshells option turned on (from the keyboard: **8**) are illustrated in Fig. A.11 for the nucleons in ^{140}Yt in the independent-particle model.

Two additional coloring schemes, gray and random (selected by consecutively typing **g**), can also be chosen. Neither has any physical significance, but they can be used to contrast with settings that emphasize the alpha, boson or quark structure of a nucleus.

Finally, within the `Quantum Code` item in the menu bar, there are options for emphasizing any specific magnitude of the selected quantum number. These can be selected by clicking the appropriate shell item in the menu panel or by typing in numerals **1**–**8** on the keyboard. Numerals **1**–**7** will highlight specific n-, j-, m-, i-, or s- quantum shells (subshells), whereas numeral **8** will toggle all of the related translucent spheres, cylinders, cones or planes associated with the selected quantum number on-and-off (Fig. A.11). For example, emphasis is drawn to the nucleons with $n = 0$ quantum numbers of ^{40}Ca in Fig. A.12a, to $n = 1$ nucleons in Fig. A.12b, and to $n = 2$ nucleons

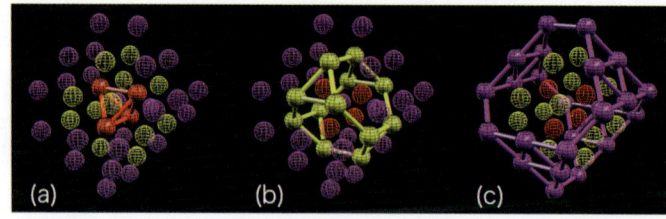

Fig. A.12. Emphasis drawn to the first three principal quantum numbers in ^{40}Ca. In (**a**), only the $n = 0$ nucleons are drawn as solid spheres, with their nearest-neighbor bonds also shown; in (**b**) only the $n = 1$ nucleons are solid with their bonds shown; and in (**c**) only the $n = 2$ nucleons and bonds are solid

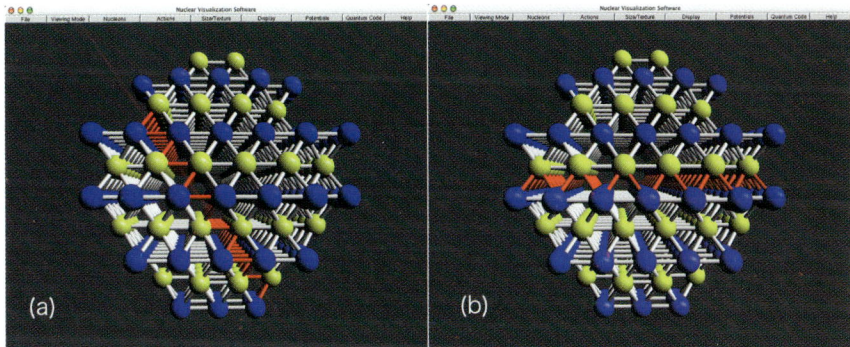

Fig. A.13. The ^{238}U nucleus with two possible fission planes shown in *red*. The probability of fission along each plane of the lattice structure can be calculated from the total electrostatic repulsion between the two fragments and the total number of two-body bonds crossing the fission plane (both of which can be displayed on screen). Oblique cuts through a nucleus (**a**) generally require breaking fewer bonds than vertical or horizontal cuts (**b**), and result in asymmetrical fission fragments

in Fig. A.12c. Similar displays can be obtained for the other four quantum numbers, j, m, i and s.

Asymmetrical Fission Fragments: `Fission`

The display of potential fission planes along various lattice planes in the fcc display mode can be done by choosing the `fission` option under the `Actions` menu bar item (or typing **f** on the keyboard). By repeating the **f** command, various possible fractures along lattice planes are shown (two distinct lattice planes are illustrated in Fig. A.13).

The fracturing of the lattice is a topic on which the fcc lattice model makes predictions that are distinct from the liquid-drop and independent-particle models. Specifically, contrary to empirical evidence, the LDM predicts *symmetrical* fission fragments ($A_{f1} \sim A_{f2}$) and the shell model produces asymmetries that are simply a function of an asymmetry parameter (β) that is adjusted *post hoc* to produce the known symmetries. In contrast, the lattice model predicts *asymmetrical* fragments as a direct consequence of: (i) the lattice structure, (ii) the Coulomb repulsion between the fragments, and (iii) the number of 2-body bonds crossing the various fission planes. The total number of bonds and Coulomb repulsion between the fragments for each fracture are displayed when the `Display/Explanatory Text/Theoretical` box is selected.

Molecular Dynamics: `Potentials`

The molecular dynamics simulation can be initiated from the `Viewing Mode` option in the menu bar or by typing **M**. Because the shape of the nuclear

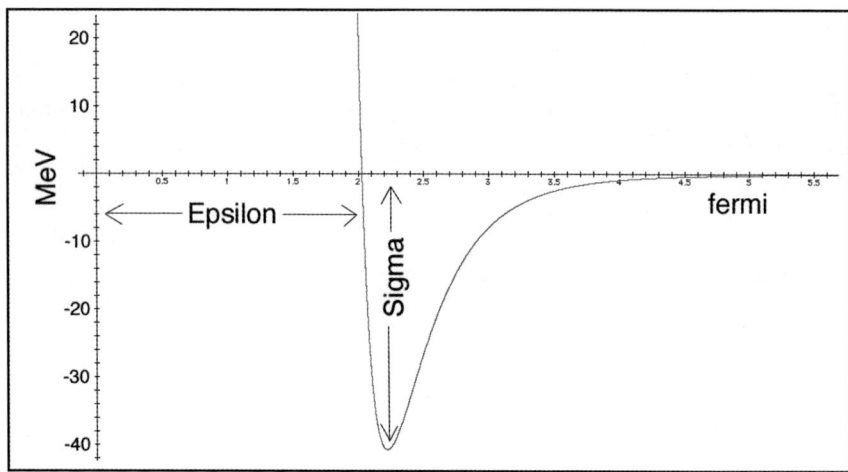

Fig. A.14. The epsilon and sigma variables for each of the nucleon-nucleon potentials can be adjusted for the molecular dynamics simulations

force acting between nucleon pairs determines the nature of the dynamics in the simulation, a variety of nuclear force parameters can be chosen under the Potentials item in the menu bar. For each of the six relevant potentials for all combinations of spin-up and spin-down protons and neutrons, the depth of the attractive portion of the potential and its radial extent can be adjusted. The potentials are all of the so-called Lennard-Jones type, as frequently used in theoretical chemistry and in many molecular dynamics simulations in nuclear theory. Each potential has two variables that determine the strength of the potential acting between the particles at a given distance. These are designated as epsilon and sigma, as illustrated in Fig. A.14. Epsilon is essentially the distance at which the potential turns from repulsive to attractive. Sigma is the maximal strength of the attractive potential.

All of the potentials have the following Lennard-Jones form:

$$P = \text{sigma}((\text{epsilon}/x)\ ^\wedge 14 - (\text{epsilon}/x)\ ^\wedge 8)x$$

Default settings (sigma = −40.0 and epsilon = 2.026) approximately reproduce empirical nuclear densities and binding energies, but small changes result in various kinds of nuclear stability and instability of possible interest. Extreme values will often result in nuclear "explosion." Default parameters for the potentials can be restored using the Default values checkbox.

Finally, the Heat setting (under Nucleons in the menu bar) determines the amount of random zero-point movement for each of the nucleons. This value is used in the simulations to set the amount of random movement in addition to the effects of the nucleon-nucleon interactions. Type **H** to increase, and **h** to decrease the temperature.

Help

Simple "help" functions can be selected from the Help item in the menu bar, or by typing @, ^, K or k on the keyboard. The at (@) mark displays information on the current version of the NVS software. The caret (^) mark displays an abbreviated list of the physics literature on the lattice models. It is not an exhaustive listing, but provides entry into the surprisingly long history of fcc and scp lattice papers in the technical literature. Typing k will display information on the current nucleon color-coding scheme and K shows the available keyboard shortcuts.

Data Files

The properties of more than 2600 isotopes are known. These include the 285 stable nuclei, some 1000 radioactive nuclei that decay toward stable states slowly enough that their properties have been measured, and more than 1300 short-lived, radioactive nuclei that have been created in particle accelerators, but have not yet been well-characterized. Depending primarily on the stability (longevity) of the isotope, various individual nuclear properties, such as spin, parity and nuclear mass, have been determined. For a smaller number of nuclei, root-mean-square charge radii, magnetic moments and quadrupole moments have also been measured. All such data are available in the physics literature (Firestone, 1996) and on web sites maintained by nuclear research institutes (e.g., The National Nuclear Data Center at the Brookhaven National Laboratories, http://www.nndc.bnl.gov). A thorough and easy-to-use database is also maintained at the Russian Joint Institute for Nuclear Research in Dubna (http://nrv.jinr.ru/nrv). This information has been incorporated into the *Nuclear Visualization Software*. Specifically, for all isotopes whose total binding energy has been measured, the following empirical data (when known) is displayed: (i) element name, (ii) number of protons, (iii) number of neutrons, (iv) total binding energy, (v) total spin, (vi) parity, (vii) RMS charge radius, (viii) magnetic moment, and (ix) quadrupole moment.

Summary

Despite the relatively small number of constituent particles, the atomic nucleus is a highly complex object, the explanation of which has demanded a surprisingly large number of theoretical models. Today, no single model is universally accepted as "the" explanation of nuclear structure, and researchers are obliged to keep many different theoretical perspectives in mind. As a tool to aid the imagination, the *Nuclear Visualization Software* may be helpful in understanding how various, seemingly contradictory nuclear structure models are in fact describing the same underlying reality.

A novel aspect of the ***NVS*** program is the emphasis drawn to the fcc/scp lattice representations of nuclei. While the features and practical uses of these models have been discussed in the physics literature, it is a quite remarkable, but not widely appreciated fact that the entire systematics of the quantum numbers assigned to nucleons (which is a direct consequence of the Schrödinger wave equation) is reproduced in an intuitive, geometrical manner within the fcc lattice (or, alternatively, within the 50%-occupied scp lattice). In other words, there is an unambiguous and precise one-to-one mapping between the known symmetries of "nuclear quantum space" and the geometry of the fcc lattice. Whether or not real nuclei maintain this geometry (in some dynamic and probabilistic fashion) is controversial, but the identity between the lattice symmetries and the n-, j-, m-, i- and s-quantum number symmetries of the Schrödinger equation is real and clearly illustrated within the **NVS** program. It is for this reason that the seemingly-contradictory character of the established models of nuclear structure theory might find unification within the fcc model – maintaining the principal strengths of independent-particle model (the entire quantum value description of nucleons, with the implied shell substructure of the nucleus) **and** those of the liquid-drop model (the constant-density nuclear core, binding energies and radii dependent on the number of nucleons, etc.). As an added bonus, various features of the cluster, boson and quark models are implied by the lattice structure, as is the nuclear skin thickness and asymmetrical fission of the actinides.

Whether or not unification of the various nuclear structure models are at hand, there are many aspects of nuclear structure and dynamics that are more easily understood when the visual mode of thinking is also employed. Particularly in light of the fact that several outstanding questions at the level of nuclear structure theory remain unanswered (What is the phase-state of nuclear matter? What is the nature and range of the nuclear force? How do nucleons aggregate in the nuclear interior?), the use of visualization techniques that are fully consistent with the quantum mechanical description of nucleon states is worth exploring.

A.2 Literature References to the Lattice Models

FCC Nuclear Model Related

Benesh, C.J., Goldman, T., & Stephenson, G.J. Valence quark distribution in A = 3 nuclei, *arXiv:nucl-th/0307038*, July, 2003.

Bevelacqua, J.J., FCC model calculations for ^{4}He excited states, *Physics Essays* 7, 389–395, 1994.

Castillejo, L., Jones, P.S.J., Jackson, A.D., Verbaarschot, J.J.M. & Jackson, A., Dense skyrmion systems, *Nuclear Physics* **A**501, 801–812, 1989.

Chao, N.C., & Chung, K.C., Tetrahedral percolation and nuclear fragmentation. *Journal of Physics* **G**17, 1851, 1991.

Chung, K.C., Nuclear fragmentation by nucleation approach. *Journal of Physics* **G**19, 1373, 1993.

A.2 Literature References to the Lattice Models

Cook, N.D., An FCC lattice model for nuclei. *Atomkernenergie* 28, 195–199, 1976.

Cook, N.D., Nuclear and atomic models. *International Journal of Theoretical Physics* 17, 21–32, 1978.

Cook, N.D., A unified theory of atomic and nuclear shell structure. *Experientia* 34, 419–420, 1978.

Cook, N.D., Quantization of the FCC nuclear theory. *Atomkernenergie-Kerntechnik* 40, 51–55, 1982.

Cook, N.D., The geometry of the atomic nucleus. *Physics Bulletin* 24, 267, 1988.

Cook, N.D., Computing nuclear properties in the FCC model. *Computers in Physics* 3, 73–77, 1989.

Cook, N.D., The attenuation of the periodic table. *Modern Physics Letters* **A**5, 1321–1328, 1990.

Cook, N.D., The problem of the mean free path of bound nucleons: implications for the nuclear force. *Modern Physics Letters* **A**5, 1531–1541, 1990.

Cook, N.D., Nuclear binding energies in lattice models. *Journal of Physics* **G** 20, 1907, 1994.

Cook, N.D., Fission of a nucleon lattice. *Bulletin of the American Physical Society* 41, 1229, 1996.

Cook, N.D., Asymmetric fission along nuclear lattice planes, *XVII RCNP Conference on Innovative Computational Methods in Nuclear Many-Body Problems*, October, 1998.

Cook, N.D., The equivalence of a 50% occupied FCC lattice and the lattice gas model, *Bulletin of the American Physical Society* 43, 22, 1998.

Cook, N.D., The nuclear symmetries software. *Bulletin of the American Physical Society* 43, 37, 1998.

Cook, N.D., Is the lattice gas model a unified model of nuclear structure? *Journal of Physics G: Nuclear and Particle Physics* 25, 1213–1221, 1999.

Cook, N.D., Asymmetric fission along nuclear lattice planes, *Proceedings of the St. Andrews Conference on Fission*, World Scientific, Singapore, pp. 217–226, 1999.

Cook, N.D., Nuclear visualization software. In, *Clustering Aspects of Nuclear Structure and Dynamics*, Nara, May 2003. (http://ribfwww.riken.go.jp/cluster8/)

Cook, N.D., The nuclear visualization software (NVS) http://www.res.kutc.kansai-u.ac.jp/~cook, 2004.

Cook, N.D., & Dallacasa, V., Face-centered solid-phase theory of the nucleus. *Physical Review* **C**36, 1883–1890, 1987.

Cook, N.D., & Dallacasa, V., The FCC nuclear model: II. Model predictions concerning nuclear radii, binding energies and excited states, *Il Nuovo Cimento* **A** 97, 184–204, 1987.

Cook, N.D., & Dallacasa, V., Nuclear RMS radii in the FCC model. *Journal of Physics* **G**13, L103, 1987.

Cook, N.D., & Dallacasa, V., A crystal clear view of the nucleus. *New Scientist,* no. 1606, March 31, 1988.

Cook, N.D., & Hayashi, T., *Proceedings of the American Physical Society Meeting,* October, 1996.

Cook, N.D., & Hayashi, T., Reproducing independent-particle model eigenstates in a close-packed lattice: The fcc model of nuclear structure. In, *Proceedings of the 17th International Symposium on Innovative Computational Methods in Nuclear Many-Body Problems,* Osaka, November, 1997.

Cook, N.D., & Hayashi, T., *Proceedings of the American Physical Society Meeting* (April, 1997)

Cook, N.D., & Hayashi, T., Lattices models for quark, nuclear structure, and nuclear reaction studies. *Journal of Physics* **G**23, 1109–1126, 1997.

Cook, N.D., & Hayashi, T., "Nuclear graphical software," *XVII RCNP Conference on Innovative Computational Methods in Nuclear Many-Body Problems* (October, 1998)

Cook, N.D., Hayashi, T., & Yoshida, N., Visualizing the atomic nucleus, *IEEE Computer Graphics and Applications,* 19(5), September/October, 1999.

Cook, N.D., & Musulmanbekov, G. The implications of nucleon substructure on nuclear structure theory. In, *Proceedings of the 8th International Conference on Clustering Aspects of Nuclear Structure and Dynamics,* Nara, Japan, November 24–29, 2003.

Cook, N.D., & Musulmanbekov, G., Nuclear structure theory reconstructed using a quark model: Liquid-drop, cluster and independent-particle model features. In, *Quarks and Nuclear Structure,* Bloomingtoon, Indiana, May, 2004 (www.qnp2004.org)

Dallacasa, V., FCC lattice model for nuclei. *Atomkernenergie-Kerntechnik* 37, 143, 1981.

Dallacasa, V., & Cook, N.D., The FCC nuclear model: I. Equivalence between the FCC lattice and nucleon eigenstates and the correspondence with the spin-orbit model. *Il Nuovo Cimento* **A** 97, 157–183, 1987.

Dimitrov, V.I., The case for a solid phase model of the nucleus. http://www.nd.edu/~vdimitro/2004.

Dyakonov, D.I., & Mirlin, A.D., *Soviet Journal of Nuclear Physics* 47, 421, 1988.

Everling, F., Binding energy systematics of 0+ states for even-even N = Z nuclides, *Proceedings of the International Workshop PINGST 2000,* p. 204–209, pingst2000.nuclear.lu.se/everling.pdf, June 2000.

Garai, J., The double tetrahedron structure of the nucleus. *arXiv.org/abs/nucl-th/0309035*

Goldman, T., Maltman, K.R., & Stephenson, G.J., *Nuclear Physics* **A** 481, 621, 1988.

Lezuo, K.J., A nuclear model based upon the close-packing of spheres. *Atomkernenergie* 23, 285, 1974.

Lezuo, K.J., Form factors for rotational levels in C^{12}, O^{16}, Si^{28} and Ca^{40}. *Zeitschrift für Naturforschung* 30a, 158, 1975; 30a, 1018, 1975.

Musulmanbekov, G., Quark correlations in nucleons and nuclei. *arXiv:hep-ph/0304262*, May, 2003.

Musulmanbekov, G., Quark and nucleon correlations in nuclei: Crystal-like nuclei? *17thInternational IUPAP Conference on Few-Body Problems in Physics*, Durham, North Carolina, June, 2003.

Musulmanbekov, G., & Al-Haidary, A., Fragmentation of nuclei at intermediate and high energies in modified cascade model. *arXiv:nucl-th/0206054*, June 2002.

Santiago, A.J., & Chung K.C., Do percolative simulations of nuclear fragmentation depend on lattice structure? *Journal of Physics* **G** 19, 349, 1993.

Yushkov, A.V., & Pavlova, N.V., *Izv. Akad. Nauk. SSSR, Ser. Fiz* 40, 826, 1976; 41, 176, 1977.

SCP and Other Lattice References

Abe, T., Seki, R., & Kocharian, A.N., **nucl-th/0312125**.

Bauer, W., Dean D., Mosel, U., & Post, U., *Physics Letters* 150**B**, 53, 1985.

Bauer, W., *Physical Review* **C**38, 1297–1303, 1988.

Bauer, W., et al., *Nuclear Physics* A452, 699–722, 1986.

Bauer, W., et al., *Revista Mexicana de Fisica* 49, 52, 1–26, 2003.

Berinde, A., et al.,*Nuclear Instruments and Methods* 167, 439–442, 1979.

Canuto, V., & Chitre, S.M., Quantum crystals in neutron stars. Invited talk at the *I.A.U. Symposium on Physics of Dense Matter*, Boulder, Colorado, Aug. 21–26, 1972.

Canuto, V., & Chitre, S.M., *International Astronomy & Astrophysics Union Symposium* 53, 133, 1974.

Canuto, V., & Chitre, S.M., *Ann. N.Y. Acad. Sci.* 224, 218, 1973.

Canuto, V., & Chitre, S.M., *Physical Review* **D**9, 1587, 1974.

Das Gupta, S., & Pan, J., *Physical Review* **C**53, 1319–1324, 1996.

Das Gupta, S., et al., *Nuclear Physics* **A**621, 897, 1997.

Das Gupta, S., et al., *Physical Review* **C**54, R2820-2, 1996; 57, 1361, 1998; 57, 1839, 1998.

Das Gupta, S., et al., *Physical Review Letters* 80, 1182, 1998.

Denisenko, K.G., Lozhkin, O.V., & Murin, Yu. A., *Soviet Journal of Nuclear Physics* 48, 264, 1988.

Dudek, J., Gozdz, A., Schunck, N., & Miskiewicz, http://chall.ifj.edu.pl/~dept2/zakopane2002/

Lee, D., Borasoy, B., & Schaefer, T., **arXiv:nucl-th/0402072**.

Lee, D. J., & Ilse C.F., **arXiv:nucl-th/0308052**

Mueller, H.-M., S. E. Koonin, R. Seki, and U. van Kolck, *Physical Review* **C** **61**, 044320 (2000) (**arxiv.org/abs/nucl-th/9910038**).

Mueller, H.-M., and R. Seki, in *Nuclear Physics with Effective Field Theory*, edited by R. Seki, U. van Kolck, and M. J. Savage (World Scientific, Singapore, 1998) pp. 191–202.

Pan, J., & Das Gupta, S., *Physical Review* **C**51, 1384–1392, 1995; 57, 1839, 1998.

Pan, J., Das Gupta, S., & Grant, M., *Physical Review Letters* 80, 1182–1184, 1998.

Park, B.Y., Lee, H.J., Vento, V., Kim, J.I., Min, D.P., Rho, M., **arxiv:hep-ph/0408010**.

Smith, J.H., *Physical Review* 95, 271, 1954.

Winans, J.G., *Physical Review* 71, 379, 1947a; 72, 435, 1947b.

A.3 Installation Notes

The *NVS* program will run on most computer systems provided that the hardware is set-up with support for OpenGL software. Newer Linux, Macintosh and Windows systems include the OpenGL graphics libraries with the operating system software, and no further preparation is needed to run the *NVS* software. Older hardware systems can often be configured with newer operating system software and/or the OpenGL software libraries, such that the *NVS* program will run, but there are limits on the backward compatibility. Specifically, the *NVS* program will not run on the older Macintoshes (pre-PowerPC machines) or Windows and Linux systems that cannot handle the OpenGL routines.

Source code for the *NVS* program is available on the enclosed CD. The entire code runs to 5,000 lines and can be recommended only to the experienced OpenGL/C programmer with a taste for spaghetti. There are, however, relatively clear-cut sections within the code where the nuclear model calculations are done. They should be of interest to physicists interested in adding new features to the *NVS* program. The code, written in ANSI-C (no C++), is provided as "Open Source" software (www.opensource.org).

The OpenGL libraries necessary for running computer graphics applications are available at no cost over the internet (http://www.opengl.org). With an appropriate C compiler, the source code with minor adjustments to the various hardware systems can be compiled and run as it is.

Function Keys

Function keys are often configured to work in conjunction with "fn," "ctrl," "option" or "alt" keys on the keyboard. If unanticipated effects are obtained using the function keys alone, try them in combination with the above special keys.

A.4 Keyboard Shortcuts

A – set to **A**lpha clustering model
B – set to **B**oson clustering model
C – display nucleons as **C**louds
D – **D**efault nucleon sequence
E – change **E**nvironmental hue
F – set to **F**CC lattice model
G – toggle various **G**raphics settings
H – increase **H**eat
I – set to the **I**ndependent-particle model
J – toggle among radial displays

K – display **K**eyboard shortcuts

L – set to the **L**iquid-drop model

M – set to **M**olecular dynamics mode
N – **N**ew (clear and default settings)

O – push last pr**O**ton to next lattice site
P – toggle the **P**icking mode
Q – set to **Q**uark model
R – toggle nuclear **R**evolution
S – set to **S**CP lattice model
T – **T**ransparency toggle
U – push last ne**U**tron to next lattice site
V – increase particle size/smoothness
W – **W**ire-frame toggle
X – toggle **X**Y plane
Y – toggle **Y**Z plane
Z – toggle **Z**X plane
1–7 – show quantum shells/subshells
9 – thinner bonds
@ – **a**bout NVS
^ – show literature references

a – set new **a**lpha cluster configuration
b – set new **b**oson cluster configuration
c – toggle among probability **c**loud densities
d – **d**efault graphics
e – **e**xplanatory text display
f – show **f**ission planes
g – **g**ray-scale and random nucleon coloring
h – decrease **h**eat
i – **i**sospin quantum number (**i**) coloring
j – total ang. mom. quantum number (**j**) coloring
k – display text on the **k**olor-coding of nucleons
l – **l**atitude and **l**ongitude stripes on nucleons
m – **m**agnetic quantum number (**m**) coloring
n – principal quantum number (**n**) coloring
o – pull last pr**o**ton back to previous lattice site
p – **p**rint information to a data file
q – set new **q**uark configuration
r – toggle nucleon **r**otation
s – **s**pin quantum number (**s**) coloring
t – reversal of **t**ransparency settings
u – pull last ne**u**tron back to previous lattice site
v – decrease particle size/smoothness
w – reversal of **w**ire-frame settings
x – toggle **x**-axis
y – toggle **y**-axis
z – toggle **z**-axis
8 – toggle translucent shells/subshells
0 – thicker bonds
* – toggle bond transparency
+ – toggle breathing modes

Function Keys

F1 – remove one proton
F3 – remove one neutron
F5 – remove one shell
F7 – make nucleons smaller
F9 – make nucleons rougher
F11 – toggle nucleon stripes
Arrow Down – rotate slower
Arrow Left – revolve slower
Page Down – thinner bonds

F2 – add one proton
F4 – add one neutron
F6 – add one shell
F8 – make nucleons larger
F10 – make nucleons smoother
F12 – toggle bond configurations
Arrow Up – rotate faster
Arrow Right – revolve faster
Page Up – thicker bonds

A.5 Keyboard Templates

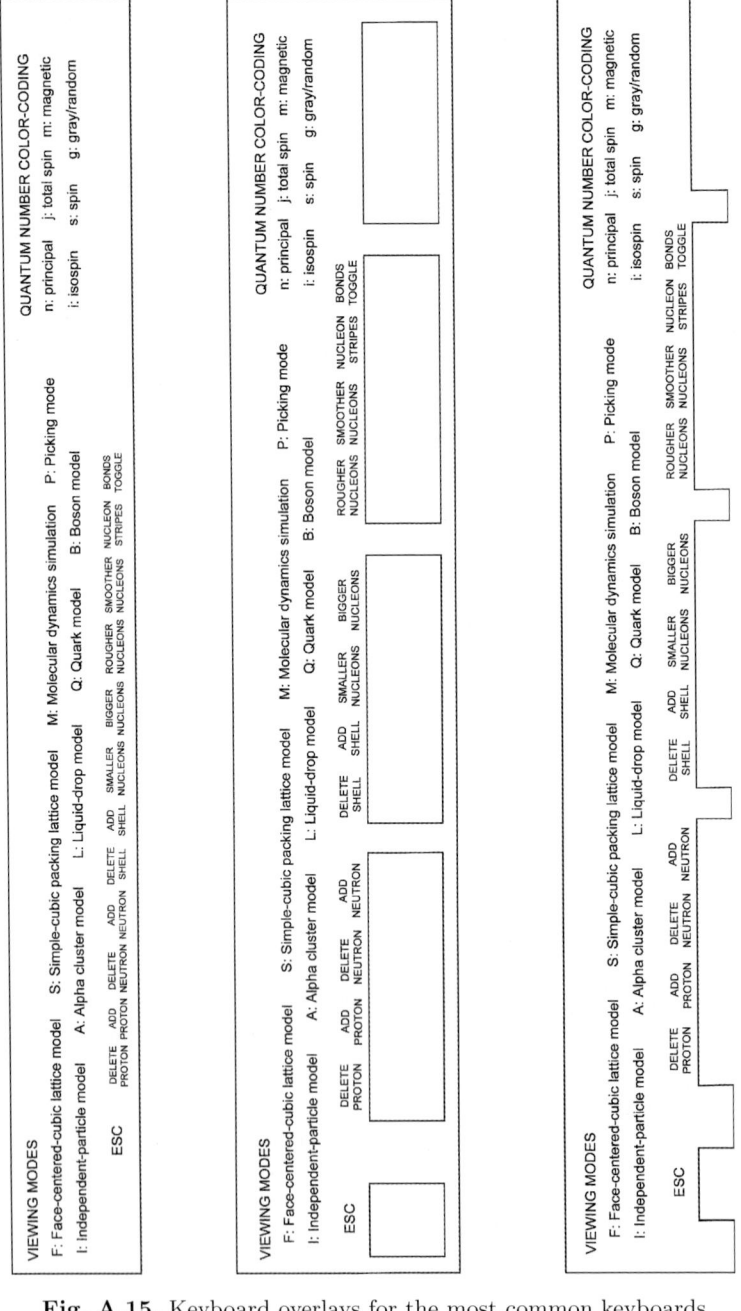

Fig. A.15. Keyboard overlays for the most common keyboards

A.6 Nuclear Model Definitions

Alpha-Particle Model: One of the earliest models of nuclear structure in which it is assumed that nucleons aggregate into 4-particle alpha clusters within the nuclear interior and/or on the nuclear surface. Synonymous with the cluster model.

Boson Model: A model of nuclear structure that assumes 1/2 spin nucleons pair to form integral-spin bosons. Several different "interacting boson models" (IBM1~IBM4) have been developed, and differ according to the assumed spin and isospin combinations.

Cluster Models: Generally synonymous with the alpha-particle model, but some versions consider ^{2}H, ^{3}H and ^{3}He, as distinct clusters, in addition to the ^{4}He cluster.

Collective Model: A development the liquid-drop model that combines liquid-drop features with a (distorted) central potential well.

Droplet Model: The droplet model and the extended droplet model are modern developments of the liquid-drop model.

Fermi Gas Model: One of the earliest models in which nuclei are thought to consist of weakly-interacting nucleons moving about as in a gas. This model forms the basis for the independent-particle model, in which distinct "orbiting" of nucleons is included.

Independent-Particle Model (IPM): The dominant theoretical paradigm in nuclear theory. The IPM is based on the Schrödinger equation that defines the energetic state of every nucleon moving under the influence of a central potential well. The state of each nucleon is uniquely determined by its five quantum numbers $(n, j, m, i$ and $s)$. Generally synonymous with the shell model.

Lattice Models (FCC and SCP): Lattice models used primarily for simulation of the high-energy fragmentation of nuclei. Nuclei are assumed to be fully- or partially-populated lattices of nucleons that interact locally with one another. Formally related to the "bond percolation" and "site percolation" lattice models.

Liquid-Drop Model (LDM): The earliest model of nuclear structure, based upon an analogy with a small liquid drop. This model has been developed as the droplet model, the extended droplet model, and various versions of the collective model. They have in common an emphasis on the strong local interactions of nucleons, and the implied volume and surface terms in the semi-empirical binding energy formula.

Molecular Dynamics (MD): Dynamic simulations in chemistry can reproduce molecular properties using realistic 2-body interactions among the constituent particles. Such techniques have been rescaled for use at the nuclear level, and are known as nuclear "molecular dynamics." The name is something of a misnomer, but means simply that a realistic nuclear force is at work in the simulations and graphical display.

Quark Model: The quark model emphasizes the interactions between quarks residing in one and the same nucleon (responsible for the properties of the individual proton or neutron) and the interactions between quarks residing in neighboring nucleons (responsible for the nuclear force and nucleon-nucleon bonding).

Shell Model: Generally synonymous with the independent-particle model. The shell model places emphasis on the closure of certain IPM shells and subshells, where nuclei are found experimentally to be especially stable and/or abundant (the "magic" nuclei).

References

Abe, T., Seki, R., Kocharian, A.N., *Physical Review* **C** 70, 014315, 2004.
Aichelin, J., & Stocker, H., *Physics Letters* **B** 163, 59, 1985.
Anagnostatos, G.S., *International Journal of Theoretical Physics* 24, 579, 1985.
Angeli, I., *Heavy Ion Physics* 8, 23, 1998.
Antonov, A.N., Hodgson, P.E., & Petrov, I. Zh., *Nucleon Momentum and Density Distributions in Nuclei*, Clarendon Press, Oxford, 1988.
Bali, G., et al., *arXiv:hep-ph/0010032*, 2000.
Barnes, T., www.qnp2004.org /T_Barnes.ppt, 2004.
Barreau, G., et al., *Nuclear Physics* **A** 432, 411, 1985.
Bauer, W., Dean, D.R., Mosel, U., & Post, U., *Physics Letters* **B** 150, 53, 1985.
Bauer, W., *Physical Review* **C** 38, 1297, 1988.
Bauer, W., Kleine Berkenbusch, M., Bollenbach, T., *Revista Mexicana de Fisica* 49, 1, 2003.
Bauer, W., Post, U., Dean, D.R., & Mosel, U., *Nuclear Physics* **A** 452, 699, 1986.
Benesh, C.J., Goldman, T., & Stephenson, G.J., *arXiv:nucl-th/0307038*, 2003.
Benhar, O., Fabrocini, A., & Fantoni, S., *Physical Review* **C** 41, 24, 1990.
Bertsch, G.F., *The Practitioner's Shell Model*, North-Holland, New York, 1972.
Bertsch, G.F., Zamich, L., & Mekjian, A., *Lecture Notes in Physics* 119, 240, 1980.
Blann, M., *Annual Reviews of Nuclear Science* 25, 123, 1975.
Blatt, J.M., & Weisskopf, V., *Theoretical Nuclear Physics*, Wiley, New York, 1952.
Bleuler, K., *Lecture Notes in Physics* 197, 1, 1984.
Blin-Stoyle, R.J., *Contemporary Physics* 1, 17, 1959.
Bobeszko, A., An fcc nuclear model (preprint, 1981).
Bohr, A., & Mottelson, B., *Nuclear Structure* (Vol. 1), Benjamin, Reading, Mass., 1969.
Bohr, N., *Nature* 137, 344, 1936.
Bohr, N., & Wheeler, J., *Physical Review* 56, 426, 1939.
Borkowski, F., et al., *Nuclear Physics* **B** 93, 461, 1975.
Bortignon, P.F., Bracco, A., & Broglia, R.A., *Giant Resonances*, Harwood, Amsterdam, 1998.
Brack, M. & Bhaduri, R.K., *Semiclassical Physics*, Addison-Wesley, Reading, Mass., 1997.

Brack, M., Damgard, J., Jensen, A.S., Pauli, H.C., Strutinsky, V.M., & Wong, C.Y., *Reviews of Modern Physics* 44, 320, 1972.
Brack, M., & Quentin, P.,*Physics Letters* **B** 76, 4484, 1974.
Brink, D.M., Friedrich, H., Weiguny, A., & Wong, C.W., *Physics Letters* **B** 33, 143, 1970.
Burcham, W.E., *Nuclear Physics: An introduction*, Longman, London, 2nd ed., 1973.
Burcham, W.E., & Jobes, M., *Nuclear and Particle Physics,* Longman, London, 1994.
Burge, E.J., *Atomic Nuclei and Their Particles*, Oxford University Press, Oxford, 1988.
Byrne, J., *Neutrons, Nuclei and Matter*, IOP Publishing, Bristol, 1994.
Caillon, J.C., & Labarsouque, J., *Physical Review* **C** 54, 2069, 1996.
Campi, X., *Journal of Physics* **A** 19, 1917, 1986.
Campi, X., *Physics Letters* **B** 208, 351, 1988.
Campi, X., & Krivine, H., *Nuclear Physics* A 620, 46, 1997.
Canuto, V., *Annual Review of Astronomy and Astrophysics* 12, 167 and 335, 1975.
Canuto, V., & Chitre, S.M., *International Astronomy and Astrophysics Union Symposium* 53, 133, 1974.
Castillejo, L., Jones, P.S.J., Jackson, A.D., Verbaarschot, J.J.M. & Jackson, A., *Nuclear Physics* **A** 501, 801, 1989.
Chao, N.C., & Chung, K.C., *Journal of Physics* **G** 17, 1851, 1991.
Cole, A.J., *Statistical Models for Nuclear Decay*, IOP Publishers, Bristol, 2000.
Condon, E.U., & Shortley, G.H., *The Theory of Atomic Spectra* , Cambridge University Press, Cambridge, 1935.
Cook, N.D., *Atomkernenergie* 28, 195, 1976.
Cook, N.D., *International Journal of Theoretical Physics* 17, 21, 1978.
Cook, N.D., *Atomkernenergie-Kerntechnik* 40, 51, 1982.
Cook, N.D., *Physics Bulletin* 24, 267, 1988.
Cook, N.D., *Computers in Physics* 3, 73, 1989.
Cook, N.D., *Modern Physics Letters* **A** 5, 1321, 1990; 5, 1531, 1990.
Cook, N.D., *Journal of Physics* **G** 20, 1907, 1994.
Cook, N.D., *Journal of Physics* **G** 25, 1, 1999.
Cook, N.D., "Asymmetric fission along nuclear lattice planes," *Second International Conference on Nuclear Fission and Neutron-Rich Nuclei* (St. Andrews, Scotland, June, 1999).
Cook, N.D., & Dallacasa, V., *Physical Review* **C** 36, 1883, 1987.
Cook, N.D., & Dallacasa, V., *Il Nuovo Cimento* **A** 97, 184, 1987.
Cook, N.D., & Dallacasa, V., *Journal of Physics* **G** 13, L103, 1987.
Cook, N.D., & Dallacasa, V., *New Scientist,* no. 1606, March 31, 1988.
Cook, N.D., & Hayashi, T., *Proceedings of the American Physical Society Meeting* 41, 1229, 1996; 43, 22, 1997.
Cook, N.D., & Hayashi, T., *Journal of Physics* **G** 23, 1109, 1997.
Cook, N.D., Hayashi, T., & Yoshida, N., *IEEE Computer Graphics and Applications* 19, 54, 1999.
Cottingham, W.N., & Greenwood, D.A., *An Introduction to Nuclear Physics*, Cambridge University Press, Cambridge, 1986.
Dacre, J., *Nuclear Physics*, Heinemann, Oxford, 1990.
Dallacasa, V., *Atomkernenergie-Kerntechnik* 37, 143, 1981.
Dallacasa, V., & Cook, N.D., *Il Nuovo Cimento* **A** 97, 157, 1987.

Das, A., & Ferbel, T.,*Introduction to Nuclear and Particle Physics*, Wiley, New York, 1994.
DasGupta, S., & Pan, J., *Physical Review* **C** 53, 1319, 1996.
DasGupta, S., et al., *Nuclear Physics* **A** 621, 897, 1997.
DasGupta, S., Pan, J., & Tsang, M.B., *Physical Review* **C** 54, R2820, 1996.
DasGupta, S., & Mekjian, A.Z., *Physical Review* **C** 57, 1361, 1998.
DasGupta, S., et al., *Physical Review* **C** 57, 1839, 1998.
DasGupta, S., et al., *Physical Review Letters* 80, 1182, 1998.
Désesquelles, P., et al., *Physical Review* **C** 48, 1828, 1993.
de-Shalit, A., & Talmi, I., *Nuclear Shell Theory*, Academic, New York, 1963.
deTar, C., *Physical Review* **D** 19, 1451, 1979.
DeVries, R.M., & DiGiacomo, N.J., *Journal of Physics* **G** 7, L51, 1981.
D'yakonov, D.I., & Mirlin, A.D., *Soviet Journal of Nuclear Physics* 47, 421, 1988.
Dymarz, R., & Kohmura, T., *Physics Letters* **B** 124, 446, 1983.
Eder, G., *Nuclear Forces: Introduction to Nuclear Physics*, 1968.
Egger, J.-P., Corfu, R., Gretillat, J., Lunke, C., Piffaretti, & Schwarz, E., *Physical Review Letters* 39, 1608, 1977.
Eisenberg, J.M., & Greiner, W., *Nuclear Theory*, North-Holland, Amsterdam, 1987.
Eisenbud, L., & Wigner, E., *Nuclear Structure*, Princeton University Press, Princeton, 1958.
Elattari, B., Richert, J., Wagner, P., & Zheng, Y.M., *Nuclear Physics* **A** 592, 385, 1995.
Elliott, J.P., "The shell model: an overview", In, *Shell Model and Nuclear Structure: where do we stand?* A. Covello, ed., World Scientific, Singapore, 1989, p. 13.
Elliott, J.P., *Nuclear Physics* **A** 507, 15c–24c, 1990.
Elliott, J.P., & Lane, A.M., "The nuclear shell model", In, *Encyclopedia of Physics*, S. Flügge, ed., Springer, Berlin, 1957, Vol. 39, p. 241.
Enge, H., *Introduction to Nuclear Physics*, Addison-Wesley, New York, 1966.
Evans, R.D., *The Atomic Nucleus*, McGraw-Hill, New York, 1955.
Fantoni, S., & Pandharipande, V., *Nuclear Physics* **A** 427 473, 1984.
Feshbach, H., *Theoretical Nuclear Physics*, Wiley, New York, 1992.
Feshbach, H., Porter, C.E., & Weisskopf, V.F., *Physical Review* 90, 166, 1953.
Feshbach, H., Porter, C.E., & Weisskopf, V.F., *Physical Review* 96, 448, 1954.
Firestone, R.B. (Ed.), *Table of Isotopes*, 8th Edition, Wiley, New York, 1996.
Fraser, J.S., & Milton, J.C.D., *Annual Review of Nuclear Science* 16, 379, 1966.
Fricke, M.P., Gross, E.E., Morton, B.J., & Zucker, A., *Physical Review* 156, 1207, 1967.
Fulmer, C.B., Ball, J.B., Scott, A., & Whiten, M.L., *Physical Review* 181, 1565, 1969.
Gadioli, E., Gadioli Erba, E., Tagliaferri, G., & Hogan, J.J., *Physics Letters* **B** 65, 311, 1976.
Gadioli, E., Gadioli Erba, E., Hogan, J.J., & Burns, K.I., *Zeitschrift für Physik* **A** 301 289, 1981.
Geltenbrot, P., Gönnenwein, F., & Oed, A., *Radiation Effects* 93, 57, 1986.
Ghirardi, G., *Sneaking a Look at God's Cards*, Princeton University Press, Princeton, 2005, pp. 111–119.
Glassgold, A.E., & Kellogg, P.J., *Physical Review* 109, 1291, 1958.
Goldberger, M.L., *Physical Review* 74, 1269, 1948.
Goldhammer, P., *Reviews of Modern Physics* 35, 40, 1963.

Goldman, T., Maltman, K.R., & Stephenson, G.J., *Nuclear Physics* **A** 481, 621, 1988.
Goldman, T., *Nuclear Physics* A532, 389c, 1991.
Gönnenwein, F., in *The Nuclear Fission Process,* Wagemans, C., ed., CRC Press, Boca Raton, 1991.
Greiner, W., *International Journal of Modern Physics* 5, 1, 1995.
Greiner, W., & Maruhn, J.A., *Nuclear Models*, Springer, Berlin, 1996.
Hahn, B.D., Ravenhall, D.G., & Hofstadter, R., *Physical Review* 101, 1131, 1956.
Halpern, I., *Annual Review of Nuclear Science* 9, 245, 1959.
Hamada, T., & Johnston, I.D., *Nuclear Physics* 34, 382, 1962.
Hamilton, J., *Endeavor* 19, 163, 1960.
Hasse, R.W., & Myers, W.D., *Geometrical Relationships of Macroscopic Nuclear Physics*, Springer, Berlin, 1988.
Hauge, P.S., Williams, S.A., & Duffey, G.H., *Physical Review* **C** 4, 1044, 1971.
Herzberg, G., *Atomic Spectra and Atomic Structure,* Dover, New York, 1944.
Heyde, K., *Basic Ideas and Concepts in Nuclear Physics*, IOP Publishing, Bristol, 1994.
Heyde, K., *The Nuclear Shell Model*, Springer, New York, 1995.
Heyde, K., *From Nucleons to the Atomic Nucleus: Perspectives in Nuclear Physics*, Springer, Berlin, 1998.
Hodgson, P.E., *Nature* 257, 778, 1975.
Hodgson, P.E., Alpha-clustering in nuclei. In, *The Uncertainty Principle and Foundations of Quantum Mechanics,* W.C. Price, ed., Wiley, New York, 1982, pp. 485–542.
Hodgson, P.E., Gadioli, E., & Gadioli Erba, E., *Introductory Nuclear Physics*, Oxford Science Publications, Oxford, 1997.
Hoffman, D.C., & Hoffman, M.M., *Annual Review of Nuclear Science*, 24, 151, 1974.
Hofstadter, R., *Reviews of Modern Physics* 28, 214, 1956.
Hofstadter, R., "Nuclear radii" in *Nuclear Physics and Technology*, vol. 2, H. Schopper, Ed., Springer, Berlin, 1967.
Hofstadter, R., & McAllister, R.W., *Physical Review* 98, 217, 1955.
Hofstadter, R., Fechter, H.R., & McIntyre, J.A., *Physical Review* 92, 978, 1953.
Hyde, E.K., *The Nuclear Properties of the Heavy Elements*, Vol. I–III, Prentice-Hall, Englewood, N.J., 1964.
Iachello, F., & Arima, A., *The Interacting Boson Model*, Springer, Berlin, 1986.
Inopin, E.V., Kinchakov, V.S., Lukyanov, V.K., & Pol, Y.S., *Annals of Physics* 118, 307, 1979.
Irvine, J.M., *Nuclear Structure Theory*, Pergamon, Oxford, 1972.
Johnson, K.E., *American Journal of Physics* 60, 164, 1992.
Jones, G.A., *The Properties of Nuclei*, Oxford University Press, Oxford, 1986.
Kaplan, I., *Nuclear Physics*, Addison-Wesley, Cambridge, Mass., 1955.
Kikuchi, K., & Kawai, M., *Nuclear Matter and Nuclear Reactions,* North-Holland, Amsterdam, 1968.
Kirson, M.W., "Some thoughts on the nuclear shell model", In: *Contemporary Nuclear Shell Models* (Eds., X.W. Pan, D.H. Feng & M. Vallieres), Springer, New York, 1997.
Koonin, S.E., "One-body nuclear dynamics", in *Nuclear Structure and Heavy-Ion Collisions,* R.A. Broglia & R.A. Ricci, eds., North-Holland, Amsterdam, 1981, pp. 233–260.

Krane, K.S., *Introductory Nuclear Physics*, Wiley, New York, 1988.
Kumar, K., *Superheavy Elements*, Hilger, London, 1989.
Lacombe, M., Loiseau, B., Richard, J.M., Vinh Mau, R., Cote, J., & Pires, O., *Physical Review* **C** 21, 861, 1980.
Landau, L.D., & Smorodinsky, Ya. *Lectures on Nuclear Theory*, Dover, New York, 1993; (first published by Plenum Press, New York, 1959).
Lawson, R.D., *Theory of the Nuclear Shell Structure*, Oxford University Press, Oxford, 1980.
Lee, D.J., *arXiv:nucl-th/0407101*, *arXiv:nucl-th/0407088*, *arXiv:nucl/th/0402072*, 2004.
Lee, D.J., & Ipsen, I.C.F., *Physical Review* **C** 68, 064003, 2003.
Lezuo, K., *Atomkernenergie* 23, 285, 1974.
Lezuo, K., *Zeitschrift für Naturforschung* 30a, 158, 1975; 30a, 1018, 1975.
Littauer, R.M., Schopper, H.F., & Wilson, R.R., *Physical Review Letters* 7, 144, 1961.
MacGregor, M.H., *Il Nuovo Cimento* **A** 36, 113, 1976.
Maltman, K., Stephenson, G.J., & Goldman, T., *Physics Letters* **B** 324, 1, 1994.
Marmier, P., & Sheldon, E., *Physics of Nuclei and Particles*, Academic, New York, 1969.
Mayer, M.G., & Jensen, J.H.D., *Elementary Theory of Nuclear Shell Structure*, Wiley, New York, 1955.
McCarthy, I.E., *Introduction to Nuclear Theory*, Wiley, New York, 1968.
Meyerhof, M., *Elements of Nuclear Physics*, McGraw-Hill, New York, 1967.
Miller, G.A., *Lecture Notes in Physics* 197, 195, 1984.
Mladjenovic, M., *The Defining Years in Nuclear Physics 1932–1960s*, IOP Publishing, Bristol, 1998.
Möller, P., & Nix, J.R., *Nuclear Physics* **A** 549, 84, 1992.
Möller, P., & Nix, J.R., *Journal of Physics* **G** 20, 1681, 1994.
Möller, P., Nix, J.R., Myers, W.D., & Swiatecki, W.J., *Atomic Data and Nuclear Data Tables* 59, 185, 1992.
Möller, P., Madland, D.G., Sierk, A.J., & Iwamoto, A., *Nature* 409, 785, 2001.
Moreau, J. & Heyde, K. in *The Nuclear Fission Process*, Wagemans, C., ed., Boca Raton, CRC Press, 1991.
Moretto, L.G.., & Wozniak, G.J., *Annual Review of Nuclear and Particle Science* 43, 379, 1993.
Morrison, D., In, *Nuclei and Particles*, E. Segre, ed., Benjamin, Reading, Mass., 1965.
Moszkowski, S.A., "Models of nuclear structure," in *Encyclopedia of Physics*, S. Flügge, ed., Springer, Berlin, 1957, vol. 39, p. 411.
Mottelson, B., "Some periods in the history of nuclear physics with particular emphasis on the compound nucleus," in *Nuclear Structure 1985*, R. Broglia, G. Hagemann & B. Herskind, eds., North-Holland, Amsterdam, 1985, pp. 3–24.
Mueller, H.-M., Koonin, S.E., Seki, R., & vanKolck, U., *Physical Review* **C** 61, 044320, 2000.
Musulmanbekov, G., *Proceedings of the Hadron Structure*, p. 266, 2002.
Musulmanbekov, G., *arXiv:hep-ph/0304262*, 2003.
Musulmanbekov, G., *AIP Conference Proceedings*, pp. 701–705, 2004.
Myers, W.D., *Droplet Model of Atomic Nuclei*, Plenum, New York, 1977.
Myers, W.D., & Swiatecki, W.J., *Nuclear Physics* 81, 1, 1966.

Nadasen, A., et al., *Physical Review* **C**23, 1023, 1981.
Nadjakov, E.G., Mainova, K.P., & Gangsky, Yu. P., *Atomic and Nuclear Data Tables* 56, 133, 1994.
Neff, T., & Feldmeier, H., *Nuclear Physics A* 713, 311, 2003.
Negele, J.W., *Comments on Nuclear and Particle Physics* 12, 1, 1983.
Negele, J.W., & Orland, H., *Quantum Many-Particle Systems*, Addison-Wesley, New York, 1988.
Negele, J.W., & Yazaki, K., *Physical Review Letters* 47, 71, 1981.
Nemeth, J., et al., *Zeitschrift für Physik* **A**325, 347, 1986.
Nicholson, J.P., (reprint) Los Alamos, 1953.
Nifenecker, H., in *Nuclear Structure*, K. Abrahams, K. Allaart & A.E.L. Dieperink, eds., Plenum, New York, 1981, p. 316.
Nilsson, S.G., et al., *Nuclear Physics* **A** 131, 1, 1969.
Nilsson, S.G., & Ragnarsson, I., *Shapes and Shells in Nuclear Structure*, Cambridge University Press, New York, 1995.
Palazzi, P., *arXiv:physics*/0301074, 2003.
Palazzi, P., http://particlez.org, 2004.
Palazzi, P., http://particlez.org, 2005.
Pan, J., & DasGupta, S., *Physical Review* **C** 51, 1384, 1995.
Pan, J., & DasGupta, S., *Physical Review* **C** 57, 1839, 1998.
Pan, J., DasGupta, S., & Grant, M., *Physical Review Letters* 80, 1182, 1998.
Pandharipande, V.R., Sick, I., & deWitt Huberts, P. K. A., *Reviews of Modern Physics* 69, 981, 1997.
Pauling, L., *Nature* 208, 174, 1965.
Pauling, L., *Physical Review Letters* 15, 499, 1965; 36, 162, 1976.
Pauling, L., *Proceedings of the National Academy of Science* 54, 989, 1965; 72, 4200, 1975; 73, 274, 1403, 1976.
Pauling, L., *Science* 150, 297, 1965.
Pauling, L., & Robinson, A.B., *Canadian Journal of Physics* 53, 1953, 1975.
Pearson, J.M., *Nuclear Physics: Energy and Matter*, Holger, London, 1986.
Petry, H.R., *Lecture Notes in Physics* 197, 236, 1984.
Pirner, H.J. & Vary, J.P., *Physical Review* **C** 33, 1062, 1986.
Povh, B., Rith, K., Scholz, C., & Zetsche, F. *Particles and Nuclei*, Springer, New York, 1995.
Powers, R.J., et al., *Nuclear Physics A* 262, 493, 1976.
Preston, M.A. & Bhaduri, R.K., *Structure of the Nucleus*, Addison-Wesley, Reading, Mass., 1975.
Reid, J.M., *The Atomic Nucleus*, Pergamon, Oxford, 1972.
Richert, J., & Wagner, P., *Physics Reports* 350, 1, 2001.
Robson, D., *Nuclear Physics* **A** 308, 381, 1978.
Rowe, D.J., *Nuclear Collective Motion*, Methuen, London, 1970.
Santiago, A.J., & Chung, K.C., *Journal of Physics* **G** 19, 349, 1993.
Schiavilla, P. V., Pandharipande, V., & Wiringa, R.,*Nuclear Physics* **A** 449, 219, 1986.
Schiffer, J.P., *Nuclear Physics* **A** 335, 339, 1980.
Seaborg, G.T., & Bloom, J.L., *Scientific American,* April, 1969, p. 57.
Seaborg, G.T., & Loveland, W., *Contemporary Physics* 28, 33, 1987.
Segre, E., *Nuclei and Particles*, Benjamin, Reading, Mass., 1965.
Seki, R., http://www.csun.edu/~rseki/collaboration, 2003.

Sick, I., & deWitt Huberts, P.K.A., *Comments on Nuclear and Particle Physics* **20**, 177, 1991.
Siemens, P.J., & Jensen, A.S., *Elements of Nuclei,* Addison-Wesley, New York, 1987.
Sitenko, A.G., & Tartakovskii, V., *Lectures on the Theory of the Nucleus*, Pergamon Press, Oxford, 1975.
Smith, C.M.H., *A Textbook of Nuclear Physics,* Pergamon, Oxford, 1965.
Smith, J.H., *Physical Review* 95, 271, 1954.
Smith, T.P., *Hidden Worlds: Hunting for quarks in ordinary matter*, Princeton University Press, Princeton, 2002.
Steuwer, R.H., "Niels Bohr and nuclear physics" in *Niels Bohr: a Centenary Volume*, A.P. French & P.J. Kennedy, eds., Harvard, Cambridge, Mass, 1985, pp. 197–220.
Stevens, P.S., The nucleus pictured by close-packed spheres (preprint, 1972).
Takahashi, K., *Progress of Theoretical Physics* 85, 779, 1991.
Tamari, V.F., "Beautiful Universe" http://home.att.ne.jp/zeta/tamari/beautiful-universe.html
Tamagaki, R., *Nuclear Physics A* 328, 352, 1979.
Townes, C.H., Foley, H.M., & Low, W., *Physical Review* 76, 1415, 1949.
Turner, M.S., 1977.
Überall, H., *Electron Scattering from Complex Nuclei*, Academic, New York, 1971.
Valentin, L., *Subatomic Physics: Nuclei and Particles*, North-Holland, Amsterdam, 1981.
Vandenbosch, R., & Huizenga, J.R., *Nuclear Fission*, Academic, New York, 1973.
vanOers, W.T.H., et al., *Physical Review* **C** 10, 307, 1974.
Walecka, J.D., *Theoretical Nuclear and Subnuclear Physics*, Cambridge University Press, New York, 1995.
Wefelmeier, W., *Zeitschrift für Physik* 107, 332, 1937.
Weisskopf, V.F., *Science* 113, 101, 1951.
Weizsäcker, C.F. von, *Zeitschrift für Physik* 96, 431, 1935.
Wheeler, J.A., "Some men and moments in the history of nuclear physics" In *Nuclear Physics in Retrospect*, R.H. Stuewer, ed., University of Minnesota Press, Minneapolis, 1979, p. 267.
Wigner, E., *Physical Review* 51, 106, 1937.
Wigner, E., "The neutron", in *Nuclear Physics in Retrospect*, R.H. Stuewer, ed., University of Minnesota Press, Minneapolis, 1979, p. 164.
Wildermuth, K., & Tang, Y.C., *A Unified Theory of the Nucleus*, Braunschweig, Vieweg, 1977.
Wilkinson, D.H., *Nuclear Physics* **A** 507, 281c–294c, 1990.
Williams, W.S.C., *Nuclear and Particle Physics*, Oxford University Press, Oxford, 1991.
Winans, J.G., *Physical Review* 71, 379, 1947; 72, 435, 1947.
Wuosmaa, A.H., Betts, R.R., Freer, M., & Fulton, B.R., *Annual Review of Nuclear and Particle Science* 45, 89, 1995.
Yang, F., & Hamilton, J.H., *Modern Atomic and Nuclear Physics*, McGraw-Hill, New York, 1996.
Yennie, D.R., Ravenhall, D.G., & Wilson, R.N., *Physical Review* 95, 500, 1954.
Yuan, H.J., Lin, H.L., Fai, G., & Moszkowski, S.A., *Physical Review* **C** 40, 1448, 1989.

Name Index

Abe, T. 203, 269
Aichelin, J. 90
Anagnostatos, G.S. 184
Angeli, I. 126
Antonov, A.N. 96, 113, 129
Arima, A. 202

Bali, G. 213
Balmer, J. 13
Barnes, T. 213
Barreau, G. 242
Bauer, W. 48, 49, 80, 81, 203
Benesh, C.J. 79, 213
Benhar, O. 207
Bertsch, G.F. 72, 95, 188
Bhaduri, R.K. 95, 115–117, 161, 170, 171
Blann, M. 90
Blatt, J.M. 5, 88, 93, 111
Bleuler, K. 48
Blin-Stoyle, R.J. 74
Bloom, J.L. 145
Bobeszko, A. 183
Bohr, A. 15, 36, 45, 63, 64, 72, 90–93, 96, 98–101, 105, 106, 109, 113, 131, 132, 155, 161, 163
Bohr, N. 10, 12, 15, 41, 44, 48, 49, 56, 57, 90, 92, 158, 160, 162
Borkowski, F. 126
Bortignon, P.F. 65, 100
Bracco, A. 52, 100
Brack, M. 144, 161, 1168–171
Brink, D.M. 66
Broglia, R.A. 100

Burcham, W.E. 95, 112, 159
Burge, E.J. 95, 163
Byrne, J. 163

Caillon, J.C. 90
Campi, X. 80, 203
Canuto, V. 82, 192, 198, 199, 212
Castillejo, L. 199
Chao, N.C. 82, 83, 203
Chadwick, N. 14, 15, 41
Chitre, S.M. 82, 192, 199, 212
Chung, K.C. 82, 83, 203
Cole, A.J. 203
Condon, E.U. 121
Cook, N.D. 39, 53, 182, 188, 196, 201, 203, 218, 219, 235, 236, 241
Cottingham,W.N. 95

Dacre, J. 95
Dallacasa, V. 39, 182, 188, 201, 218, 219
Damgard, J. 153, 169, 172
Das, A. 58
DasGupta, S. 81, 203, 206–208
Désesquelles, P. 203
de-Shalit, A. 5, 95, 114
DeVries, R.M. 90, 103, 104, 111
deWitt Huberts, P. K. A . 207
DiGiacomo, N.J. 90, 103, 104, 111
Duffey, G.H. 223, 224
Dyakonov, D.I. 82
Dymarz, R. 90

Eder, G. 112
Egger, J.-P. 218

Name Index

Einstein, A. 10, 49
Eisenberg, J.M. 95
Eisenbud, L. 95
Elattari, B. 82
Elliott, J.P. 48, 75, 111, 141
Enge, H. 90, 95, 107
Evans, R.D. 95

Fall, J.B. 110
Fantoni, S. 207
Fechter, H.R. 126
Feldmeier, H. 67
Ferbel, T. 58
Fermi, E. 15, 158
Feshbach, H. 90, 100, 106
Firestone, R.B. 134, 265
Fraser, J.S. 151, 155
Fricke, M.P. 90
Fulmer, C.B. 90

Gadioli Erba, E. 114
Gadioli, E. 90, 114
Gamow, G. 56
Geltenbort, P. 156
Ghirardi, G. 247
Glassgold, A.E. 90
Gönnenwein, F. 157, 173, 244
Goldberger, M.L. 90, 97
Goldhammer, P. 67
Goldman, T. 78, 79, 213
Grant, M. 270
Greenwood, D.A. 95
Greiner, W. 5, 50, 144, 146
Gross, E.E. 135

Hahn, B.D. 15, 99, 151, 158
Halpern, I., 151
Hamada, T. 143
Hamilton, J.H. 60, 91, 121, 160, 219
Hasse, R.W. 58, 64
Hauge, P.S. 223, 224
Haxel, O. 15
Hayashi, T. 53, 196, 203
Heisenberg, W. 15
Herzberg, G. 20, 121
Heyde, K. 101, 102, 151, 155, 159, 162
Hodgson, P.E. 67, 91, 92, 96, 113, 114, 129
Hoffman, D.C. 154, 155

Hoffman, M.M. 154, 155
Hofstadter, R. 48, 124, 126, 127, 132, 133, 137, 218
Huizenga, J.R. 160
Hyde, E.K. 160

Iachello, F. 202
Inopin, E.V. 223
Irvine, J.M. 95
Iwamoto, A. 166

Jackson, A.D. 69
Jensen, A.S. 88, 90, 94
Jensen, J.H.D. 15, 43, 72, 95, 161, 166
Jobes, M. 95, 159
Johnson, K.E. 56
Johnston, I.D. 143
Jones, G.A. 76, 102
Jones, P.S.J. 97, 130

Kaplan, I. 95
Kellogg, P.J. 90
Kirson, M.W. 141
Kohmura, T. 90
Koonin, S.E. 100
Krane, K.S. 15, 73, 76, 153, 161, 167, 180
Krivine, H. 203
Kumar, K. 147

Labarsouque, J. 90
Lacombe, M. 143
Landau, L.D. 5, 125–129, 131
Lane, A.M. 75, 111
Lawson, R.D. 95, 159
Lee, D. 203
Lezuo, K. 183
Littauer, R.M. 132, 198, 212, 213
Loveland, W. 145

MacGregor, M.H. 68, 69, 70, 118, 219, 223
Maltman, K.R. 79, 213
Marmier, P. 97
Maruhn, J.A. 5, 50, 51, 95, 144
Mayer, M.G. 15, 43, 72, 161, 166
McCarthy, I.E. 97, 159
McIntyre, J.A. 126
Meyerhof, M. 112

Miller, G.A. 48
Milton, J.C.D. 151, 155
Mirlin, A.D. 82
Mladjenovic, M. 41, 160
Möller, P. 62, 144, 147, 171
Moreau, J. 151, 155, 162
Moretto, L.G. 83
Morrison, D. 89, 90
Moszkowski, S.A. 46, 47
Mottelson, B., 15, 36, 45, 63, 64, 90, 92, 93, 96, 98–101, 105, 106, 131, 155, 161, 163
Mueller, H.-M. 203, 269, 270
Musulmanbekov, G. 79, 213
Myers, W.D. 31, 32, 58, 61, 63, 64

Nadasen, A. 108, 117
Neff, T. 67
Negele, J.W. 90, 99, 104–106
Nifenecker, H. 161
Nilsson, S.G. 75, 77–79, 95, 148, 158, 169, 170, 181
Nix, J.R. 62, 147

Orland, H. 106
Palazzi, P. 211
Pan, J. 203, 206–208
Pandharipande, V. R. 207, 208
Pauli, H.C. 113
Pauli, W. 20, 68, 94, 102, 103, 105, 109–121
Pauling, L. 24, 26, 68, 70, 71
Pearson, J.M. 91
Petrov, I.Zh. 96, 113, 129
Petry, H.R. 48
Porter, C.E. 90, 100
Povh, B. 30, 92, 129, 138, 159
Powers, R.J. 220
Preston, M.A. 95, 115–117, 161, 170

Quentin, P. 144

Ragnarsson, I. 77–79, 95, 148, 158, 170, 181
Rainwater, J. 15
Ravenhall, D.G. 126
Reid, J.M. 91
Richert, J. 80, 81, 203
Robinson, A.B. 70

Robson, D. 48, 78, 79
Rowe, D.J. 95
Rutherford, E. 14, 15, 41, 49, 123
Rydberg, B. 13

Santiago, A.J. 82, 203
Schiavilla, P. V. 207
Schiffer, J.P. 90, 102
Schrödinger, E. 15
Seaborg, G.T. 15, 145
Segre, E. 32, 33, 89, 113
Seki, R. 203
Sheldon, E. 97
Shortley, G.H. 121
Sick, I. 207
Siemens, P.J. 88, 90, 94
Sitenko, A.G. 129
Smith, C.M.H. 91
Smith, J.H. 48
Smorodinsky, Ya. 5, 125–131
Steuwer, R.H. 49
Stevens, P.S. 183
Stocker, H. 90
Strutinsky, V.M. 161, 169, 170
Swiatecki, W.J. 31, 32, 63

Takahashi, K. 198, 203
Talmi, I. 5, 95, 114
Tamagaki, R. 198, 203
Tamari, V.F. 211
Tang, Y.C. 67
Tartakovskii, V. 129
Townes, C.H. 32
Turner, M.S. 119

Überall, H. 134

Valentin, L. 95
Vandenbosch, R. 160
vanOers, W.T.H. 90

Wagner, P. 80, 81, 203
Walecka, J.D. 95
Wefelmeier, W. 184
Weisskopf, V.F. 5, 44, 88, 90, 93, 94, 100, 108, 109, 111, 119
Wheeler, J.A. 15, 41, 48, 56, 158, 160, 162
Wigner, E. 48, 72, 95, 183

Wildermuth, K. 67
Wilkinson, D.H. 48, 77
Williams, W.S.C. 113, 130, 131, 159
Wilson, R.N. 126
Winans, J.G. 239
Wiringa, R. 168
Wozniak, G.J. 83

Wuosmaa, A.H. 67

Yang, F. 60, 121, 160, 219
Yazaki, K. 104–106
Yennie, D.R. 126
Yoshida, N. 53
Yuan, H.J. 106, 107
Yukawa, H. 15

Subject Index

alpha particle model 6, 18, 42, 46, 50–53, 55, 56, 65–67, 222–225
angular momentum 189, 190, 225–233
atomic structure 9, 11, 12, 19–27
azimuthal quantum number 20, 191–194

binding energy 31, 32, 58–62, 225–239
boiling point 24, 27
boson model 51

Chernobyl 4
cluster model 6, 18, 42, 46, 50–53, 55, 56, 65–71, 118
collective model 46, 47, 50–53, 56–65
compound nucleus 15, 47, 52, 72, 90–93, 109, 158
Coulomb force 58–60, 69

density, nuclear 123–134, 214–216

electronegativity 24, 26
exclusion principle 94, 97, 108–122

fcc lattice model 82, 83
Fermi density curve 17, 18, 133–137
Fermi gas 46, 47, 96
fission 57, 118, 151–175, 239–246
fission fragments 152–174, 239–246

Hiroshima 4

independent-particle model 6, 46, 50–53, 55, 56, 72–76, 177–181
ionization energy 23, 26

isotones 28–30
isotopes 28–30

lattice-gas 53, 204–208
lattice models 7, 18, 48–50, 52, 53, 80–84, 175–246
liquid-drop model 6, 18, 41, 46, 47, 50–53, 55–57, 135–137, 148, 158, 162

magnetic moment 74, 75
magic numbers 27–40, 72–76, 187, 188
mean free path 87–122,
melting point 25, 27

neutron 16, 41, 197, 198
nuclear force 130, 131, 132, 141–150
nuclear potential-well 75–79, 168–172
nuclear power 3, 7

optical model 45–47, 105, 106

pairing effect 59–62
parity 193, 194
Pauli principle (see exclusion principle)
percolation model 53, 80–84
periodic table 11, 12, 22, 145
proton 16, 17, 197, 198

quadrupole moment 32–35
quantum numbers 19, 20, 39, 182–193
quarks 48, 78–80, 198, 200, 211–214

radius, atomic 21–26

radius, nuclear 15, 17, 35–38, 62–64, 123–140, 220–222
radius, nucleon 17, 123–129

Schmidt lines 74, 75
Schrödinger equation 6, 12, 19, 21, 27, 38, 40, 177–181
scp lattice model 80–84
Segre chart 27, 28
separation energy 34–36
shell model 18, 43–47, 50–53, 99, 104, 106, 111, 148
shells, atomic 12, 19–27, 43, 44
shells, nuclear 27–38, 43–45, 50, 72–75, 183–188

skin, nuclear 89, 103, 104, 133–137, 216–220
Skyrme potential 106, 107
spheron model 70, 71
spin-orbit coupling 72–73, 189–191
super-heavy nuclei 141–150, 233–235
symmetry effect 61, 62

Three Mile Island 4

uncertainty principle 49, 56, 57
unification 5–8, 208, 209

vaporization 25, 27

DATE DUE

JUN 1 4 2010	
RECEIVED	

BRODART, CO.   Cat. No. 23-221-003